Lecture Notes in Mathematics 1520

Subseries:
LOMI and Euler International
Mathematical Institute, St. Petersburg

Yu. G. Borisovich Yu. E. Gliklikh (Eds.)

Global Analysis – Studies and Applications V

Springer-Verlag
Berlin Heidelberg New York
London Paris Tokyo
Hong Kong Barcelona
Budapest

Editors

Yuriĭ G. Borisovich
Yuriĭ E. Gliklikh
Department of Mathematics
Voronezh State University
394693, Voronezh, Russia

Consulting Editor

A. M. Vershik
Department of Mathematics and Mechanics
St. Petersburg State University
198904, Petrodvorets, St. Petersburg, Russia

The articles in this volume are translations of the articles that appeared originally in Russian in the books "Algebraicheskie voprosy analiza i topologii" and "Nelineinye operatory v global'- nom analize" published by Voronezh University Press in 1990 and 1991.

Mathematics Subject Classification (1991): 58-02, 47A56, 55N, 57T, 58B15, 58C06, 58C30, 58D05, 58F, 58G03, 58G17, 58G32, 58H10, 68U05, 81P20

ISBN 3-540-55583-8 Springer-Verlag Berlin Heidelberg New York
ISBN 0-387-55583-8 Springer-Verlag New York Berlin Heidelberg

Printed in Germany

Typesetting: Camera ready by author/editor
Printing and binding: Druckhaus Beltz, Hemsbach/Bergstr.
46/3140-543210 - Printed on acid-free paper

PREFACE

This Lecture Notes volume (a sequel to 1108, 1214, 1334 and 1453) continues the presentation to English speaking readers of the Voronezh University Press series "Novoe v global'nom analize" (New Developments in Global Analysis) which is based on the seminars on Global Analysis in Voronezh University and Annual Voronezh Winter Mathematical Schools. Here we publish the articles selected from two issues of the series entitled "Algebraic questions of Analysis and Topology" (1990) and "Nonlinear Operators in Global Analysis" (1991), so the articles are divided into two chapters whose scopes are centred at the above topics. Some papers (e.g. the paper by Yu.E.Gliklikh of the first chapter and the paper by B.Yu.Sternin and V.E.Shatalov of the second one) are revised and expanded specially for this edition. That is why we break the tradition and do not indicate the year of the original publication of the chapters in the Contents.

We are indebted to the members of the editorial board of the series (A.T.Fomenko, A.S.Mishchenko, S.P.Novikov, M.M.Postnikov, A.M.Vershik et al.) and to the sponsor of the Russian issue of 1991, Voronezh Construction and Technology Bureau of Radio Communications (Voronezh, ul. Elektrosignalnaya, 1) headed by V.Ya.Shadchnev.

Yu.G.Borisovich
Yu.E.Gliklikh

October, 1991

CONTENTS

CHAPTER 1

CHAPTER 2

STOCHASTIC ANALYSIS, GROUPS OF DIFFEOMORPHISMS AND LAGRANGIAN DESCRIPTION OF VISCOUS INCOMPRESSIBLE FLUID

Yu.E.Gliklikh
Department of Mathematics
Voronezh State University
394693, Voronezh, USSR

The aim of this paper is to describe some constructions of stochastic differential geometry in the form convenient for the specialists in global analysis and ordinary differential geometry not familiar with stochastics. The interest in this machinery is caused by its applications to mathematical physics. We consider Ito equations on Riemannian manifolds (finite and infinite dimensional) and study the relations of their solutions with the Nelson's (forward and backward) mean derivatives. In particular we correct some misunderstandings in [9].

I would like to express my thanks to David Elworthy for very useful discussions on the subject when I was visiting Warwick University.

1. Some preliminaries from stochastic analysis.

Let F be a separable Hilbert space, R^n be the n-dimensional Euclidean space, $t \in [0, \ell]$. Let $w(t)$ be a Wiener process in R^n defined on a probability space $(\Omega, \mathcal{F}, \mathbb{P})$ and adapted to the non-decreasing family of complete σ-algebras $\mathcal{B}_t$. Consider a stochastic function A(t), non-anticipating with respect to $\mathcal{B}_t$ such that for every $t \in [0, \ell]$ A(t) is a bounded linear operator $A(t) : R^n \longrightarrow F$. We shall use two types of stochastic integrals of A(t) over $w(t)$. The first of them is the Ito integral which can be defined (for $A(t)$ satisfying some additional conditions) as an integral of Riemann type as follows (see [3]). Consider a sequence of partitions q of $[0,\ell]$, $q=(0=t_0<t_1<\dots<$ $<t_q=\ell)$, then

$$\int_0^\ell A(t)dw(t) = \lim_{\text{diam } q\to\infty} \sum_{i=0}^{q-1} A(t_i)(w(t_{i+1}) - w(t_i)). \qquad (1)$$

(if the limit exists) is called the Ito integral. The anticipating integral (see e.g. [10]) is constructed as the following limit (if it exists)

$$\int_0^\ell A(t)d_*w(t) = \lim_{\text{diam } q\to\infty} \sum_{i=o}^{q-1} A(t_{i+1})(w(t_{i+1})-w(t_i)) \qquad (2)$$

Note that, generally speaking, (1) is not equal to (2). Obviously the above integrals are connected by the formula

$$\int_0^\ell A(t)d_*w(t) = \int_0^\ell dA(t)dw(t) + \int_0^\ell A(t)dw(t) \qquad (3)$$

where $\int_0^\ell dA(t)dw(t)$ is a second order integral (see e.g. [3,6,10])

$$\int_0^\ell dA(t)dw(t) = \lim_{\text{diam } q\to\infty} \sum_{i=0}^{q-1} (A(t_{i+1})-A(t_i))(w(t_{i+1})-w(t_i)) \qquad (4)$$

Consider a smooth vector field $a(t,X)$ on F and a smooth field of linear operators $A(t,X)$,

$$A(t,X) : R^n \longrightarrow F$$

where $t\in[0,\ell]$, $X\in F$. Ito stochastic differential equation is an equality of the form

$$\xi(t) = \xi_0 + \int_0^t a(t,\xi(t))dt + \int_0^t A(t,\xi(t))dw(t) \qquad (5)$$

where the first integral in the right-hand side is an ordinary integral. Equation (5) is usually written in a differential form

$$d\xi(t) = a(t,\xi(t))dt + A(t,\xi(t))dw(t). \qquad (6)$$

<u>Definition 1.</u> (see e.g. [11]. We say that Ito equation (5) has a

strong solution if for any realization of the Wiener process $w(t)$ in R^n (adapted to the corresponding $\mathcal{B}_t$) there exists a stochastic process $\xi(t)$ in F, which is non-anticipating with respect to $\mathcal{B}_t$, such that for every t a.s. (5) is satisfied.

Everywhere below we consider the realization of $w(t)$ as the coordinate process $w(t,\omega) = \omega(t)$ on the probability space $(\tilde{\Omega}, \tilde{\mathcal{F}}, \nu)$, where $\tilde{\Omega} = C^0([0,\ell], R^n)$ is the Banach space of continuous mappings $\omega : [0,\ell] \to R^n$, $\tilde{\mathcal{F}}$ is the σ-algebra generated by cylindrical sets and ν is the Wiener measure; here $\tilde{\mathcal{B}}_t$ is the σ-algebra generated by cylindrical sets with bases over $[0,t]$ and completed by all sets of zero measure ν.

Recall that if $\varphi : F \to F_1$ is a smooth mapping (e.g. a change of coordinates in F), then

$$d\varphi(\xi(t)) = \varphi' \circ a(t,\xi(t))dt + \frac{1}{2}\operatorname{tr}\varphi''(A(t,\xi(t)),A(t,\xi(t))dt + \varphi' \circ A(t,\xi(t))dwt(t) \tag{7}$$

where φ' and φ'' are the first and the second derivatives, $\operatorname{tr}\varphi''(A,A) \doteq \sum_{i=1}^{n} \varphi''(Ae_i, Ae_i)$ for an arbitrary orthonormal basis $e_1,\dots,e_n$ in R^n. (7) is a variant of Ito formula (see [11]).

Using formulas (3) and (4), we can express a solution $\xi(t)$ to (5) with smooth $A(t,X)$ in terms of the anticipating integral (2)

$$\xi(t) = \xi_0 + \int_0^t a(t,\xi(t))dt - \int_0^t \operatorname{tr}A'(A(t,\xi(t)))dt + \int_0^t A(t,\xi(t))d_*w(t) \tag{8}$$

(see e.g. [6,10]) where the bilinear operator $A'(t,X)(\cdot,\cdot) : F \times R^n \to F$ is the derivative of $A(t,X)$, $\operatorname{tr}A'(A(t,X)) = \sum_{i=1}^{n} A'(t,X)(A(t,X)e_i,e_i)$.

Note that the non-tensor terms $\operatorname{tr}\varphi''(A,A)$ and $\operatorname{tr}A'(A)$ appear in (7) and (8) as a consequence of the following property of $w(t)$: $\int_0^t (dw,dw) = \int_0^t \operatorname{tr}\Psi dt$ for a bilinear map $\Psi : R^n \times R^n \to F$ (see e.g. [2, 11]).

Any stochastic process $\xi(t)$ in F defines three families of σ-subalgebras in $\mathcal{F}$: "the past" $\mathcal{P}_t^\xi$, generated by the inverse images of Borel sets in F with respect to the mappings $\xi(s) : \Omega \to F$ for $s \le t$, "the future" $\mathcal{F}_t^\xi$, generated by the same method for $s \ge t$, and "the present" ("now") $\mathcal{N}_t^\xi$, generated analogously by $\xi(t)$ itself. These families are assumed to be completed with all sets of zero probability.

Note that for the above realization of $w(t)$ as the coordinate process we have $\mathcal{P}_t^w = \tilde{\mathcal{B}}_t$. It should be also mentioned that for a strong solution $\xi(t)$ to (5) $\mathcal{P}_t^{\xi} = \mathcal{P}_t^{w}$.

Denote by $E(\cdot\,|\,\mathcal{B})$ the conditional expectation with respect to σ-subalgebra $\mathcal{B}$, and let $E_t^{\xi}(\cdot) = E(\cdot\,|\,\mathcal{N}_t^{\xi})$. Following Nelson (see e.g. [13,14]) for a stochastic process $\xi(t)$ in F we define the "mean forward derivative" by

$$D\xi(t) = \lim_{\Delta t \to +0} E_t^{\xi}\left(\frac{\xi(t+\Delta t) - \xi(t)}{\Delta t} \right) \tag{9}$$

and the "mean backward derivative" by

$$D_*\xi(t) = \lim_{\Delta t \to +0} E_t^{\xi}\left(\frac{\xi(t) - \xi(t-\Delta t)}{\Delta t} \right) \tag{10}$$

where $\Delta t \to +0$ means that $\Delta t \to 0$ and $\Delta t > 0$. It should be pointed out that, generally speaking, $D\xi(t) \neq D_*\xi(t)$ (if $\xi(t)$ had smooth sample trajectories, these derivatives would coincide) and by definition $D\xi(t)$ and $D_*\xi(t)$ are vectors (i.e. under changes of coordinates they transform via ordinary tensor law).

It is obvious that for a solution $\xi(t)$ to (5) $D\xi(t) = a(t,\xi(t))$ (see [7]). Note that there exists a vector field $a_*(t,X)$ on F such that $D_*\xi(t) = a_*(t,\xi(t))$.

The vector $v = \frac{1}{2}(D\xi(t) + D_*\xi(t))$ is called the current velocity of $\xi(t)$; the vector $u = \frac{1}{2}(D\xi(t) - D_*\xi(t))$ is called the osmotic velocity of $\xi(t)$.

Let $F = R^n$ and consider in R^n equation (5) with $A = \sigma I$, where I is the identity operator, σ is a real constant. In this case the osmotic velocity u of the solution $\xi(t)$ to (5) is described as follows (see [13,14]). Let $\rho(t,x)$ be the probability density of $\xi(t)$ with respect to the Lebesgue measure on $[0,\ell] \times R^n$. Then $u(t,x) = \operatorname{grad} \log_e \sqrt{\rho(t,x)}$.

Let $Y(t,x)$ be a C^2- smooth vector field on R^n. Then for a solution $\xi(t)$ to (5) on R^n with $A = \sigma I$ we can define the "mean forward" and "mean backward derivatives" of Y by

$$DY(t,\xi(t)) = \lim_{\Delta t \to +0} E_t^{\xi}\left(\frac{Y(t+\Delta t, \xi(t+\Delta t)) - Y(t,\xi(t))}{\Delta t} \right),$$

$$D_*Y(t,\xi(t)) = \lim_{\Delta t \to +0} E_t^{\xi}\left(\frac{Y(t,\xi(t)) - Y(t-\Delta t, \xi(t-\Delta t))}{\Delta t} \right). \tag{11}$$

It is easy to obtain the formulas (see [13,14])

$$DY = \left(\frac{\partial}{\partial t} + a\cdot\nabla + \frac{\sigma^2}{2}\Delta\right)Y,$$

$$D_*Y = \left(\frac{\partial}{\partial t} + a_*\cdot\nabla - \frac{\sigma^2}{2}\Delta\right)Y, \qquad (12)$$

where $\nabla = \left(\frac{\partial}{\partial x^1}, \dots, \frac{\partial}{\partial x^n}\right)$, Δ is the Laplacian, the dot denotes the scalar product in R^n.

2. Some calculations.

It is obvious that $Dw(t) = 0$ for a Wiener process $w(t)$ in R^n because $w(t)$ is a martigale.

Lemma 1. For $t \in (0, \ell)$ the following equalities are valid:

$$D_*w = \lim_{\Delta t \to +0} E_t^w\left(\frac{w(t) - w(t-\Delta t)}{\Delta t}\right) = \frac{w(t)}{t};$$

$$D^w \frac{w(t)}{t} = \lim_{\Delta t \to +0} E_t^w\left(\frac{\frac{w(t+\Delta t)}{t+\Delta t} - \frac{w(t)}{t}}{\Delta t}\right) = -\frac{w(t)}{t^2};$$

$$D_*^w \frac{w(t)}{t} = \lim_{\Delta t \to +0} E_t^w\left(\frac{\frac{w(t)}{t} - \frac{w(t-\Delta t)}{t-\Delta t}}{\Delta t}\right) = 0.$$

Proof. $D\,w(t) = 0$ because $w(t)$ is a martingale. Thus $D_*w(t)_x = -2u^w(t,x)$, where $u^w(t,w(t))$ is the osmotic velocity of $w(t)$. According to [13,14] $u^w(t)_x = \mathrm{grad}\,\log_e\sqrt{\rho^w(t,x)}$ where the density $\rho^w(t,x)$ is well known (see e.g. [11]): $\rho^w(t,x) = (2\pi t)^{-n/2}\cdot \exp(-x^2/2t)$. Direct calculations show: $\mathrm{grad}\,\log_e\sqrt{\rho^w} = \frac{1}{2}\cdot\frac{x}{t}$, so $D_*^w w(t) = \frac{w(t)}{t}$. Then it is easy to see that

$$D^w \frac{w(t)}{t} = \frac{d}{dt}\left(\frac{1}{t}\right)\cdot w(t) + \frac{1}{t}D\,w(t) = -\frac{w(t)}{t^2};$$

and $D_*^w \frac{w(t)}{t} = \frac{d}{dt}\left(\frac{1}{t}\right)\cdot w(t) + \frac{1}{t}D_*w(t) = 0$, Q.E.D.

It should be noted that $\frac{w(t)}{t}$ is ill-defined at $t = 0$. But the following statement holds:

Lemma 2. The integral $\int_0^t \frac{w(\tau)}{\tau}d\tau$ exists a.e. for all $t \in [0, \ell]$.

Indeed, using the formula for $\rho^W(t,x)$ by standard calculations we can obtain $E(\int_0^t \|\frac{w(\tau)}{\tau}\| d\tau) < C\cdot\sqrt{t}$, where E denotes the expectation, the constant $C>0$ depends only on the dimension n. So, the result follows from the classical Chebyshev's inequality.

Let $\xi(t)$ be a solution to (5). Introduce ξ-backward derivative of $w(t)$ by the formula

$$D_*^{\xi} w(t) = \lim_{\Delta t\to +0} E_t^{\xi}\left(\frac{w(t)-w(t-\Delta t)}{\Delta t}\right).$$

Definition 2. The process $w_*^{\xi}(t) = -\int_0^t D_*^{\xi} w(\tau)d\tau + w(t)$ will be called the backward Wiener process with respect to $\xi(t)$.

We should emphasize that $w_*^{\xi}(t)$ is determined for a given solution $\xi(t)$; from lemma 1 it follows that $w_*^w(t) = -\int^t (w(\tau)/\tau)d\tau + w(t)$ but it is a complicated problem to calculate $w_*^{\xi}(t)$ for more general processes $\xi(t)$. It is obvious that $D_*^{\xi} w_*^{\xi}(t)=0$ and $D^{\xi} w_*(t) = -D_*^{\xi} w(t)$. Note that these equalities are not valid if D_*^{ξ} and D^{ξ} are replaced by D_* and D respectively. For example, it is shown by M.Yor et al. that $w_*^w(t)$ is a Wiener process with respect to $\mathcal{P}_t^{w_*^w}$.

Lemma 3. Let F, $a(t,X)$ and $A(t,X)$ be as in section 1 and $\xi(t)$ be a strong solution to (5), $t\in[0,\ell]$. Then (i) for $t\in[0,\ell)$ $D^{\xi}\int_0^t A(\tau,\xi(\tau))dw(\tau) = 0$. (ii) for $t\in(0,\ell]$ $D_*^{\xi}\int_0^t A(\tau,\xi(\tau))dw(\tau) =$ $= -\operatorname{tr}A'(A(t,\xi(t)) + A(t,\xi(t))\circ(D_*^{\xi}w(t))$.

Proof. The statement (i) is obvious because $\int_0^t A(\tau,\xi(\tau))dw(\tau)$ is a martingale with respect to $\mathcal{P}_t^{\xi}$.

Let us prove (ii). Using (8) and the properties of conditional expectation we obtain

$$D_*\int_0^t A(\tau,\xi(\tau))dw(\tau) = D_*\left(-\int_0^t \operatorname{tr}A'(A(\tau,\xi(\tau)))d\tau + \int_0^t A(\tau,\xi(\tau))d_*w(\tau)\right) =$$

$$= -\operatorname{tr}A'(A(t,\xi(t))) + \lim_{\Delta t\to +0} E_t^{\xi}\left(A(t,\xi(t))\circ\left(\frac{w(t)-w(t-\Delta t)}{\Delta t}\right)\right) =$$

$$= -\operatorname{tr}A'(A(t,\xi(t))) + A(t,\xi(t))\circ\lim_{\Delta t\to +0} E_t^{\xi}\left(\frac{w(t)-w(t-\Delta t)}{\Delta t}\right). \text{ Q.E.D.}$$

3. Stochastic equation on smooth manifold.

In what follows in this section we consider a smooth manifold M modelled on a separable Hilbert space F. Let us fix a certain connection H on M (i.e. the so called "horizontal" distribution on TM) and denote its exponential map by $\exp_m : T_mM \longrightarrow M$ (see e.g. [2, 7]).

Let $a(t,m)$ be a vector field on M, $A(m)$ be a field of linear operators $A(m) : R^n \longrightarrow T_mM$, $w(t)$ be a Wiener process in R^n. Consider a class of stochastic processes $\widetilde{(a(t,m), A(m))}$ in the tangent space T_mM which consists of solutions to stochastic differential equations

$$X(t+\tau) = \int_t^{t+\tau} \mathring{a}(s,X(s))ds + \int_t^{t+\tau} \mathring{A}(s,X(s))dw(s) \qquad (13)$$

in T_mM, where $\mathring{a}(s,X)$ and $\mathring{A}(s,X)$ are Lipschitz, vanish outside a certain neighbourhood of the origin in T_mM, and are such that $\mathring{a}(\cdot,0) = a(t,m)$ and $\mathring{A}(\cdot,0) = A(m)$. Note that the solutions of (13) are strong.

The expression (see [2])

$$d\xi(t) = \exp_{\xi(t)}\widetilde{(a(t,\xi(t)), A(\xi(t)))} \qquad (14)$$

is called the Ito equation in the form of Ya.I.Belopol'skaya - Yu.L. Daletskiĭ. It means that the process $\xi(t+\tau)$ for $\tau > 0$ belongs to the class $\exp_{\xi(t)}\widetilde{(a(t,\xi(t)), A(\xi(t)))}$ until it leaves a certain neighbourhood of $\xi(t)$.

Fix a certain chart (U,φ) on M and let $\Gamma_m(X,Y)$ be a local connector of the connection H in this chart (see e.g. [2, 5, 7]). The local "coordinate" description of (14) in (U,φ) has the form

$$d\xi(t) = a(t,\xi(t)) - \frac{1}{2}\,\mathrm{tr}\,\Gamma_{\xi(t)}(A(t,\xi(t)),A(t,\xi(t)))dt + A(t,\xi(t))dw(t) \qquad (15)$$

(see e.g. [2,7]). Comparing the rule of transformation for $\Gamma_m(X,Y)$ under the change of coordinates (see e.g. [5]) with (7) one can easily prove that (15) is a covariant formula.

Note that equation (14) is compatible with mappings of manifolds. Let $f : M \longrightarrow N$ be a C^2- mapping and let there be a connection on the manifold N with the exponential map $\exp^N$ such that $f(\exp\circ X) = \exp(Tf\circ X)$ for each $X \in TM$. One can easily show that for $\xi(t)$ on M satisfying (15) the process $f(\xi)$ on N satisfies the equation

$$df(\xi(t)) = \exp^N_{f(\xi(t))}(Tf\circ a(t,\xi(t)),\widetilde{\ } Tf\circ A(\xi(t))).$$

A more detailed description and justification of this construction can be found in [2].

According to formulas (9) and (10) one can correctly define mean derivatives $D\xi(t)$ and $D_*\xi(t)$ for a solution $\xi(t)$ of an Ito equation on M. We recall that there exist vector fields such that the mean derivatives are presented as a superposition of those vector fields and $\xi(t)$. We shall describe these vector fields for solutions to (14) below (cf. section 1).

The construction of mean derivatives of vector fields should be generalized. Denote by $K : TTM \longrightarrow TM$ the connection map (connector) of the connection H(see e.g. [5,7]). Let $Y(t,m)$ be a C^2-smooth vector field on M, $\xi(t)$ be a solution to an Ito equation on M. By analogy with the constructions of ordinary differential geometry we define the covariant mean forward $\mathbb{D}Y$ and mean backward $\mathbb{D}_*Y$ derivatives by the formulas

$$\mathbb{D}Y(t,\xi(t)) = K\circ \lim_{\Delta t\to +0} E^{\xi}_t\left(\frac{Y(t+\Delta t,\xi(t+\Delta t)) - Y(t,\xi(t))}{\Delta t}\right),$$

$$\mathbb{D}_*Y(t,\xi(t)) = K\circ \lim_{\Delta t\to +0} E^{\xi}_t\left(\frac{Y(t,\xi(t)) - Y(t-\Delta t,\ (t-\Delta t))}{\Delta t}\right). \qquad (16)$$

An equivalent definition of $\mathbb{D}Y$ and $\mathbb{D}_*Y$ in terms of parallel displacement along $\xi(t)$ is described e.g. in [14]) .

<u>Theorem 1.</u> Let $\xi(t)$ be a solution to (14). Then $D\xi(t)$ exists and $D\xi(t) = a(t,\xi(t))$.

<u>Proof.</u> Fix a point $m\in M$ and consider the vector $D\xi(t)_m \in T_mM$. In the normal chart of the point m (where $\Gamma_m(X,Y) = 0$) by the methods of section 1 it is easy to calculate that $D\xi(t)_m = a(t,m)$. Since there are vectors of T_mM in both sides of the last equality, it is valid in any chart. Q.E.D.

Now for the given fields $a(t,m)$ and $A(t,m)$ we construct an equation of Ito type on M such that for its solution $\xi(t)$ $D_*\xi(t) = a(t,\xi(t))$. Denote by ∇A the covariant derivative of A with respect to the connection H, which is the field of bilinear operators $\nabla A(t,m)(\cdot,\cdot) : T_mM\times R^n \longrightarrow T_mM$. Consider the field $\nabla A(t,m)(A(t,m)\cdot,\cdot): R^n\times R^n \longrightarrow T_mM$ and the vector field $tr\nabla A(A)(t,m) = \sum_{i=1}^{n}\nabla A(t,m)(Ae_i,e_i)$ where $e_1,\ldots,e_n$ is an orthonormal frame in R^n.

Let us introduce the following equation on M

$$d\xi(t) = \exp_{\xi(t)}\widetilde{(a(t,\xi(t)) + \mathrm{tr}\nabla A(A)(t,\xi(t)) - A(t,\xi(t))\circ D_*^{\xi} w(t), A(t,\xi(t)))} \qquad (17)$$

where $\widetilde{(a(t,\xi(t)) + \mathrm{tr}\nabla A(A)(t,\xi(t)) - A(t,\xi(t))\circ D_*^{\xi} w(t), A(t,\xi(t)))}$ denotes the class of stochastic processes in $T_{\xi}M$ which consists of the solutions to equations

$$X(\tau) = \int_0^{\tau} \mathring{a}(s,X(s))ds + \int_0^{\tau} \mathrm{tr}\mathring{A}'(\mathring{A}(s,X(s)))ds - \int_0^{\tau} \mathring{A}(s,X(s))\circ D_*^{\xi} w(s)ds + \int_0^{\tau} \mathring{A}(s,X(s))dw(s) \qquad (18)$$

where $\mathring{a}(s,x)$ and $\mathring{A}(s,X)$ are the same as in (13) with ad additional condition that $\mathring{A}(s,X)$ is smooth, $\mathring{A}' : T_mM \times R^n \longrightarrow T_mM$ is the ordinary derivative of $\mathring{A}$ as in (8).

Let us express (17) in a local chart by analogy with formula (15) for equation (14). It is clear that under the action of $\exp_m$ on T_mM the vectors, tangent to T_mM, are transformed by the action of the derivative $\exp_m'$. We use the formula $(\exp_m'\mathring{A})' = \exp_m''(\cdot\,,\mathring{A}\cdot) + \exp_m'\mathring{A}'(\cdot,\cdot)$ as well as the fact that the derivatives of $\exp_m$ at the zero of the space T_mM are expressed in the form: $\exp_m'(0) = I$, $\exp_m''(0)(\cdot,\cdot) = -\Gamma_m(\cdot,\cdot)$ (see e.g. [2]). Thus, in particular, under the action of $\exp_m$ the vector $\mathrm{tr}\mathring{A}'(s,0)(\mathring{A}\cdot,\cdot)$ turns into the vector $\exp_m'(0)\mathrm{tr}\mathring{A}'(s,0)(\mathring{A}\cdot,\cdot) = \mathrm{tr}A'(t,m)(A\cdot,\cdot) + \mathrm{tr}\Gamma_m(A\cdot,A\cdot) = \mathrm{tr}\nabla A(A)(t,m)$ (this explains the notations in (17)), and $\mathring{A}(s,0)$ turns into $A(t,m)$. So we obtain the expression of (17) in the local chart in the form:

$$d\xi(t) = a(t,\xi(t))dt + \mathrm{tr}\nabla A(A)(t,\xi(t))dt - A(t,\xi(t))\circ(D_*^{\xi}w(t))dt - \frac{1}{2}\,\mathrm{tr}\,\Gamma_{\xi(t)}(A,A)dt + A(t,\xi(t))dw(t). \qquad (19)$$

<u>Theorem 2.</u> Let $\xi(t)$, $\xi(0) = m_o$, be a strong solution to (17) for t belonging to some interval $[0,\ell]$. Then for $t\in(0,\ell]$ $D_*\xi(t)$ exists and $D_*\xi(t) = a(t,\xi(t))$.

<u>Proof.</u> For the sake of simplicity consider $\xi(t)$ in the normal chart of the point m_o. Let $\xi(\tau) = \exp_{m_o} X(\tau)$ where

$$X(\tau) = \int_0^{\tau} \hat{a}(s,X(s))ds + \int_0^{\tau} \mathrm{tr}\hat{A}'(\hat{A})(s,X(s))ds - \int_0^{\tau} \hat{A}(s,X(s))\circ(D_* w(s))\,ds + \int_0^{\tau} A(s,X(s))dw(s)$$

is a process in $T_{m_o}M$ which exists by definition of solution to equation (17). Since $D_*\xi(t)$ is a vector, it is sufficient to prove the equality $D_*X(\tau) = a(\tau,X(\tau))$ which follows from lemma 3(ii). Q.E.D.

Let us replace $\mathrm{tr}\nabla A(A)(t,m)$ in (18) by $\mathrm{tr}A'(t,m)(A\cdot,\cdot)+$ $+\mathrm{tr}\,\Gamma_m(A\cdot,A\cdot)$ (see above) and then according to formula (8) and definition 2

$$\int_0^t \mathrm{tr}A'(s,\xi(s))(A\cdot,A\cdot)ds-\int_0^t A(s,\xi(s))\circ(D_*^{\xi}w(s))ds+\int_0^t A(s,\xi(s))dw(s)$$

by $\int_0^t A(s,\xi(s))))d_*w_*(s)$. So (18) takes the form

$$\xi(t)-\xi(t-\tau)=\int_{t-\tau}^t a(s,\xi(s)))ds+\tfrac{1}{2}\int_{t-\tau}^t \Gamma_{\xi(s)}(A,A)ds+\int_{t-\tau}^t A(s,\xi(s))d_*w_*(s) \tag{20}$$

Consider a class of stochastic processes $(a(t,m)\overset{*}{,}A(t,m))$ in the tangent space T_mM which consists of solutions to (backward) stochastic equations

$$X(t)-X(t-\tau)=\int_{t-\tau}^t \hat{a}(s,X(s))ds+\int_{t-\tau}^t \hat{A}(s,X(s))d_*w_*(s)$$

where $\hat{a}$ and $\hat{A}$ are analogous to the corresponding terms in (13) with an additional assumption on $\hat{A}$ to be smooth. The expression

$$d_*\xi(t)=\exp_{\xi(t)}(a(t,\xi(t))\overset{*}{,}A(t,\xi(t))) \tag{21}$$

means that the process $\xi(t-\tau)$ for any $\tau>0$ belongs to the class $\exp_{\xi(t)}(a(t,\xi(t))\overset{*}{,}A(t,\xi(t)))$ until it leaves a certain neighbourhood of $\xi(t)$. It is easy to see that (20) is the coordinate description of (21) and thus (21) is in some sense equivalent to (17).

A certain different description of forward and backward stochastic equations on manifolds, based on formulas in local coordinates, close to (15) and (20), is given e.g. in [14].

4. Parallel displacement along stochastic processes and Ito development

In this section we consider only finite dimensional Riemannian manifolds of dimension n and Wiener processes determined in R^n.

Suppose in addition that the Riemannian manifold M is complete. Fix a point $m_o\in M$. Let a stochastic process $a(t)$ be given in $T_{m_o}M$ which a.e. has continuous trajectories and which is non-anticipating with respect to a certain non-decreasing family $\mathcal{B}_t$ of complete σ-algebras, $t \in [0,\ell]$. Let a Wiener process w(t) be given in $T_{m_o}M$; it is defined on the same probability space as a(t) is, and is adapted to $\mathcal{B}_t$.

Fix $k \in \mathbb{Z}$ and consider a subdivision $0=t_o^{(k)} < t_1^{(k)} < \ldots < t_{N(k)}^{(k)} = \ell$ of $[0,\ell]$ into the intervals of length 2^{-k}. Define a process $\xi^{(k)}(t)$ on M as follows. Let the geodesics start at the point m_o with the initial direction $a(0)t_1^k + w(t_1^{(k)})$ for t $[0,t_1^{(k)}]$. Translate the vectors a(t) and w(t) by Riemannian parallel displacement along the above geodesics to the point t_1, and then let the geodesics start at those points with the initial directions $a(t_1^{(k)})(t_2^{(k)} - t_1^{(k)}) + (w(t_2^{(k)}) - w(t_1^{(k)}))$ up to the moment $t_2^{(k)}$, then analogously up to the moment $t_3^{(k)}$, etc.

It is evident that the process $\xi^{(k)}(t)$ has piece-wise geodesic sample trajectories. Under some conditions (existence of a uniform, in a certain sense, atlas on M, boundedness of the process a(t), etc.) $\xi^{(k)}(t)$ is well-defined on $[0,\ell]$ and converges a.e. uniformly on $[0,\ell]$ to a certain process (see [7]). According to [7] this process will be called Ito development $R_I\xi(t)$ of the stochastic process $z(t) = \int_0^t a(\tau)d\tau + w(t)$ in $T_{m_o}M$.

Now we describe some results on the existence of Ito developments, necessary for further use.

Definition 3. If the development $R_I w(t)$ of Wiener process w(t) in $T_m M$ is well-defined for $t \in [0,\infty)$ and some point $m_o \in M$, the Riemannian manifold M is called stochastically complete.

Ordinary completeness of Riemannian manifold is necessary for its stochastic completeness. The following sufficient condition follows from the papaer by Yau (see [6]).

Theorem (Yau). Let the Ricci curvature of a complete Riemannian manifold M be bounded below by some $\alpha > -\infty$. Then M is stochastically complete.

From the construction of Ito development it is easy to see that R_I is an extension of the inverse classical Cartan development from the set of piece-wise smooth curve (beginning at zero) onto almost all continuous curves in $T_{m_o}M$ with respect to the measure μ_z corresponding to z(t).

Theorem 3. Let the process a(t) satisfy the condition

$$\mathbb{P}\left(\int_0^\ell \|a(\tau)\|^2 d\tau < \infty\right) = 1 \qquad (22)$$

and the manifold M be stochastically complete. Then for $z(t) = \int_0^t a(\tau)d\tau + w(t)$ $R_I z(t)$ is well-defined a.e. when $t \in [0, \ell]$.

<u>Proof.</u> It follows from (22) that the measure μ_z is absolutely continuous with respect to Wiener measure (see e.g. [12]). Let $\xi^{(k)}(t)$ be a sequence of processes with piece-wise geodesic trajectories, which a.e. uniformly on $[0, \ell]$ with respect to Wiener measure converges to $R_I\, w(t)$. Then the same sequence a.e. with respect to μ_z uniformly converges to the development of the coordinate process of the probability space $(\tilde{\Omega}, \tilde{\mathcal{F}}, \mu_z)$, i.e. to $R_I z(t)$; $\tilde{\Omega}$ is the space of continuous curves $\omega: [0, \ell] \longrightarrow T_{m_0} M$, $\tilde{\mathcal{F}}$ is generated by cylindrical sets . Q.E.D.

The Riemannian parallel displacement along the development $R_I z(t)$ is well-defined as the a.e. uniform limit of parallel displacements along $\xi^{(k)}(t)$ in the classical sense (recall that the trajectories of $\xi^{(k)}(t)$ are piece-wise smooth). Denote by $\Gamma_{t,s}$ the operator of parallel displacement from $\xi(t)$ along $\xi(\cdot)$ to $\xi(s)$.

<u>Theorem 4</u> [7]. The development $\xi(t) = R_I\, z(t)$ of the process $z(t) = \int_0^t a(\tau)d\tau + w(t)$ satisfies the following Ito equation in the form of Ya.I.Belopol'skaya - Yu.L.Daletskiĭ

$$d\xi(t) = \exp_{\xi(t)}(\Gamma_{0,t}a(t)dt + \Gamma_{0,t}dw(t))$$

From the definition it follows that the parallel displacement along $\xi(t)$ conserves the scalar products (and norms, consequently) of vectors because the classical Riemannian parallel displacement has this property.

5. Constructions on groups of diffeomorphisms and applications to hydrodynamics.

Let $\mathbb{T}^n$ be a flat n-dimensional torus, i.e. the Riemannian metric $\langle\, ,\rangle$ on T^n is obtained from the Euclidean metric in R^n after the factorization with respect to the integral lattice. First we briefly describe the necessary geometrical objects on the manifolds of diffeomorphisms of T^n . These constructions for the general compact oriented Riemannian manifold M can be found in e.g. [4,7].

Let $s > \frac{n}{2} + 1$. Denote by $D^s(\mathbb{T}^n)$ the set of all C^1-diffeomorphisms of M belonging to the Sobolev class H^s. Recall that when $s > \frac{n}{2} + K$, $K > 0$, the space of Sobolev maps H^s is continuously imbedded in the space of C^K maps.

It is possible to define the structure of C^∞-smooth Hilbert manifold on $D^s(\mathbb{T}^n)$ (see [4,7]). Here the tangent space $T_e D^s(\mathbb{T}^n)$ at the point $e = id$ is a separable Hilbert space $H^s(T\mathbb{T}^n)$ of all H^s-vector fields on M (the scalar product in $H^s(T\mathbb{T}^n)$ is naturally generated by the Riemannian metric $\langle \, , \rangle$ on $\mathbb{T}^n$) and the tangent space $T_\eta D^s(\mathbb{T}^n)$ at the point $\eta \in D^s(\mathbb{T}^n)$ consists of all mapping $Y : \mathbb{T}^n \longrightarrow T\mathbb{T}^n$ such that $\pi Y = \eta$, where $\pi : TM \longrightarrow M$ is a natural projection (i.e. $Y = X \circ \eta$ where $X \in H^s(T\mathbb{T}^n) = T_e D^s(\mathbb{T}^n)$).

$D^s(\mathbb{T}^n)$ is a topological group, where the superposition $\circ$ is involved as a multiplication. For each $\eta \in D^s(\mathbb{T}^n)$ the right-hand translation $R_\eta : D^s(\mathbb{T}^n) \longrightarrow D^s(\mathbb{T}^n)$, $R_\eta \theta = \theta \circ \eta$, is a C^∞-smooth mapping with the derivative $TR_\eta X = X \circ \eta$, $X \in TD^s(\mathbb{T}^n)$. The left-hand translation $L_\eta \theta = \eta \circ \theta$ is only continuous on $D^s(\mathbb{T}^n)$, but when η is of the class H^{s+1}, L_η is C^1-smooth with the derivative $TL_\eta X = T\eta \circ X$, $X \in TD^s(\mathbb{T}^n)$.

Obviously one can define right-invariant vector fields on $D^s(\mathbb{T}^n)$. Let $\bar{X}$ be such a field and X be a vector of this field belonging to $T_e D^s(\mathbb{T}^n)$. The following property of $\bar{X}$ is very important for us: $\bar{X}$ is C^k - smooth on $D^s(\mathbb{T}^n)$ iff the vector field X on M belongs to the class H^{s+k}, where $k = 1, 2, \ldots, \infty$, $H^\infty = C^\infty$. Any right-invariant vector field $\bar{X}$ (C^1-smooth in the general case and continuous, when $s > \frac{1}{2}n + 2$) has the flow on $D^s(\mathbb{T}^n)$. The integral curve $\eta(t)_e$ beginning at e is the flow of X on $\mathbb{T}^n$; $\eta(t)_\varphi = \eta(t)_e \circ \varphi$, $\varphi \in D^s(\mathbb{T}^n)$.

Denote by $D^s_\mu(\mathbb{T}^n)$ the submanifold in $D^s(\mathbb{T}^n)$ consisting of all diffeomorphisms which preserve the form of Riemannian volume on M. $D^s_\mu(\mathbb{T}^n)$ is also a subgroup in $D^s(\mathbb{T}^n)$. $T_e D^s_\mu(\mathbb{T}^n)$ is the space of all zero-divergence H^s vector fields on M, $T_\eta D^s_\mu(\mathbb{T}^n) = \{ Y = X \circ \eta \mid X \in T_e D^s_\mu(\mathbb{T}^n), \eta \in D^s_\mu(\mathbb{T}^n)\}$. All the properties of right(left)-hand translations, right-invariant vector fields etc. mentioned for $D^s(M)$ are valid for $D^s_\mu(\mathbb{T}^n)$.

Let $\eta \in D^s(\mathbb{T}^n)$, $X, Y \in T_\eta D^s(\mathbb{T}^n)$. Determine the scalar product $(\, ,)_\eta$ in $T_\eta D^s(\mathbb{T}^n)$ by the formula

$$(X, Y)_\eta = \int_M \langle X(m), Y(m) \rangle_{\eta(m)} \mu(dm) \tag{23}$$

where $\mu(dm)$ is the form of Riemanniam volume. Using (23) for all $\eta \in D^s(M)$ we define the Riemannian metric on $D^s(M)$. Obviously this metric introduces the topology of the functional space $L_2 = H^0$ in the tangent spaces, which is weaker than the initial topology H^s.

The restriction of (23) to $TD^s_\mu(\mathbb{T}^n)$ is weakly Riemannian metric on $D^s_\mu(\mathbb{T}^n)$ which is evidently right-invariant.

Consider the connector $K : TT\mathbb{T}^n \longrightarrow T\mathbb{T}^n$ of the Levi-Civita connection of the metric $\langle \, , \rangle$ (see e.g. [5,7]). Recall that the covariant derivative $\nabla_a b$ of the Levi-Civita connection for vector fields a and b on $\mathbb{T}^n$ is defined by the formula $\nabla_a b = K \circ Tb(a)$.

For vector fields X,Y on $D^s(\mathbb{T}^n)$ define the covariant derivative $\overline{\nabla}_X Y$ by the formula

$$\overline{\nabla}_X Y = K \circ TY(X). \tag{24}$$

One can easily see that at each $\eta \in D^s(M)$ $TY(X)_\eta$ is a mapping of $\mathbb{T}^n$ into $TT\mathbb{T}^n$, so (20) defines $\overline{\nabla}_X Y$ correctly. It is shown in [4] that $\overline{\nabla}$ is covariant derivative of the Levi-Civita connection of the metric (,) on $D^s(\mathbb{T}^n)$. The geodesic pulverization $\bar{Z}$ of this connection is described as follows:

$$\bar{Z}(X) = Z \circ X \tag{25}$$

for $X \in TD^s(\mathbb{T}^n)$, where Z is the geodesic pulverization of the Levi-Civita connection on $\mathbb{T}^n$ (i.e. the vector field on $T\mathbb{T}^n$). One can easily obtain from (25) the following statement: $\bar{Z}$ is $D^s(\mathbb{T}^n)$-right-invariant and C^∞-smooth on $TD^s(\mathbb{T}^n)$.

Denote by $P_e : T_eD^s(\mathbb{T}^n) \longrightarrow T_eD^s_\mu(\mathbb{T}^n)$ the $(,)_e$-orthogonal projection of vector fields onto their zero-divergence components, so that all gradients are projected by P_e into zero (see e.g. [4]). Consider the mapping $P : TD^s(\mathbb{T}^n)|_{D^s_\mu(\mathbb{T}^n)} \longrightarrow TD^s_\mu(\mathbb{T}^n)$ determined for each $\eta \in D^s_\mu(\mathbb{T}^n)$ by the formula

$$P_\eta = TR_\eta \circ P_e \circ TR_\eta^{-1}.$$

It is obvious that P is $D^s_\mu(\mathbb{T}^n)$-right-invariant. There is an important and rather complicated result (see [4]): P is a C^∞-smooth mapping. Note the important property of P_e: for every $Y \in T_eD^s(\mathbb{T}^n)$ we have

$$P_e(Y) = Y + \text{grad } p \tag{26}$$

where p is a certain H^{s+1}-function on $\mathbb{T}^n$ unique to within the constants.

According to the standard construction of differential geometry now we may define the connector $\tilde{K}$ and the covariant derivative $\tilde{\nabla}$ of the Levi-Civita connection on $D^s_\mu(\mathbb{T}^n)$ by the formulas

$$\tilde{K} = P \circ K, \tag{27}$$

$$\widetilde{\nabla}_X Y = P \circ \overline{\nabla}_X Y = \widetilde{K} \circ TY(X), \qquad (28)$$

where X,Y are vector fields on $D^s_\mu(\mathbb{T}^n)$.

The geodesic pulverization S of this connection is a vector field on $TD^s_\mu(\mathbb{T}^n)$ of the form

$$S = TP \circ \overline{Z} \qquad (29)$$

It evidently follows from (25) and (29) that S is D^s_μ -right invariant and C^∞-smooth on $TD^s_\mu(\mathbb{T}^n)$. Denote by $\widetilde{\exp}$ the corresponding exponential map of a neighbourhood of the zero section in $TD^s_\mu(\mathbb{T}^n)$ onto $D^s_\mu(\mathbb{T}^n)$; $\widetilde{\exp}$ is $D^s_\mu(\mathbb{T}^n)$-right-invariant, C^∞-smooth and covers some neighbourhood of each point in $D^s_\mu(\mathbb{T}^n)$ (see [4,7]).

It is well-known that all tangent spaces to $\mathbb{T}^n$ are naturally isomorphic to R^n. Denote by $A(m) : R^n \longrightarrow T_m\mathbb{T}^n$, $m \in \mathbb{T}^n$, this natural isomorphism. Thus the field A of linear isomorphisms of R^n onto tangent spaces to $\mathbb{T}^n$ is constructed. Obviously for each given $y \in R^n$ the vector field $A\circ y : \mathbb{T}^n \longrightarrow T\mathbb{T}^n$ on $\mathbb{T}^n$ is constant (i.e. one may imagine that the same vector y is applied at every point of $\mathbb{T}^n$) and consequently $A\circ y$ is a C^∞-smooth zero-divergent vector field. Moreover, the constant vector field $A\circ y$ is harmonic because evidently $d(A\circ y) = 0$.

Thus the field A may be considered as a linear operator $\widetilde{A}(e) : R^n \longrightarrow T_eD^s_\mu(\mathbb{T}^n)$. For $\eta \in D^s_\mu(\mathbb{T}^n)$ denote by $\widetilde{A}(\eta) : R^n \longrightarrow T_\eta D^s_\mu(\mathbb{T}^n)$ the operator determined by the formula $\widetilde{A}(\eta)\circ y = [\widetilde{A}(e)\circ y]\circ\eta = [A\circ y]\circ\eta$. So the field $\widetilde{A}$ of linear operators mapping R^n into tangent spaces to $D^s_\mu(\mathbb{T}^n)$ is constructed. Obviously $\widetilde{A}$ is right-invariant. Since the field A on $\mathbb{T}^n$ is C^∞-smooth, the right-invariant field $\widetilde{A}$ on $D^s_\mu(\mathbb{T}^n)$ is C^∞-smooth.

The construction of the field $\widetilde{A}$ is a variant of the general construction of [6].

For each $X \in TD^s_\mu(\mathbb{T}^n)$ consider the Levi-Civita connection $\widetilde{H}_X$ at X. Define on $TD^s_\mu(\mathbb{T}^n)$ the field of operators $\widetilde{A}^T(X) = T\pi^{-1}\widetilde{A}(\pi X)|_{\widetilde{H}_X} : R^n \longrightarrow T_X TD^s_\mu(\mathbb{T}^n)$.

Using the connection $\widetilde{H}$ we can construct the connection $\widetilde{H}^T$ on the manifold $TD^s_\mu(\mathbb{T}^n)$ (i.e. a special horizontal distribution on $TTD^s_\mu(\mathbb{T}^n)$) as it is described in [2,5]. Denote by $\widetilde{\exp}^T : TTD^s_\mu(\mathbb{T}^n) \longrightarrow TD^s_\mu(\mathbb{T}^n)$ the exponential map of this connection (of course, $\widetilde{\exp}^T$ is defined only on a neighbourhood of zero section in $TTD^s_\mu(\mathbb{T}^n)$). This map has the following important property: for any $Y \in TTD^s_\mu(\mathbb{T}^n)$ $\pi\,\widetilde{\exp}^T Y = \widetilde{\exp}\, T\pi Y$ (see [2,5]).

Lemma 4. (i) $\widetilde{\nabla}\widetilde{A} = 0$; (ii) $\widetilde{\nabla}^T\widetilde{A}^T = 0$, where $\widetilde{\nabla}$ is the covariant derivative (24) on $D^s_\mu(\mathbb{T}^n)$, $\widetilde{\nabla}^T$ is the covariant derivative of the connection $\widetilde{H}^T$ on $TD^s_\mu(\mathbb{T}^n)$.

Proof. According to (24) $\widetilde{\nabla}\widetilde{A} = P\circ K\circ T\widetilde{A}$. By construction $K\circ T\widetilde{A}$ is the right-invariant field of bilinear operators obtained by right-hand translations of the bilinear operator $K\circ TA$ in $T_eD^s_\mu(\mathbb{T}^n)$. Since $A(m)$ has constant coordinates with respect to the natural coordinate system on $\mathbb{T}^n$ and K is the connector of Euclidean connection on $\mathbb{T}^n$, obviously $K\circ TA = 0$ which proves (i).

Since $\widetilde{\nabla}^T\widetilde{A}^T$ is a $D^s_\mu(\mathbb{T}^n)$-right-invariant field on $TD^s_\mu(\mathbb{T}^n)$, it is sufficient to prove (ii) at the points $(e,Y)\in T_eD^s_\mu(\mathbb{T}^n)$. Consider a chart on $D^s_\mu(\mathbb{T}^n)$ including e. Using the corresponding coordinate systems over this chart we obtain $\widetilde{A}^T(e,Y)(x)=(\widetilde{A}(e)\circ x,-\Gamma_e(Y,\widetilde{A}(e)\circ x)$,

$$\widetilde{\nabla}^T\widetilde{A}^T(e,Y)\circ(x,\mathring{Y})=\widetilde{A}^{T\prime}(e,Y)(x,\mathring{Y})+\Gamma^T_{(e,Y)}(\widetilde{A}^T(e,Y)\circ x,\mathring{Y})$$

where $x\in R^n$, $\mathring{Y} = (Y_1, Y_2)\in T_{(e,Y)}TD^s_\mu(\mathbb{T}^n)$, $\widetilde{A}^{T\prime}$ is the derivative of $\widetilde{A}^T$ calculated in the coordinates mentioned above, $\Gamma^T_{(e,Y)}$ is the local connector of $\widetilde{H}^T$ which is expressed in terms of the local connector Γ_e of $\widetilde{H}$ by the formula: $\Gamma^T_{(e,Y)}(Z_1,Z_2,Y_1,Y_2) = (\Gamma_e(Z_1,Y_1)$, $d\Gamma_e(Y,Z_1,Y_1) + \Gamma_e(Z_2,Y_1) + \Gamma_e(Z_1,Y_2))$ (see [2,5]). For the sake of simplicity suppose that the above chart on $D^s_\mu(\mathbb{T}^n)$ is the normal chart of $\widetilde{\exp}$ at e, i.e. $\Gamma_e = 0$ and, consequently, $\widetilde{A}'(e) = \widetilde{\nabla}A(e) = 0$. Then it is easy to calculate that $\widetilde{\nabla}^T\widetilde{A}^T(e,Y)\circ(x,\mathring{Y}) = (0,-d\Gamma_e(Y,\widetilde{A}(e)\circ x,Y_1))+ (0,d\Gamma_e(Y,\widetilde{A}(e)\circ x, Y_1)) = 0$. Q.E.D.

Fix a real constant $\sigma > 0$. It is obvious that for a given vector field on $D^s_\mu(\mathbb{T}^n)$ (on $TD^s_\mu(\mathbb{T}^n)$) we may consider the stochastic equations of type (14) and (17) involving the exponential map $\widetilde{\exp}$ and the operator field $\sigma\widetilde{A}$ (the exponential map $\widetilde{\exp}^T$ and the operator field $\widetilde{A}^T$ respectively) as well as a Wiener process $w(t)$ in R^n. Lemma 4 means that the relations between these equations of (14) and (17) types are rather simple.

In the rest of the section we try to describe the motion of perfect incompressible fluid on $\mathbb{T}^n$ by analogy with the modern Lagrangian description of perfect incompressible fluid [1,4]. We show that the stochastic differential geometry on $D^s_\mu(\mathbb{T}^n)$ plays the same basic role for the viscous fluid as the ordinary differential geometry on $D^s_\mu(\mathbb{T}^n)$ for perfect fluid. The first version of this approach was announced in [8].

Let $F\in T_eD^s_\mu(\mathbb{T}^n)$, $\widetilde{F}$ be the corresponding right-invariant vector field on $D^s_\mu(\mathbb{T}^n)$, $\widetilde{F}^\ell$ be its natural vertical lift on $TD^s_\mu(\mathbb{T}^n)$. Let $\widetilde{\xi}(t)$ be a strong solution of Ito equation on $D^s_\mu(\mathbb{T}^n)$ of the form

$$d\tilde{\xi}(t) = \widetilde{\exp}_{\tilde{\xi}(t)}(\widetilde{a(t,\tilde{\xi}(t)), \sigma\tilde{A}(\tilde{\xi}(t))}) \tag{30}$$

for some right-invariant vector field $\tilde{a}(t,g)$ ($a(t,m)$ is the corresponding vector field on $\mathbb{T}^n$) and there exists a right-invariant vector-field $\tilde{u}(t,g)$ such that

$$D_*\tilde{\xi}(t) = u(t,\tilde{\xi}(t)) \tag{31}$$

and for the process $\tilde{u}(t,\tilde{\xi}(t))$ in $TD^s_\mu(\mathbb{T}^n)$

$$D_*\tilde{u}(t,\tilde{\xi}(t)) = S(\tilde{u}(t,\tilde{\xi}(t)) + \tilde{F}^\ell(t,\tilde{u}(t,\tilde{\xi}(t)) \tag{32}$$

where S is the geodesic pulverization (29). Using the connector $\tilde{K}$(27) define $\tilde{\mathbb{D}}_*D_*\tilde{\xi}(t)$ according to (16). From (32) it follows that

$$\tilde{\mathbb{D}}_*D_*\tilde{\xi}(t) = F(t,\tilde{\xi}(t)). \tag{33}$$

Thus (33) is a stochastic analogue of the classical Newton's law of dynamics and in the case of F=0 the process $\xi(t)$ becomes a certain stochastic analogue of a geodesic curve on $D^s_\mu(\mathbb{T}^n)$.

Denote by $u(t)$ the vector in $T_eD^s_\mu(\mathbb{T}^n)$ (i.e. the divergence-free H^s-vector field on $\mathbb{T}^n$) corresponding to $\tilde{u}(t,g)$.

<u>Theorem 5.</u> Let for $\tilde{\xi}(t)$ and $\tilde{u}(t,g)$ (31) and (32) hold. Then $u(t)$ as a vector field on $\mathbb{T}^n$ satisfies the classical Navier-Stokes equation with the viscosity coefficient $\sigma^2/2$:

$$\frac{\partial}{\partial t}u - \frac{1}{2}\sigma^2\Delta u + (u\cdot\nabla)u + \operatorname{grad} p = F. \tag{34}$$

<u>Proof.</u> According to the usual machinery of differential geometry we can describe the vector $S(\tilde{u}(t,\tilde{\xi}(t)))$ as follows

$$S(\tilde{u}(t,\tilde{\xi}(t))) = \lim_{\Delta t\to+0} E_t^{\tilde{u}(\tilde{\xi})}\left(\frac{\tilde{u}(t,\tilde{\xi}(t)) - \tilde{u}(t,\tilde{\xi}(t-\Delta t))}{\Delta t}\right) -$$

$$- \tilde{K}\circ\lim_{\Delta t\to+0} E_t^{\tilde{u}(\tilde{\xi})}\left(\frac{\tilde{u}(t,\tilde{\xi}(t)) - \tilde{u}(t,\tilde{\xi}(t-\Delta t))}{\Delta t}\right)$$

Since S, $\tilde{F}$, $\tilde{u}$ etc. are right-invariant, from (32) and the last formula it follows that

$$\frac{\partial}{\partial t}u = TR^{-1}_{\tilde{\xi}(t)}\left(-\tilde{K}\circ\lim_{\Delta t\to+0} E_t^{\tilde{u}(\tilde{\xi})}\left(\frac{u(t,\tilde{\xi}(t))-\tilde{u}(t,\tilde{\xi}(t-\Delta t))}{\Delta t}\right)\right)\in T_eD^s_\mu(\mathbb{T}^n)$$

It is easy to see that the process $\tilde{\tilde{\xi}}(t)$ may be considered as a stochastic flow $\xi(t)$ on $\mathbb{T}^n$, $\xi(0,m)=m$, with the diffusion $A \circ A^*$ and $D_*\xi(t) = u(t)$. Using formulas (22) and (23) and applying the second formula (12) for differentiation along $\xi(t)$ on $\mathbb{T}^n$, one can easily show that

$$TR^{-1}_{\tilde{\xi}(t)}(-\tilde{K}\circ \lim_{\Delta t \to +0} E_t^{\tilde{u}(\tilde{\xi})}(\frac{\tilde{u}(t, \tilde{\xi}(t)) - \tilde{u}(t, \tilde{\xi}(t-\Delta t))}{\Delta t})) =$$

$$= -P_e((u(t)\cdot\nabla)u(t) - \frac{1}{2}\sigma^2\Delta u(t)) = -((u(t)\cdot\nabla)u(t) - \frac{1}{2}\sigma^2\Delta u(t) + \operatorname{grad} p)$$

which proves (34). Q.E.D.

References

1. Arnold V. Sur la géométrie différentielle des groupes de Lie de dimension infinie et ses applications a l'hydrodynamique des fluides parfaits. In: Ann. Inst. Fourier, 1966, t.16, N 1.
2. Belopol'skaya Ya.I. and Dalecky Yu.L. Stochastic equations and differential geometry. Kluwer, 1989.
3. De Witt-Morette C. and Elworthy K.D. A stepping stone to stochastic analysis. In: New stochastic methods in physics, Physics Reports, 1981, vol.77, N 3.
4. Ebin D.G., Marsden J. Groups of diffeomorphisms and the motion of an incompressible fluid. Annals of Math., 1970, vol.92, N 1, p.102-163.
5. Eliasson H.I. Geometry of manifolds of maps. J. Diff. Geometry, 1967, vol.1, N 2.
6. Elworthy K.D. Stochastic differential equations on manifolds. Cambridge University Press, 1982 (London Mathematical Society Lecture Notes Series, vol.70).
7. Gliklikh Yu.E. Analysis on Riemannian manifolds and problems of mathematical physics. Voronezh University Press, 1989.(in Russian)
8. Gliklikh Yu.E. Stochastic differential geometry of the groups of diffeomorphisms and the motion of viscous incompressible fluid. Fifth International Vilnius conference on probability theory and mathematical statistics, Abstracts of communications, 1989, vol.1, p.173-174.
9. Gliklikh Yu.E. Infinite-dimensional stochastic differential geometry in modern Lagrangian approach to hydrodynamics of viscous incompressible fluid. In: "Constantin Caratheodory: an International Tribute" (Th.M.Rassias, ed.), World Scientific, 1991.vol.1.
10. Ito K. Extension of stochastic integrals. In: Proc. of Intern. Symp. SDE (Kyoto, 1976). New York, 1978.
11. Korolyuk V.S., Portenko N.I., Skorohod A.V. et al. Handbook in probability theory and mathematical statistics. Moscow, Nauka, 1985 (in Russian).
12. Liptser R.S. and Shiryayev A.N. Statistics of random processes. Springer, 1984.
13. Nelson E. Dynamical theories of Brownian motion. Princeton University Press, 1967.
14. Nelson E. Quantum fluctuations. Princeton University Press, 1985.

FROM TOPOLOGICAL HOMOLOGY: ALGEBRAS WITH DIFFERENT PROPERTIES OF HOMOLOGICAL TRIVIALITY

A.Ya.Helemskiĭ
Department of Mechanics and
Mathematics
Moscow State University
119899, Moscow, USSR

"Topological homology" is a short title for the homological theme in the theory of Banach and polynormed algebras. It is of comparatively recent usage, in Romania and somewhere else (cf. [1]). It seems that this title correctly reflects the essence of the matter: the continuity is very important in definitions of homological characteristics of "algebras of analysis", and the results themselves sound in the language of algebra, topology and analysis.

As to the heritage from algebra, the area has obtained two principal impulses from "pure" homology. The first one, which has caused its very emergence, was connected with the discovering by Hochschild of cohomology groups of associative algebras and of applications of these groups to extensions and derivations [B, nos.206-209] [1]. The second impulse is connected with the birth of the "full" homological algebra of Cartan, Eilenberg and MacLane which has given, among its numerous achievements, the opportunity to investigate cohomology groups as well as some new important concepts from the point of view of the unified notion of derived functor [B, nos.4-5]. Let us remark also that recently Connes [2] and Tzygan [3] have invented the cyclic (co)homology, which can happen, in retrospective, to be a new important impulse coming from algebra. Anyhow, the first swallow - and we hope that it is really a messenger of the spring - has already come flying [4].

Nowadays the topological homology, which does exist a quarter of a century, has become considerably larger, and it does not appear possible to describe its modern state with some degree of completeness on about twenty pages. Our aim is more modest: it is to give a most

[1] To avoid a formidable list of literature we often use references like [B, no.N] which means the item number N in the bibliography of the book [B]

general notion about only one of its several main sujets. We shall discuss some classes of Banach algebras which can be distinguished by their properties of triviality of their cohomology according to different classes of their coefficients - and, more generally, by properties of projectivity, injectivity or flatness of different classes of (bi)modules over these algebras.

So, we warn the reader that the whole series of inner problems of topological homology (Tor functors, homological dimensions, many conditions of the projectivity and of the flatness, many results on cohomology with some concrete coefficients) as well as a big circle of questions concerning applications (multioperator functional calculus, some problems of the complex analytic geometry, derivations, automorphisms, perturbations of algebras and of representations,...) are left out of the limits of this paper. Even our discussion of homologically trivial algebras almost does not involve here results on polynormed algebras which are not Banach, and also such topic as strongly amenable C^*-algebras in the sense of Johnson.

As to books and big articles which completely or partially concern problems of topological homology, we note [B] , [5] (B, nos.7,15,152, 155, 177, 202-204], [6, p.229-274] .

1. Standard complexes and cohomology groups.

The oldest concept of topological homology is the most simple and natural "Banach" analogue of Hochschild groups, which was introduced in 1962 by Kamowitz [B, no.107] . The definition, which slightly generalizes the original one, is as follows.

Let A be a Banach algebra, and let X be a Banach A-bimodule (see, e.g., [B] about these things). Denote by $C^n(A,X)$; $n = 1,2,\dots$ the space of continuous n-linear operators $f : A\times\dots\times A \longrightarrow X$ (we call them n-cochains), and let us take $C^0(A,X)$ as X. Now let us consider the so-called standard complex

$$0 \to X \to C^1(A,X) \to \dots \to C^n(A,X) \to C^{n+1}(A,X) \to \dots \qquad (\widetilde{C}(A,X))$$

where the "coboundary operator" δ^n is given by

$$\delta^n f(a_1,\dots,a_{n+1}) = a_1\cdot f(a_2,\dots,a_{n+1}) + \sum_{k=1}^{n}(-1)^k f(a_1,\dots,a_k a_{k+1},\dots, a_{n+1}) + (-1)^{n+1} f(a_1,\dots,a_n)\cdot a_{n+1}$$

for n 0 and by $\delta^0 x(a) = a\cdot x - x\cdot a$

(it is easy to verify that $\delta^{n+1}\delta^{n} = 0$). Let us also take $Z^n(A,X)$ as Ker δ^n and $B^n(A,X)$ as Im δ^{n-1}.

Definition 1. The n-th cohomology of the complex $\widetilde{C}(A,X)$ (in other words, the factorspace $Z^n(A,X)/B^n(A,X)$) is called usual, or continuous, n-dimensional cohomology group of Banach algebra A with coefficients in Banach A-bimodule X, and is denoted by $\mathcal{H}^n(A,X)$.

It is worth noting that $Z^n(A,X)$ but not, generally speaking, $B^n(A,X)$ is a Banach space, and therefore $\mathcal{H}^n(A,X)$ is a complete prenormed space.

Behold the first - simplest and, as we shall see, the most rigid - condition of the homological triviality.

Definition 2. A Banach algebra A is called contractible if $\mathcal{H}^1(A,X) = 0$ for all Banach A-bimodules X.

Running ahead (see Theorem 21), we note that the contractibility of A implies $\mathcal{H}^n(A,X) = 0$ for all X and for all $n > 0$.

We have no possibility to dwell on some details of different interpretations and applications of groups $\mathcal{H}^n(\cdot\ ,\ \cdot)$; as to $n = 2,3$, see, e.g., [B, nos. 94, 107,97,155,158], [B], [6, p.253-260]. As to $n = 1$, it is easy to observe that an element of $Z^1(A,X)$ ("1-dimensional cocycle") is just a continuous operator $D : A \longrightarrow X$ which satisfies an identity $D(ab) = a\cdot D(b) + D(a)\cdot b$, that is the so-called continuous derivation of A with the range X. At the same time an element of $B^1(A,X)$ ("1-dimensional coboundary") is a derivation of the special form $D(a) = a\cdot x - x\cdot a$ with fixed $x \in X$ - that is the so-called inner derivation. Thus $\mathcal{H}^1(A,X)$ is the space of continuous derivations "modulo inner ones". Since automorphisms of Banach algebras, which are close to identity, are exponents of derivations (see [B, no.1], [5]) it implies the importance of $\mathcal{H}^1(\cdot,\ \cdot)$ for some problems of the mathematical apparatus of quantum mechanics (see, e.g., [B, no.132], [7]).

Let us return to the pioneering paper of Kamowitz. It does contain also the first non-trivial result on the computation of the cohomology in some concrete situation.

Theorem 1. [B, no.107]. Let Ω be a compact set, X be a symmetric (that means $a\cdot x = x\cdot a$) Banach bimodule over $C(\Omega)$. Then $\mathcal{H}^1(C(\Omega),X) = \mathcal{H}^2(C(\Omega),X) = 0$.

Question 1 (which is the oldest open question in the entire topological homology). Is it true, in the assumptions of Theorem 1, that $\mathcal{H}^n(C(\Omega),X) = 0$ for every, or at least for some, $n \geqslant 3$?

As to the other natural question - can one dispense with the symmetry condition? - it was answered after some time (70-72) in the negative. The following general theorem is true.

Theorem 2. Let A be a commutative Banach algebra. Then

(I) [B, no.194] A is contractible iff $A = \mathbf{C}^m$ (with coordinatewise multiplication) for some m .

(II) [B, nos.200,202] If $\mathcal{H}^2(A,X) = 0$ for all X, then the spectrum of A is finite (and hence if, in addition, A is semisimple, then $A = \mathbf{C}^m$).

The reader with some algebraic experience could certainly presage the first part of this theorem. At the same time the second part of the theorem bears the imprint of some "pathological" properties of Banach geometry: let us recall for comparison that two-dimensional Hochschild groups of the algebra of polynomials vanish with all coefficients. However, it is already impossible to replace $\mathcal{H}^2(\cdot,\cdot)$ by $\mathcal{H}^3(\cdot,\cdot)$ in theorem 2; see Theorem 26 below.

In the beginning of the seventies Kadison and Ringrose [B, no.127] and, independently, Johnson [B, no.3] have indicated a class of coefficients which was much more tractable than in the general case.

Suppose that a Banach A-bimodule X is dual to some Banach space X_*. Then it is easy to show that the following two properties of X are equivalent: (I) for every $a \in A$ operators $X \to X : x \mapsto a\cdot x, x\cdot a$ are weak* continuous, and (II) X_* has a structure of a Banach A-bimodule such that the operations in X have the form $\langle a\cdot x, x_* \rangle = \langle x, x_*\cdot a \rangle$ and $\langle x\cdot a, x_* \rangle = \langle x, a\cdot x_* \rangle$. A bimodule X with these properties is called dual, and X_* (whose structure is uniquely defined) is called its predual bimodule. As obvious parts of this definition, the notions of a dual right (respectively, left) Banach A-module with predual left (right) Banach A-module appear.

Studying the cohomology with dual coefficients, Johnson has introduced a new property of homological triviality which provides, as we shall see below, a much larger class, than contractible algebras.

Definition 3 [B, no.3]. A Banach algebra A is called amenable (or amenable-after-Johnson, when we shall need precision) if $\mathcal{H}^1(A,X) = 0$ for every dual A-bimodule X.

As a matter of fact, this condition implies $\mathcal{H}^n(A,X) = 0$ for all dual X and for all $n > 0$. This result, as well as the origin of the term "amenable", will be clarified later in Sect.3 and 4. Now we shall proceed to another variant of the notion of cohomology groups which was introduced by Kadison and Ringrose in 1971 for the special case of operator algebras and which reflected essentially the operator framework.

Let **A** be an operator C*-algebra in a Hilbert space H - in other words, C*-subalgebra of the C*-algebra B(H) of all bounded operators in H. A dual A-bimodule X is called normal provided the operators A X : a a·x,x·a are normal (that is, continuous respectively to the ultraweak topology in A and weak* topology in X) for every x X. For example, the ultraweak closure A of A, as well as B(H) and, moreover, every von Neumann algebra which contains A, are normal A-bimodules. At the same time the dual A-bimodule A* is not normal, generally speaking.

A cochain $f : A\times\ldots\times A \longrightarrow X$, where X is a dual A-bimodule, is called normal provided it is normal in each variable; all 0-dimensional cochains are also proclaimed normal. Normal n-cochains form a closed subspace in $C^n(A,X)$ which we shall denote by $C^n_w(A,X)$. If X is normal, these subspaces form, as it is easy to see, a subcomplex in C(A,X) which we shall denote by $C_w(A,X)$.

Definition 4 [B, no.127] . The n-th cohomology of $C_w(A,X)$ is called normal n-dimensional cohomology group of operator C*-algebra A with coefficients in normal A-bimodule X.

A special property of homological triviality corresponds to this definition; it was distinguished by Connes in 1976 [B, no.113] .

Definition 5. Operator C*-algebra A is called amenable-after-Connes[1] if $\mathcal{H}^1_w(A,X) = 0$ for every normal X.

(Again one can replace the condition on 1-dimensional cohomology by the corresponding condition on n-dimensional cohomology for all $n>0$. But this fact is more difficult to prove than in the cases of contractibility and of amenability-after-Johnson; see Theorem 21 below).

There exists an important connection between both types of amenability, which was actually observed by Johnson, Kadison and Ringrose (JKR) (cf. [B, no.99]).

Theorem 3. If an operator C*-algebra A is amenable-after-Johnson, then its ultraweak closure $\bar{A}$ is amenable-after-Connes.

Here is a simple proof. We must establish that every normal derivation $D : \bar{A} \longrightarrow X$, where X is a normal A-bimodule, has the form $D(a) = a\cdot x - x\cdot a$ for some x X. Since D and X are normal, it is sufficient to check such an equality for all a A. But the latter follows from the fact that $D|_A$ is a derivation of an amenable algebra

[1] Connes himself, having concentrated on von Neumann algebras, used the term "amenable as von Neumann algebra". Some authors say also "ultraweak amenability" [B, no.130] or "normal amenability" [8] .

with the range in a dual bimodule.

Apart from an independent interest, the normal cohomology provides an effective tool for the computation of the "usual" one. The following theorem, which is due to Johnson, Kadison and Ringrose, is a basis for these applications.

<u>Theorem 4</u> ([B, no.99] ; see also [9]). Let A be an operator C*-algebra, X be a normal A-bimodule. Then $\mathcal{H}^n_w(A,X) = \mathcal{H}^n(A,X)$ for all $n \geqslant 0$.

As a matter of fact, this theorem was proved by JKR under the additional assumption that X is also a normal A-bimodule. Later (see Theorem 17) we shall outline a possible approach to the proof which permits one to dispense with this assumption. However, another theorem of JKR is indispensible for all knwon proofs of Theorem 4 and has an obvious independent value:

<u>Theorem 5</u> [B, no.99] . Let $A_1,\dots,A_n$ be operator C*-algebras, and let E be a dual Banach space, and let $F_o : A_1\times\dots\times A_n \longrightarrow E$ be a polylinear operator, which is norm continuous and normal on each variable. Then there exists (obviously unique) extension $F : \overline{A}_1\times\dots\times\overline{A}_n \longrightarrow E$ of F_o which preserves the polylinear operator norm of F_o and is also normal on each variable.

Both cited results imply

<u>Theorem 6</u>. Let A be an operator C*-algebra, and let X be a normal A-bimodule. Then $\mathcal{H}^n_w(A,X) = \mathcal{H}^n_w(A,X) = \mathcal{H}^n(A,X) = \mathcal{H}^n(A,X)$.

Indeed, Theorem 5 implies that complexes $\widetilde{C}_w(A,X)$ and $\widetilde{C}_w(\overline{A},X)$ are isomorphic; hence, their cohomologies coincide. Thus we have the first equality; the rest are provided by Theorem 4, which is considered for A and $\overline{A}$.

As an application to a very important class of algebras, JKR (see idem) have obtained the following result. We recall that a von Neumann algebra is called hyperfinite, if it is an ultraweak closure of some family of its finite-dimensional *-subalgebras, which is directed by inclusion.

<u>Theorem 7</u>. Let A be a hyperfinite von Neumann algebra, and let X be a normal A-bimodule. Then $\mathcal{H}^n(A,X) = \mathcal{H}^n_w(A,X) = 0$ for all n 0. In particular, every such algebra (e.g. B(H)) is amenable-after-Connes.

It was already mentioned in the introduction that the results on cohomology with concrete coefficients are for the most part outside of the scope of this paper. Nevertheless we shall state several open questions and comment on them.

<u>Question 2</u>. Is it true that $\mathcal{H}^n(A,A) = 0$ for all $n \geqslant 1$ and for

every (not necessary hyperfinite, compare Theorem 7) von Neumann algebra A?

(The well-known theorem of Kadison and Sakai [B, nos.124, 160] establishes the positive answer for $n = 1$).

<u>Question 3</u>. Let A be an operator C*-algebra in a space H. Is it true that $\mathcal{H}^1(A,B(H)) = 0$?

This question was raised and studied by Christensen 6, p.261-274 who has established several sufficient conditions of the positive answer; it is really so, in particular, if A has a cyclic vector.

<u>Remark</u>. Christensen (for $n = 1$) and afterwards Christensen, Effros and Sinclair (CES) [10] have obtained several results on the triviality of groups $\mathcal{H}^n(A,B(H))$ by reducing usual cochains to the so-called completely bounded ones. This way has led CES to the definition of a new and apparently very important version of cohomology - "completely bounded cohomology groups of an operator C*-algebra with coefficients in completely bounded bimodules" (denoted by $\mathcal{H}^n_{cb}(\cdot,\cdot)$). They have managed to prove in the cited paper that $\mathcal{H}^n_{cb}(A,B(H))=0$ for all A (compare Question 3) and for all $n>0$. As a corollary, they have obtained the vanishing of $\mathcal{H}^n(A,B(H))$; $n>0$ in the assumption that A is a properly infinite von Neumann algebra.

As to another standard bimodule over an operator C*-algebra A, acting in H - namely, the bimodule (and algebra) K(H) of compact operators in H - recently Popa has proved that $\mathcal{H}^1(A,K(H)) = 0$ for every von Neumann algebra A. Thus he has given an answer to an old question of Johnson.

Now let us give special attention to the A-bimodule A* over an arbitrary Banach algebra A. Two reasons have aroused an interest to it. The first is connected with the following recent result of Bade, Curtis and Dales (BCD).

<u>Theorem 8</u> [11] . Let A be a commutative Banach algebra with $\mathcal{H}^1(A,A^*) = 0$. Then $\mathcal{H}^1(A,X) = 0$ for every symmetric Banach A-bimodule X.

Indeed, since X and A* are symmetric, we must take non-zero $D \in Z^1(A,X)$ and construct non-zero $\widetilde{D} \in Z^1(A,A^*)$. The condition easily provides that $A = A^2$; therefore there exists $a \in A$ with $D(a^2) = 2aD(a) \neq 0$. We are only to put $\widetilde{D} : a \mapsto f : f(b) = g(bD(a))$ where $g \in X^*$ is such that $g(D(a^2)) \neq 0$.

Let us recall Theorem 1. Later it was strengthened by Johnson [B, no.3] who has replaced $C(\Omega)$ by arbitrary amenable algebra in its condition. This result and the very definition of amenability justify the following terminology.

Definition 6. A Banach algebra A (not necessarily commutative) is called[1] weakly amenable if $\mathcal{H}^1(A,A^*) = 0$.

A non-trivial theorem of Haagerup [B, no.187, Sect.4] claims that every C*-algebra is weakly amenable; and Christensen and Sinclair have shown [4] that if our C*-algebra has no non-zero continuous traces, then $\mathcal{H}^n(A,A^*) = 0$ for all $n>0$. BCD in [11] have displayed several interesting examples of weakly amenable, as well as amenable, algebras. In particular, they have proved that the Beurling algebra $l_1(w)$ with the weight $w(n) = (1 + |n|^{\alpha})$; $n \in \mathbb{Z}$ is amenable for $\alpha = 0$, is not amenable but is weakly amenable for $0 < \alpha < \frac{1}{2}$ and is not weakly amenable for $\alpha \geq \frac{1}{2}$. Several interesting characterizations of commutative weakly amenable algebras are obtained by Groenbaek in [12].

The second source of the interest to groups $\mathcal{H}^n(A,A^*)$ is their connection with the cyclic homology and cohomology, which were introduced by Tzygan [3] and Connes [2]. The Banach version of cyclic cohomology, which will be defined below, is a particular case of a general concept of the cyclic cohomology of a cyclic Banach space (that is, of a functor from the standard "cyclic category" Λ [13] to the category of Banach spaces).

Following [4], let us consider the space $A^{(n+1)*}$ of (n+1)-linear functionals on A, which can be naturally identified with $C^n(A,A^*)$. We shall call f $A^{(n+1)}$ (as well as the corresponding cochain) cyclic, if $f(a_1,\ldots,a_n,a_o) = (-1)^n f(a_o,a_1,\ldots,a_n)$ for all variables. It is easy to check that subspaces $CC^n(A,A^*)$ in $C^n(A,A^*)$, consisting of cyclic cochains, form a subcomplex in $\widetilde{C}(A,A^*)$. Its n-th cohomology is called n-dimensional cyclic cohomology of Banach algebra A and is denoted by $\mathcal{H}C^n(A)$ (or $\mathcal{H}^n_\lambda(A)$).

Christensen and Sinclair, applying their results on groups $\mathcal{H}^n(A,A^*)$ and an analogue of the exact sequence of Connes [2] (which connects $\mathcal{H}C^n(A)$ and $\mathcal{H}^n(A,A^*)$), have established the following

Theorem 9 [4]. If A is an amenable C*-algebra, then $\mathcal{H}C^n(A)$ vanishes for all odd n and coincides with the space of continuous traces on A for all even n. If A is an arbitrary C*-algebra without non-zero continuous traces, then $\mathcal{H}C^n(A) = 0$ for all $n \geq 0$.

---

1) After a suggestion of Johnson, cf. [11]

2. The "full" homology enters. Expression of cohomology groups via Ext.

The majority of the already presented results was obtained by the methods, based on the investigation of the equation $\delta^n g = f;\ n \geqslant 1$ in the corresponding version of the standard complex. We refer to these methods as to "direct" ones; the reader can obtain a good account of the corresponding algebraic and analytic technics (which is often very refined and sophisticated) from the papers, say, [B, nos. 3, 95,98,155] [10] [4].

However, Theorem 2 is an exception; this result and many of those which will be presented later are obtained by methods of "full" homology (see Introduction). These methods save us from being tied to standard complexes, and use instead the specially chosen resolutions and long exact sequences.

The carrying over of these ideas and methods to "algebras of analysis" was done, essentially, in papers [B, nos.194-200] and, independently, in [B, nos.177,7]. The approach was actually the same, and it has used ideas of "relative homology" due to Hochschild and others [B, nos.209,5,220]. We proceed to main definitions.

Let A be a Banach algebra. We shall denote by A_+ its unitization, by A^{op} its opposite algebra, by A^{env} its enveloping algebra $A_+ \widehat{\otimes} A_+^{op}$. The category of Banach left (right, bi-) modules will be denoted by A-mod (mod-A, A-mod-A), and that of Banach spaces will be denoted by Ban. We recall that a right A-module is a left A^{op}-module, and A-bimodule is a left A^{env}-module; therefore one can automatically carry over everything, which will be said about left modules, to other types of modules.

For X,Y A-mod we shall denote by $_A h(X,Y)$ the Banach space of continuous morphisms from X to Y. The co(contra)variant functor of morphisms, which is defined by fixation of X(Y), will be denoted by $_A h(X,?)(_A h(?,Y))$: A-mod $\longrightarrow$ Ban.

All our categories are additive (though not abelian); therefore one can speak about their complexes, in particular about splittable (=contractible) ones. Note that even in Ban, because of the abundance of uncomplemented subspaces, the splittability of a complex is a much stronger property, than its exactness (as of a complex of linear spaces). That makes reasonable the following

Definition 7. A complex in A-mod is called admissible if it splits as a complex in Ban (that is, it has a contracting homotopy consisting of continuous operators). A morphism in A-mod is called admissible if it can participate in some admissible complex.

Now we shall introduce several classes of modules, which are considered as "homologically best". We shall restrict ourselves to left modules: the transfer to other types is obvious. There are many roads which lead to each of these classes; here is one of the shortest.

Definition 8. $P \in A$-mod is called projective if every admissible epimorphism $\sigma : X \to P$ has a right inverse morphism (in A-mod). J A-mod is called injective, if every admissible monomorphism $i : J \quad Y$ has a left inverse morphism. F A-mod is called flat, if its dual module (see Sect.1) F^* is injective (in mod-A).

The equivalent definition of projectivity is that the functor ${}_A h(P,?)$ preserves the exactness of admissible complexes; that of injectivity is that the functor ${}_A h(?,J)$ has the same property; that of flatness is that the functor $? \mathop{\widehat{\otimes}}_A F$: mod-A $\to$ Ban has the same property.

There is an important class of projective modules which consists of so-called free modules: these have the form $A_+ \otimes E$; $E \in$ Ban with the operation $a \cdot (b \otimes x) = ab \otimes x$. Every P A-mod is projective iff it is a retract (direct module summand) of a free module. The analogous role for injective modules is played by so-called cofree modules: these have the form $B(A_+,E)$; E Ban (here and below $B(\cdot,\cdot)$ is a symbol of a space of all continuous operators) with the operation $[a \cdot f](b) := f(ba)$.

One old and rather concrete result about the projectivity is as follows.

Theorem 10 [B, no.196] Let A be a unital commutative Banach algebra with the spectrum Ω, and let I be its closed ideal with the hull h(I). Then for the projectivity of the A-module I it is necessary, and when $A = C(\Omega)$ it is also sufficient, that the set $\Omega \setminus h(I)$ be paracompact.

And here is a result on the injectivity, this time a recent one.

Theorem 11 [9]. Let A be an operator C^*-algebra, and let $A^{n^*}_\alpha$; $n > 1$ be the space of continuous n-linear functionals on A, which are normal on some (arbitrary chosen) variables. Then $A^{n^*}_\alpha$ is an injective Banach A-bimodule.

As to flatness, it is easy to see that every projective module is certainly flat. The converse is false: one can at least compare Theorem 10 with the following result.

Theorem 12 [B, no.198] . If a left closed ideal in a Banach algebra A has a right b.a.u.[1)], then it is a flat left Banach A-module.

[1)] b.a.u. means, as usual, "bounded approximate unit".

Proceeding to cyclic modules (= to factormodules of A_+), we shall present the following "almost criterion" of their flatness.

Theorem 13 [B, no.205] . Let I be a closed left ideal in A_+. Then for the flatness of A_+/I it is sufficient, and when $I^\perp = \{f : A_+^* : f|_I = 0$ has a Banach complement in A_+ (e.g., I itself has such a complement in A_+) it is also necessary that I does have a right b.a.u.

Let us note for comparison, that I, which has a Banach complement in A_+, is projective iff it has a (usual) right unit.

Remark. The role of b.a.u. in both theorems is connected with the following well known fact: A has a right b.a.u. iff A^{**} with the Arens multiplication has a right unit. One can use the latter for the construction of $I = (A_+/I)^*$ as a retract of certainly injective A_+^*.

We return to geberal definitions and constructions.

Definition 9. Let A be a Banach algebra, and let X,Y A-mod. A complex

$$0 \leftarrow X \xleftarrow{\varepsilon} P_0 \leftarrow P_1 \leftarrow \dots \quad (0 \leftarrow X \leftarrow \mathcal{P}) \text{ over } X \text{ (respectively,}$$

$$0 \rightarrow Y \rightarrow J_0 \rightarrow J_1 \rightarrow \dots \quad (0 \rightarrow Y \rightarrow \mathcal{J}) \text{ under } Y) \text{ is called}$$

projective (respectively, injective) resolution of X(Y), if this complex is admissible, and all P_k are projective (J_k are injective); $k \geqslant 0$.

Projective resolution is a simultaneous form of writing down the following process: X is presented (which is always possible) as a factormodule of a projective one (P_0), the natural projection (ε) being admissible; then the same procedure is applied to the kernel of ε , and so on. On some step the kernel of the corresponding natural projection can happen to be projective itself, and then a resolution will appear, which will be fringed by zeroes from the right. It is remarkable, that the number of such a step depends only on X itself, but not on a concrete choice of projective modules and quotient maps. This number is called homological dimension of X, and it is its most important characteristic. As to results concerning homological dimension, see the book [B] and its list of references, and also [15][16] .

The following theorem presents, in the "Banach packing" one of the principal achievements of homological algebra.

Theorem 14. Let $0 \leftarrow X \leftarrow \mathcal{P}$ ($0 \rightarrow Y \rightarrow \mathcal{J}$) be a projective (respectively, injective) resolution of a left Banach module X(Y). Then for all $n \geqslant 0$ the n-th cohomology of the complex

$$0 \rightarrow {}_A h(P_0, Y) \rightarrow {}_A h(P_1, Y) \rightarrow \dots \qquad ({}_A h(\mathcal{P}, Y))$$

does coincide, up to a topological isomorphism, with the n-th cohomology of the complex

$$0 \longrightarrow {}_Ah(X,J_0) \longrightarrow {}_Ah(X,J_1) \longrightarrow \dots \qquad ({}_Ah(X,\mathcal{J}))$$

and both of them do not depend on the particular choice of resolutions $\mathcal{P}$ and/or $\mathcal{J}$.

"Definition" 10. The (common) n-th cohomology of complexes ${}_Ah(\mathcal{P},Y)$ and ${}_Ah(X,\mathcal{J})$ is denoted by $\mathrm{Ext}^n_A(X,Y)$.

By analogy, one can define Ext for right modules and for bimodules; the latter will be denoted by $\mathrm{Ext}_{A-A}^{\;n}(\cdot,\cdot)$.

Theorem 14 easily implies

Theorem 15. (I) X is projective iff $\mathrm{Ext}^1_A(X,Y) = 0$ for all Y iff $\mathrm{Ext}^n_A(X,Y) = 0$ for all Y and $n>0$. (II) Y is injective iff $\mathrm{Ext}^1_A(X,Y) = 0$ for all X iff $\mathrm{Ext}^n_A(X,Y) = 0$ for all X and $n>0$.

As to applications of Ext and their interpretation as the container of obstructions to the lifting and extension of morphisms in particular, see e.g., [5, Ch.VII]. We shall discuss here another thing - their connection with the cohomology. In the following theorem for an arbitrary Banach algebra A we consider A_+ and A_+^* as bimodules, and for an operator C^*-algebra A we consider also the bimodule $\overline{A}_*$, which is predual to $\overline{A}$ (it is actually the subbimodule of A^* consisting of normal functionals).

Theorem 16. (I) $\mathcal{H}^n(A,X) = \mathrm{Ext}_{A-A}^{\;n}(A_+,X)$; $n \geqslant 0$ for any A and for any $X \in A\text{-mod-}A$.

(II) In addition, $\mathcal{H}^n(A,X) = \mathrm{Ext}^n_{A-A}(X_*,A_+^*)$; $n \geqslant 0$ for any X_* and for any dual $X \in A\text{-mod-}A$ with the predual X_*.

(III) [9] $\mathcal{H}^n_w(A,X) = \mathrm{Ext}^n_{A-A}(X_*,A_*)$; $n \geqslant 0$ for any operator C^*-algebra A and for any normal $X \in A\text{-mod-}A$ with the predual X_*.

The proof of (I) is founded on the computing of the indicated Ext with the help of the Banach analogue of the known "bimodule bar-resolution", which has the form $0 \longleftarrow A_+ \longleftarrow A_+ \widehat{\otimes} A_+ \longleftarrow \dots \longleftarrow A_+ \widehat{\otimes} A \widehat{\otimes} \dots \widehat{\otimes} A \widehat{\otimes} A_+ \longleftarrow$ $\longleftarrow \dots$ $(\mathbb{B}(A_+))$; then the complex ${}_Ah_A(\mathbb{B}(A_+),X)$, where ${}_Ah_A(\cdot,\cdot)$ is a notation of a space of bimodule morphisms, happens to be isomorphic to $\widetilde{C}(A,X)$. (II) is a corollary of (I) and of the general formula $\mathrm{Ext}^n_A(Y,Z^*) = \mathrm{Ext}^n_{A^{op}}(Z,Y^*)$ (with A^{env} as A), which gives the known "conjugate associativity" ${}_Ah(Y,Z^*) = (Z \widehat{\otimes}_A Y)^* = {}_{A^{op}}h(Z,Y^*)$ in the case of $n = 0$. (III) is recently obtained; in its proof Ext is calculated with the help of some special injective resolution of the bimodule A_*. This resolution has the form $0 \to \overline{A}_* \to A^2_* \to A^3_* \to \dots$, where A^n_* consists of n-linear continuous functionals on A, which are normal on each variable (cf. Theorem 11). The functor ${}_Ah_A(X_*,?)$, being applied to this resolution, provides - by virtue of normality of X - a complex, which is isomorphic to $\widetilde{C}_w(A,X)$.

As one of the applications of this theorem, one can obtain a rather simple proof of Theorem 4 in its following slightly strengthened form.

Theorem 17 [9]. Let A be an operator C*-algebra, and let X be a dual A-bimodule with the predual X_*, such that at least one of operators $A \to X : a \mapsto a\cdot x$ and $a \mapsto x\cdot a$ is normal for all $x \in X$. Then $\mathrm{Ext}^n_{A-A}(X_*,A^*) = \mathrm{Ext}^n_{A-A}(X_*,\bar{A}_*)$; $n \geq 0$.

The proof uses the injectivity of the bimodule A^{n*} of all continuous n-linear functionals on A and of its subbimodule A_1^{n*} formed by functionals which are normal on the first variable (Theorem 11). By computing the first Ext with the help of the resolution of A^* formed by A^{n*}, n = 1,2,..., and by computing the second one with the help of the resolution of A_* formed by A_1^{n*} ; n = 1,2,..., we come to taking the n-th cohomology of two isomorphic complexes.

In some important cases one can express the cohomology also via left-module Ext. Recall that for X,Y A-mod one can consider B(X,Y) as a bimodule with operations $[a\cdot\varphi](x) := a\cdot(\varphi(x))$ and $[\varphi\cdot a](x) = \varphi(a\cdot x)$.

Theorem 18 [B, no.194]. $\mathcal{H}^n(A,B(X,Y)) = \mathrm{Ext}^n_A(X,Y)$; $n \geq 0$ for all $X,Y \in$ A-mod.

Here is one of useful applications (compare Question 3).

Theorem 19 [B, no.106] Let A be a Banach algebra of operators in a Banach space E, which contains all finite-dimensional operators. Then $\mathcal{H}^n(A,B(E)) = 0$ for all $n > 0$.

Indeed, when we take A-module E (with $a\cdot x := a(x)$), we see that B(E) is just B(E,E) (see above). Therefore, by virtue of Theorems 19 and 15, it is sufficient to establish that E is projective. But it is a retract of the free module A_+; this can be shown, with fixed x_0 E and $f_0 \in E^*$; $\langle f_0, x_0\rangle = 1$, with the help of morphisms $A_+ \to E : a \mapsto a(x_0)$ and $E \to A : x \mapsto \langle f_0, \cdot\rangle x$.

The same equality $\mathcal{H}^n(A,B(E)) = 0$, for some other $A \subseteq B(E)$ than in this theorem, is valid provided E is injective, or even flat when E is a Hilbert space. Let us remark that Golovin [16] has established the flatness of a Hilbert space H as a nodule over a nest operator algebra in H.

Remark. The cyclic Banach cohomology of a cyclic Banach space E (cf. the end of Sect.1) can be expressed, up to a topological isomorphism) via some Ext similar to that which was introduced by Connes in [13]. Actually, $\mathcal{H}C^n$ E) = $\mathrm{Ext}^n_{\ell_1(\Lambda)}(\mathbb{C}^{\natural}, E)$ where $\ell_1(\Lambda)$ is the ℓ_1 -

space with the basis formed by morphisms of the cyclic category and with the multiplication generated by a composition law of these morphisms. Here E and the "constant cyclic object" $\mathbb{C}^{\natural}$ are considered as left Banach $l_1(\Lambda)$-modules.

3. Homologically trivial algebras: general results.

Suppose that we know that this or that class of modules and/or bimodules over a given Banach algebra contains only projective, or injective, or flat (bi)modules. What can be said about such an algebra? Now we shall discuss several most important conditions of this kind. In the beginning we shall show that the conditions of the vanishing of cohomology, which are expressed in definitions 2,3 and 5, are particular cases of such conditions.

Theorem 20. (I) A Banach algebra A is contractible iff A_+ is projective in A-mod-A iff A is unital, and A is projective in A-mod-A.

(II) A Banach algebra A is amenable iff A_+ is flat (= A_+^* is injective) in A-mod-A iff A has a b.a.u., and A is flat (= A^* is injective) in A-mod-A.

(III) 9 An operator C*-algebra is amenable-after-Connes iff A_* is injective in A-mod-A.

As to the assertions (I) and (II), the first equivalence is an immediate corollary of Theorems 14 and 15. As to a b.a.u. in (II) (the unital case in (I) is simpler), the module A_+/A is flat in A-mod and in mod-A provided A is amenable (see Theorem 25 below). Therefore Theorem 13, being considered for A and A^{op}, implies that A has both one-sided and hence a two-sided b.a.u.

The assertion (III), which was obtained recently, is more delicate. Proving "only if", we cannot, as in (II), suppose that $\mathrm{Ext}^1_{A-A}(Y,A_*)$ for all Y A-mod-A: Theorem 16 (III) provides this equality only for prenormal (=predual to normal) bimodules. In particular, the injective bimodule A_*^2 (see above) appears not to be prenormal (cf. [17]) - and only this property would be sufficient for a presentation of A_* as a retract of A_*^2 . Nevertheless one can get the desired presentation with the help of some special subbimodule A_o^2 of A_*^2 which was invented by Haagerup and Effros. It happens to be prenormal, and some non-trivial lemma of Effros (cf. [17, lemma 2.3]) implies that there exists a morphism from A_*^2 to A_o^2 , which is identity map on a bimodule identified with A_*.

Theorems 20 and 15 directly imply the assertion which was mentioned in Sect.1:

Theorem 21. Let A be a contractible Banach algebra (respectively, amenable Banach algebra, amenable-after-Connes operator C^*-algebra). Then $\mathcal{H}^n(A,X)$ (respectively, $\mathcal{H}^n(A,X)$, $\mathcal{H}^n_w(A,X)$) vanishes for all (respectively, all dual, all normal) A-bimodules X and $n>0$.

Since some standard bimodules (free, cofree, mentioned in Theorem 11,...) are certainly projective or injective, it is sufficient for the testing of the discussed properties of algebras to know, whether some particular morphism is a retraction or coretraction (that is, whether it has right or left inverse in A-mod-A). Indeed, let us consider $\pi_+ : A_+\widehat{\otimes}A_+ \longrightarrow A_+$, $\pi : A\widehat{\otimes}A \longrightarrow A : a\otimes b \longmapsto ab$ and also their dual and bidual morphisms. Apart from this, for an operator C^*-algebra let us consider $\pi_* : A_* \longrightarrow A^2_*$ (which sends, like π^*, f to g : g(a,b) : = f(ab)) and also $(\pi_*)^*$.

Theorem 22 (cf. [B][B, no.95][17][9]). (I) A is contractible iff π_+ is a retraction. (Ia) A unital A is contractible iff π is a retraction. (II) A is amenable iff π^*_+ is a coretraction iff π^{**}_+ is a retraction. (IIa) A with a b.a.u. is amenable iff π^* is a coretraction iff π^{**} is a retraction. (III) An operator C^*-algebra A is amenable-after-Connes iff π_* is a coretraction iff $(\pi_*)^*$ is a retraction.

All these criteria are rather effective and they have many consequencies. As an example, here is a simple proof of the known theorem (cf. Connes [B, no.114], Haagerup [B, no.187], Effros [17]) concerning a connection between both types of the amenability.

Theorem 23. An operator C^*-algebra A is amenable-after-Johnson iff its enveloping von Neumann algebra A^{**} is amenable-after-Connes.

"Only if" is just a special case of Theorem 3. In order to prove "if" with the help of Theorem 22(III), we observe that $\pi_* : (A^{**})_* \longrightarrow (A^{**})^2_*$ has a right inverse morphism of A^{**}- , and hence of A-bimodules. However, by virtue of Theorem 5, π_* is the same thing as $\pi^* : A^* \longrightarrow A^{2*}$, the latter checking the amenability-after-Johnson in Theorem 22 (IIa).

So, the contractibility and both amenabilities are just properties of projectivity, injectivity or flatness of one particular bimodule connected with A : A_+ or, according to the situation, A^*_+ or A_* . The projectivity or flatness of all left A-modules is another important typical property of a given A. As to the connection of such properties, we know the following.

Theorem 24. If A is contractible, then all one-sided (left and right) A.modules are projective. If A is amenable(-after-Johnson), then all one-sided A-modules are flat.

Let us comment on the second assertion (the first one is simpler). The dual of every X A-mod is isomorphic to ${}_A h(X, A_+^*)$ with the operation $[\varphi \cdot a](x) := [\varphi(x)](a)$. Theorem 22(II) implies that the latter is a retract of ${}_A h(X, A_+^{*2})$, which is isomorphic to the free right Banach module $B(A_+, X^*)$.

<u>Questions 4 and 5</u>. Let all left Banach modules over a Banach algebra A be projective (respectively, flat). Does it imply that A is contractible (respectively, amenable)?

<u>Remark</u>. There exist non-contractible algebras with projective one-sided modules in pure algebra: the field of rational functions provides an example.

It may seem that the projectivity (or flatness) of all A-bimodules is a far stronger property than the projectivity (flatness) of A_+ alone. But it is not so.

<u>Theorem 25</u>. All bimodules over a contractible (respectively, amenable) algebra are projective (flat).

The reason is that the operation of taking a tensor product of algebras, and hence of taking the enveloping algebra preserves the properties of the contractibility and amenability. As to the contractability, one can easily deduce it from Theorem 22 ((I) and (II)). The proof in the amenability case is somewhat more complicated (see Johnson [B, no.3], and also [B, no.204]).

On the contrary, the following class of algebras differs from the already mentioned ones.

<u>Definition 11</u>. A Banach algebra A is called biprojective, if it is projective as an A-bimodule.

The following theorem served as the initial stimulus for the study of non-unital (that is, non-contractible; cf. Theorem 20(I)) biprojective algebras.

<u>Theorem 26</u> [B, nos.200,202] . If A is biprojective, then $\mathcal{H}^n(A, X) = 0$ for all X A-mod-A and $n \geq 3$.

In the spirit of Theorem 22, to check the biprojectivity it is convenient to use the following

<u>Theorem 27</u> [B] . A is biprojective iff $\pi : A \hat{\otimes} A \longrightarrow A$ is a retraction.

As an example, it is not difficult to prove, following this way, that the algebra $A = E \hat{\otimes} E^*$, where E Ban , with the multiplication $(x\, x^*)(y\, y^*) := \langle x^*, y \rangle x \otimes y^*$ is biprojective.

At last, in order to complete the picture, we shall also introduce biflat algebras - those A which are flat as A-bimodules. By virtue of

Theorem 20 (II), amenability is just biflatness plus existence of a b.a.u.

The following diagram shows the relations among classes of algebras discussed above [1].

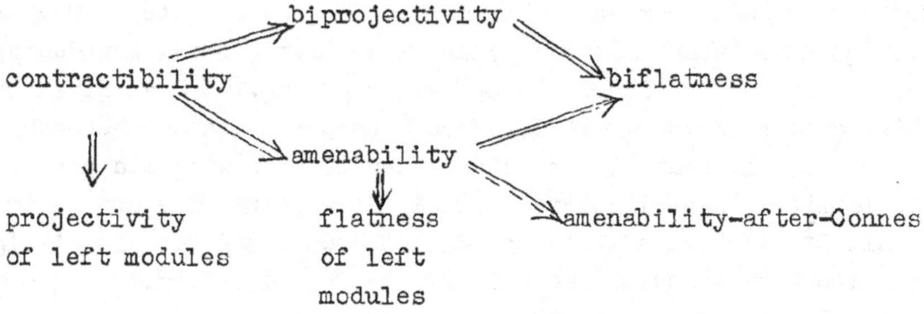

None of these logical arrows can be reversed, with the possible exception of vertical ones (cf. Questions 4 and 5). B(H) is amenable-after-Connes, but (see below) not -after-Johnson. K(H) provides an example (see below) of an amenable non-biprojective algebra, and the algebra of nuclear operators N(H), which is isomorphic to $H \hat{\otimes} H^*$ (see above) gives an example of a biprojective non-amenable algebra.

4. Conditions of the homological triviality in concrete classes of algebras.

As to the property of contractibility, the algebra of n n matrices M_n provides a classical example inherited from "pure" homology. For such an A the known morphism $a \mapsto a\sum_{k=1}^{n} e_{1k} \otimes e_{1k}$ [B, no.5], or ρ: $a \mapsto \frac{1}{n} a \sum_{k,l=1}^{n} e_{kl} \otimes e_{kl}$ (here $_k$ is a matrix unit) serves as a right inverse to in Theorem 22 (Ia). The second mentioned morphism has the advantage of having norm 1 [B, no.199] after the identification of M_n with B(H); dimH = n .

The easy corollary is that every finite-dimensional semisimple algebra, being a finite direct sum of matrix algebras, is contractible. Up to this moment it is unknown whether there exist other contractible Banach algebras. More generally, let us consider the following hierarchy of properties of a Banach algebra: 1) A is semisimple and finite-dimensional ⟹ 2) A is contractible ⟹ 3) all left A-modules are projective ⟹ 4) all irreducible A-modules are projective.

[1] the arrow ⟹ means: for operator C*-algebras only.

Question 6. Are all properties 1)-4) equivalent? At least can some of the indicated logical arrows be reversed?

As to commutative algebras, the positive answer was known long ago [B, no.194]. The principal difficulty in the general case is that maximal ideals are not bound to have Banach complements. One can surmount it under some assumptions "about good geometry". Taylor [B, no.177] has obtained 2) $\Longrightarrow$ 1) assuming that A has a bounded approximation property. Afterwards Selivanov [B, no.165] has given the positive answer to the whole Question 6 assuming that either A/RadA or all irreducible A-modules have the (usual) approximation property. This implies, in particular, that all four properties are equivalent for all L^1-algebras of locally compact groups and for all C*-algebras. As to the latter, quite recently Lazar, Tsui and Wright [18] have established the equivalence of "1)" to the following property: $\mathcal{H}^1(A, B) = 0$ for every C*-algebra B which contains A.

Remark. Outside the framework of Banach structures the question, similar to the 6th one, certainly has a negative answer. As examples, the algebra of all functions on an arbitrary set (with the topology of pointwise convergence) and the algebra of distributions on a compact Lie group are contractible (Taylor [B, no.7]).

As to biprojective algebras, they form a conspicuously larger class than the contractible ones. Apart from the already menstioned in Sect.3 $E\hat{\otimes}E^*$(and N(H)), say, c_0 and ℓ_1 also belong to this class. Both have $a \mapsto a\sum_{k=1}^{\infty} p^k \otimes p^k$, where $p^k = (0,\dots,0,1,0,\dots)$ is the k-th "basis vector", as a right inverse to π in Theorem 27. As to group algebras, there is

Theorem 28 [B]. Let G be a locally compact group. Then $L^1(G)$ is biprojective iff G is compact iff C $L^1(G)$-mod with the operation $f\cdot z = (\int_G f(s)ds)z$ is projective.

(The "comultiplication" $f \mapsto g$: $g(s,t) := f(st)$ becomes a right inverse to π after the identification of $L^1(G)\hat{\otimes}L^1(G)$ with $L^1(G\times G)$).

Nevertheless Selivanov has shown [B, no.166] that the biprojectivity is such a strong assumption on a Banach algebra, that one can give, at least for a semisimple algebra with the approximation property, almost complete description of its structure. The theorem of Selivanov, in its slightly simplified form, claims that every algebra in the indicated class is a topological direct sum of algebras of the "type of $E\hat{\otimes}E^*$". (Thus ℓ_1 and c_0, on the one hand, and N(H), on the other hand, can be considered as mutually opposite extreme cases of the general construction of the algebras under discussion). The "only if" part of the following theorem is one of the corollaries.

Theorem 29 [B, nos.199,166]. A C*-algebra is biprojective iff it is c_0-sum of an (arbitrary) family of full matrix C*-algebras (in other words, its spectrum is discrete, and its irreducible representations are finite-dimensional).

As a particular case of this theorem, B(H) and K(H) are not biprojective.

Now let us proceed to the most popular class of homologically defined algebras - to amenable ones. The very term is motivated by the discovery of Johnsom [B, no.3] of the connection with the time-honoured concept of amenable locally compact group. This connection, with some later addition, is the following.

Theorem 30. Let G be a locally compact group. Then $L^1(G)$ is amenable iff G is amenable iff C mod-$L^1(G)$ with the operation $z \cdot f := (\int_G f(s)ds)z$ is flat.

One of the proofs is exposed in [B]. We shall only clarify why, in the simplest case of a discrete G, the amenability (= existence of left-invariant mean $M : L^\infty(G) \to \mathbb{C}$) of such a group implies the existence of a left inverse to π^* in Theorem 22 (IIa). Since $L^1(\cdot)^* = L^\infty(\cdot)$, π^* can be identified with $L^\infty(G) \to L^\infty(G\times G) : f \mapsto u : u(s,t) = f(st)$. Left-invariance of M provides the coincidence of operators $\zeta_k : L^\infty(G\times G) \to L^\infty(G) : u \mapsto f(s) := M(\bar{v}_k(s))$; $k = 1,2$, where $\bar{v}_1(s) : t \mapsto u(st,t^{-1})$ and $\bar{v}_2(s) : t \mapsto u(t,t^{-1}s)$. But ζ_1 is a morphism of left, and ζ_2 - that of right $L^1(G)$-modules. Therefore their common value is a morphism of bimodules, which is, because of $M(=1) = 1$, a left inverse to π^*.

... In conclusion, let us discuss the results about the amenability in the class of C*-algebras. Kadison and Ringrose (KR) [B, no.128] has established in 1971 the amenability of all uniformly hyperfinite C*-algebras (K(H), in particular), and of all $C_0(\Omega)$ as well. It became apparent a little bit later that actually every approximatively-dimensional C*-algebra, that is such an A, which is the uniform closure of the union of some family A_ν; $\nu \in \Lambda$ of its finite-dimensional sub-C*-algebras, directed by inclusion, is amenable. Following [B], let us comment on this fact. Every $\pi_\nu^* : A_\nu^* \to (A_\nu \hat{\otimes} A_\nu)^*$ has a left inverse ρ_ν in A_ν-mod-A_ν of the norm 1 (cf. the beginning of this section). If we take $\varphi \in (A \hat{\otimes} A)^*$, restrict it on $A_\nu \hat{\otimes} A_\nu$, then apply ρ_ν and extend the resulting functional, preserving its norm, to all A, we shall construct a map (not necessarily operator!) $\zeta_\nu : (A \hat{\otimes} A)^* \to A^*$. It is not difficult to prove, with the help of theorems of Alaoglu and Tichonoff that ζ_ν; $\nu \in \Lambda$ converges in the corresponding topology to some ζ, which must be now a morphism of A-bimodules.

It is the latter, which will be a left inverse to π^*, being required in Theorem 22 (IIa).

Since hyperfinite von Neumann algebras are ultraweak closures of approximatively finite-dimensional operator C*-algebras, we obtain, taking into account Theorem 3, the amenability-after-Connes of these algebras (the fact mentioned earlier in Theorem 7).

Almost simultaneously with KR, Johnson [B, no.3], who also used the technics connected with amenable groups, has proved that every postliminal C*-algebra, in particular, the same K(H) and $C_0(\Omega)$, is amenable. (Later Sheinberg [B, no.214] has shown that the only amenable uniform algebras are $C_0(\Omega)$).

The first example of non-amenable C*-algebra was indicated by Bunce in 1976: it is the reduced C*-algebra of every free group. This is a corollary of some general theorem of Bunce [B, no.72], where it was shown, in particular, that the amenability of $C_r^*(G)$ for a discrete G is equivalent to the amenability of G.

However, the problem concerning the general characterization of amenable C*-algebras has remained open until 1982. Its solution has required the simultaneous consideration of the problem of a characterization of amenable-after-Connes von Neumann algebras. The main role was played by "injective" von Neumann algebras, which were discovered by Connes in 1976 in his well-known paper [B, no.113]; thus were called those $A \subseteq B(H)$, for which there exists a projector $P \in B(B(H))$ with the range A and with the norm 1. Connes has proved that this property is equivalent to the hyperfiniteness and to several other important properties of a von Neumann algebra.

Afterwards it became clear that the indicated class of algebras is closely connected with a very important class of C*-algebras, which were discovered a little bit later. We mean so-called nuclear algebras. By analogy with the nuclearity of Grothendieck, Lance [B, no.134] called a C*-algebra A nuclear, if there existed only one C*-norm in $A \otimes B$ for any C*-algebra B (there are many such norms in the general case). One can come to nuclear C*-algebras by many ways, which appear to be quite different. We need the following important characterization (Choi and Effros [19]): A is nuclear iff its enveloping von Neumann algebra A** is injective (in the sense of Connes). Now we are able to formulate the main result concerning both types of amenability.

Theorem 31. (I) A C*-algebra A is amenable-after-Johnson iff it is nuclear.

(II) A von Neumann algebra A is amenable-after-Connes iff it is injective (or, equivalnetly, hyperfinite).

The second assertion was proved before the first one, in 1978. This deep result was established by Connes [B, no.114] (however, under the assumption of A being separable, which was removed by Elliot [B, no. 222]). The argument essentially used Theorem 7 (by Johnson, Kadison and Ringrose) which was formulated in Sect.1. Now we, "grown wise with experience", see that (I) is an immediate corollary of (II) combined with Theorem 23 and the theorem of Choi and Effros mentioned above. Nevertheless, actually the whole Theorem 31 was completed earlier than the simple proof of "if" part of Theorem 23 became known. Indeed, Connes (see idem), aware of only the "only if" part at that time, has deduced from that part that the amenability-after-Johnson does imply the nuclearity (and as a by-product he proved that B(H), being, as Wasserman has already proved [B, no.77], non-nuclear, is certainly not amenable-after-Johnson). As to the remaining implication "nuclearity $\Longrightarrow$ amenability", this was a subject matter of a difficult theorem, proved in 1982 by Haagerup [B, no.187].

References

1. Putinar M. On analytic modules: softness and quasicoherence. Complex analysis and applications, 1985, Publ. House of the Bulgarian Acad. Sci. Sofia, 1986, 534-547.

2. Connes A. Non-commutative differential geometry, Parts I and II, I.H.E.S. 62 (1985), 157-360.

3. Tzygan B.L. Homology of matrix Lie algebras over rings and Hochschild homology, Uspekhi Mat. Nauk 38 (1983), 217-218 (in Russian).

4. Christensen E., Sinclair A.M. On the vanishing of $H^n(A,A^*)$ for certain C*-algebras, Pacific J. Math. 137 (1989), 55-63.

B. Helemskii A.Ya. The Homology of Banach and Topological Algebras. Kluwer, Dordrecht, 1989.

5. Helemskiĭ A.Ya. Banach and polynormed algebras: the general theory, representations, homology. Nauka, Moscow, 1989 (in Russian) - to be translated into English, Oxford Univ. Press, London, 1991.

6. Operator algebras and applications. Proc. of Symp. in Pure Math., v.38, Part II. Kadison R.V., ed. Providence, 1982.

7. Christensen E., Evans D.E. Cohomology of operator algebras and quantum dynamical semigroups, J. London Math. Soc. 20 (1979).

8. Effros E.G. Advances in quantized functional analysis. Proc. ICM, 1986, v.2, 906-916.

9. Helemskiĭ A.Ya. Homological algebra background of the "amenability-after-Connes": injectivity of the predual bimodule, Mat. Sb. 180, no.12, 1680-1690 (in Russian).

10. Christensen E., Effros E.G., Sinclair A.M. Completely bounded multilinear maps and C*-algebraic cohomology, Invent. Math. 90 (1987), 279-296.

11. Bade W.G., Curtis P.C., Dales H.G. Amenability and weak amenability for Beurling and Lipshitz algebras, Proc. London Math. Soc. (3) 55 (1987), 359-377.

12. Groenbaek N. A characterization of weakly amenable algebras, Studia Math. XCIV (1989) 149-162.

13. Connes A. Cohomologie cyclique et foncteur Ext^n, C.R.Acad. Sci. Paris, serie I, 296 (1983), 953-958.

14. Pugach L.I. Homological properties of functional algebras and analytic polydiscs in their maximal ideal spaces, Rev. Roumaine Math. Pure and Appl. 31 (1986), 347-356 (in Russian).

15. Ogneva O.S. Coincidence of homological dimensions of Frechet algebra of smooth functions on a manifold with the dimension of the manifold, Funct. anal. i pril. 20 (1986), 92-93 (in Russian).

16. Golovin Yu.O. Homological properties of Hilbert modules over nest operator algebras, Mat. Zametki 41 (1987), 769-775 (in Russian).

17. Effros E.G. Amenability and virtual diagonals for von Neumann algebras, J. Funct. Anal. v.78 (1988), 137-153.

18. Lazar A.J., Tsui S.-K., Wright S. A cohomological characterization of finite-dimensional C*-algebras, J. Operator Theory 14 (1985)

19. Choi M.-D., Effros E.G. Nuclear C*-algebras and injectivity: the general case, Indiana Univ. Math. J. 26(1977), 443-446.

DUALITY IN STABLE SPENCER COHOMOLOGIES

V.V.Lychagin, L.V.Zil'berglеit
All-Union Correspondence Institute
of Civil Engineering
S.Kalitnikovskaya, 30
109807, Moscow, USSR

In the present paper we continue the study of the methods for the calculation of stable Spencer cohomologies started in [3].

Spencer cohomologies, from the point of view of the theory of sheaves, are the cohomologies of a manifold with coefficients in the sheaf of solutions of this system of differential equations. This is at least so if the stable Spencer complex is locally exact. In this case ordinary duality theorems from the theory of sheaves give duality theorems in stable Spencer cohomologies. A posteriori it appears that dual cohomologies are Spencer cohomologies of a certain (generally speaking, another) system of differential equations.

Here we propose a more direct way of proving such duality theorems. We use the methods of differential calculus and homological algebra, which fact allows us, in particular, to by-pass the question of a local exactness of Spencer complex. The basic results of the work are two duality theorems: theorems 2 and 3. These theorems, applied to the differential equations given by the classical differential operators, lead to the dualities of Poincare and Serre.

1. Green's formula.

Green's formula is the basis for the mechanism of obtaining duality theorems. For our purpose it is sufficient to use Green's formula for first-order differential operators.

At first we introduce the necessary notations. Let M be a smooth compact oriented manifold of dimension n. We denote by $\Gamma(\alpha)$ a module of smooth sections of a vector bundle $\alpha : E(\alpha) \rightarrow M$. If $\alpha = \mathbb{1} : M \times R$ M is a trivial linear bundle, then the section module of this bundle, which is naturally identified with the algebra of smooth functions on M, is denoted by F. We denote by $\pi_k : J^k(\alpha) \to M$ the bundle of k-jets of a bundle α and by $\pi_{k,k-1} : J^k(\alpha) \longrightarrow J^{k-1}(\alpha)$ the natural projection. The section module of the bundle π_k is denoted by $\mathcal{J}^k(\alpha)$. Let $\mathrm{Diff}_k(\alpha, \beta)$ be a module of linear differential operators of not higher than k-th order acting from the sections of a bundle α into the sections of a bundle β. Each operator $\nabla \in \mathrm{Diff}_k(\alpha,\beta)$ corresponds to homomorphism φ_∇, such that the diagram

$$\begin{array}{ccc} \Gamma(\alpha) & \xrightarrow{\quad\nabla\quad} & \Gamma(\beta) \\ & \searrow^{j_k} & \uparrow \varphi_\nabla \\ & & \mathcal{J}^k(\alpha) \end{array}$$

is commutative.

Later on we shall deal mainly with first-order operators. Recall that the symbol of the operator $\nabla \in \mathrm{Diff}_1(\alpha,\beta)$ is the morphism of vector bundles

$$\sigma(\nabla) : \alpha \otimes \tau^* \longrightarrow \beta$$

such that

$$\sigma(\nabla)\,(a(x) \otimes d_x f) = \nabla(fa)(x),$$

where $f \in F$, $f(x) = 0$, $a \in \Gamma(\alpha)$, $\tau^* : T^*M \longrightarrow M$ is a cotangent bundle.

For every differential I-form $\lambda \in \Gamma(\tau^*)$ let us denote by $\sigma_\lambda(\nabla)$ the value of a morphism $\sigma(\nabla)$ on form λ, i.e. such a morphism of vector bundles $\sigma_\lambda(\nabla) : \alpha \longrightarrow \beta$, that $\sigma_\lambda(\nabla)(a) = \sigma(\nabla)(a \otimes \lambda)$ for any section $a \in \Gamma(\alpha)$.

We call the bundle $\alpha^t = \mathrm{Hom}(\alpha, \Lambda^n \tau^*)$ dual to the bundle α. Denote by

$$\langle\ ,\ \rangle : \Gamma(\alpha) \otimes \Gamma(\alpha^t) \longrightarrow \Lambda^n \tau^*$$

a natural pairing between the sections of bundles α and α^t.

For every morphism $s \in \mathrm{Hom}(\alpha,\beta)$ denote by $s^t \in \mathrm{Hom}(\beta^t, \alpha^t)$ a dual morphism defined, as usually, by the relation:

$$\langle sa, b^t \rangle = \langle a, s^t b^t \rangle,$$

where $a \in \Gamma(\alpha)$, $b^t \in \Gamma(\alpha^t)$.

<u>Proposition 1</u>. Let $\nabla \in \mathrm{Diff}_1(\alpha,\beta)$, $\square \in \mathrm{Diff}_1(\beta^t, \alpha^t)$. Consider a mapping $\gamma(\nabla, \square)$ associating the pair (a, b^t), $a \in \Gamma(\alpha)$, $b^t \in \Gamma(\beta^t)$ with n-form $\langle \nabla(a), b^t \rangle - \langle a, \square(b^t) \rangle$.

This mapping is a first-order differential operator, acting from the bundle $\alpha \otimes \beta^t$ into the bundle $\Lambda^n \tau^*$ iff the equality

$$\sigma_\lambda^t(\nabla) + \sigma_\lambda(\Box) = 0$$

is satisfied for all covectors $\lambda \in \Gamma(\tau^*)$.

If the conditions of the proposition are fulfilled, the symbol of the operator $\gamma(\nabla, \Box)$ defines the homomorphism $w_\nabla : \alpha \otimes \beta^t \longrightarrow \Lambda^{n-1} \tau^*$, acting in the following way:

$$\sigma_\lambda(\gamma(\nabla,\Box))(a \otimes b^t) = \langle \sigma_\lambda(\nabla)a, b^t \rangle = \lambda \wedge w_\nabla(a, b^t).$$

This formula shows that the symbols of the operators $\gamma(\nabla, \Box)$ and $d \circ w_\nabla$ coincide. So, the operators are distinguished by a homomorphism. Consequently, replacing operator $\Box$ by the zero-order operator we may say that the operators γ and $d \circ w_\nabla$ coincide. The operator $\Box$, obtained in this way, we denote by ∇^t and call it a dual operator to operator ∇. These arguments show that holds the following

Proposition 2. For every operator $\nabla \in \mathrm{Diff}_1(\alpha, \beta)$ there exists a unique dual operator $\nabla^t \in \mathrm{Diff}_1(\beta^t, \alpha^t)$ defined by Green's formula:

$$\langle \nabla(a), b^t \rangle - \langle a, \nabla^t(b^t) \rangle = dw_\nabla(a, b^t),$$

where $a \in \Gamma(\alpha)$, $b^t \in \Gamma(\beta^t)$.

2. Hodge - Spencer Theory.

In this section we present the well-known Hodge-Spencer theory in the form convenient for us. The basic result of this section is Theorem 1 which establishes the isomorphism between cohomologies of an elliptic complex and the complex dual to it.

Further M is a compact oriented Riemannian manifold with metric g. Each metric g_α in bundle α defines the isomorphism

$$g_\alpha : \alpha \longrightarrow \alpha^*$$

where $\alpha^* = \mathrm{Hom}(\alpha, \mathbb{1})$ is a conjugate bundle.

Define the operator

$$*_{\alpha} \; : \Lambda^{j} \tau^{*} \otimes \alpha \longrightarrow \Lambda^{n-j} \tau^{*} \otimes \alpha^{*}$$

for all $j = 0,1,\dots,n$ by the relation

$$*_{\alpha}(w \otimes a) = \; * w \otimes g_{\alpha}(a),$$

where $w \in \Lambda^{j}(M) = \Gamma(\Lambda^{j} \tau^{*})$, $a \in \Gamma(\alpha)$ and $* : \Lambda^{j} \tau^{*} \to \Lambda^{n-j} \tau^{*}$ is Hodge operator, corresponding to metric g.

Note that $*_{\alpha^{t}} = *_{\alpha}^{-1}$.

If bundles α and β are provided with metrics g_{α} and g_{β}, then each differential operator $\nabla \in \mathrm{Diff}_1(\alpha, \beta)$ may correspond, as usually, to the operator ∇^{*} $\mathrm{Diff}_1(\beta, \alpha)$, conjugate with respect to the metric

$$(\nabla(a), b)_{\beta} = (a, \nabla^{*}(b))_{\alpha}$$

where $a \in \Gamma(\alpha)$, $b \in \Gamma(\beta)$ and, for example, $(a_1, a_2)_{\alpha} = \int_M \langle a_1, *_{\alpha} a_2 \rangle$.

The following commutative diagram shows the connection between two types of conjunctions of differential operators:

$$\begin{array}{ccc}
\Gamma(\beta) & \xrightarrow{\;*_{\beta}\;} & \Gamma(\beta^{t}) \\
\nabla^{*} \big\downarrow & & \big\downarrow \nabla^{t} \\
\Gamma(\alpha) & \xrightarrow{\;*_{\alpha}\;} & \Gamma(\alpha^{t})
\end{array}$$

The diagram's commutativity is a corollary to Green's formula:

$$(\nabla(a), b)_{\beta} = \int_M \langle \nabla(a), *_{\beta} b \rangle = \int_M \langle a, \nabla^{t}(*_{\beta} b) \rangle =$$
$$= \int_M \langle a, *_{\alpha}(*_{\alpha}^{-1} \nabla^{t} *_{\beta} b) \rangle = (a, *_{\alpha}^{-1} \nabla^{t} *_{\beta} b)_{\alpha} .$$

From the above constructions the transition to a dual operator is functorial and the transition to a conjugate one is artificial, since the latter is connected with the choice of a metric. The duality theorems, which have a natural character, establish an isomorphism between the cohomologies of this complex and the dual one. This fact explains the importance of the transition from operator ∇ to operator ∇^{t}. The transition to the conjugate operators has a purely technical character and it is necessitated by the historical reasons, i.e. Hodge-Specncer theory. Let us pass on to exact definitions.

Consider a complex of first-order differential operators

$$0 \rightarrow \Gamma(\alpha_0) \xrightarrow{\nabla_1} \Gamma(\alpha_1) \xrightarrow{\nabla_2} \Gamma(\alpha_2) \longrightarrow \dots \longrightarrow \Gamma(\alpha_N) \rightarrow 0 \quad (1)$$

Then, passing on to dual bundles, we obtain the complex:

$$0 \longleftarrow \Gamma(\alpha_0^t) \xleftarrow{\nabla_1^t} \Gamma(\alpha_1^t) \xleftarrow{\nabla_2^t} \Gamma(\alpha_2^t) \longleftarrow \dots \longleftarrow \Gamma(\alpha_N^t) \longleftarrow 0 \quad (2)$$

If, besides, the bundles are provided with a metric, then the transition to conjugate operators yields the complex

$$0 \longleftarrow \Gamma(\alpha_0) \xleftarrow{\nabla_1^*} \Gamma(\alpha_1) \xleftarrow{\nabla_2^*} \Gamma(\alpha_2) \longleftarrow \dots \longleftarrow \Gamma(\alpha_N) \longleftarrow 0 \quad (3)$$

The following commutative diagram establishes the connection between complexes (2) and (3):

$$\begin{array}{ccccccccc}
0 \longleftarrow & \Gamma(\alpha_0) & \xleftarrow{\nabla_1^*} & \Gamma(\alpha_1) & \xleftarrow{\nabla_2^*} & \Gamma(\alpha_2) & \longleftarrow \dots \longleftarrow & \Gamma(\alpha_N) & \longleftarrow 0 \\
 & \downarrow *_{\alpha_0} & & \downarrow *_{\alpha_1} & & \downarrow *_{\alpha_2} & & \downarrow *_{\alpha_N} & \\
0 \longleftarrow & \Gamma(\alpha_0^t) & \xleftarrow{\nabla_1^t} & \Gamma(\alpha_1^t) & \xleftarrow{\nabla_2^t} & \Gamma(\alpha_2^t) & \longleftarrow \dots \longleftarrow & \Gamma(\alpha_N^t) & \longleftarrow 0
\end{array}$$

Recall that complex (1) is called elliptical, if the corresponding symbolic complex

$$0 \longrightarrow \Gamma(\alpha_0) \xrightarrow{\sigma_\lambda(\nabla_1)} \Gamma(\alpha_1) \xrightarrow{\sigma_\lambda(\nabla_2)} \Gamma(\alpha_2) \longrightarrow \dots \longrightarrow \Gamma(\alpha_N) \longrightarrow 0 \quad (4)$$

is acyclic for any non-zero covector $\lambda \in \Gamma(\tau^*)$.

The ellipticity of complex (1) is equivalent to the ellipticity of the dual complex (2) or the conjugate complex (3). This follows from the fact that the symbolic complex of complex (2) is obtained from the symbolic complex (4) by applying the functor Hom($\quad, \Lambda^n \tau^*$).

The proof of the first duality theorem, given below, is based on the following two facts: firstly, the basic result of Hodge-Spencer theory, which states that in each class of cohomologies of an elliptic complex given on a compact oriented Riemannian manifold there exists exactly one harmonic representative. Secondly, Proposition 3, which connects Laplace operators of complex (1) and the dual complex (2).

For complex (1) the Laplace operators

$$\Delta_j : \Gamma(\alpha_j) \longrightarrow (\alpha_j)$$

have the form:

$$\Delta_j = \nabla_j^* \circ \nabla_j + \nabla_{j-1} \circ \nabla_{j-1}^*$$

while for the dual complex (2)

$$\Delta_j^t : \Gamma(\alpha_j^t) \longrightarrow \Gamma(\alpha_j^t)$$

$$\Delta_j^t = \nabla_j^t \circ (\nabla_j^t)^* + (\nabla_{j-1}^t)^* \circ \nabla_{j-1}^t$$

Proposition 3. The following diagram

$$\begin{array}{ccc} \Gamma(\alpha_j) & \xrightarrow{\ *_{\alpha_j}\ } & \Gamma(\alpha_j^t) \\ \Delta_j \downarrow & & \Delta_j^t \downarrow \\ \Gamma(\alpha_j) & \xrightarrow{\ *_{\alpha_j}\ } & \Gamma(\alpha_j^t) \end{array}$$

is commutative.

Proof. Let us transform the expression for operators Δ_j and Δ_j^t:

$$\Delta_j = *_{\alpha_j}^{-1} \nabla_j^t *_{\alpha_{j+1}} \nabla_j + \nabla_{j-1} *_{\alpha_{j-1}}^{-1} \nabla_{j-1}^t *_{\alpha_j}$$

$$\Delta_j^t = \nabla_j^t *_{\alpha_{j+1}^t}^{-1} \nabla_j *_{\alpha_j^t} + *_{\alpha_j^t}^{-1} \nabla_{j-1} *_{\alpha_{j-1}^t} \nabla_{j-1}^t =$$

$$\nabla_j^t *_{\alpha_{j+1}} \nabla_j *_{\alpha_j}^{-1} + *_{\alpha_j} \nabla_{j-1} *_{\alpha_{j-1}}^{-1} \nabla_{j-1}^t$$

Therefore hold the equalities

$$*_{\alpha_j} \Delta_j = \nabla_j^t *_{\alpha_{j+1}} \nabla_j + *_{\alpha_j} \nabla_{j-1} *_{\alpha_{j-1}}^{-1} \nabla_{j-1}^t *_{\alpha_j} =$$

$$\Delta_j^t *_{\alpha_j} .$$

Denote by H^k the cohomologies of complex (1) in the term $\Gamma(\alpha_k)$ and by H_k the cohomologies of the dual complex (2) in the term $\Gamma(\alpha_k^t)$. Then holds

Theorem 1. On a compact oriented manifold the operators $*_{\alpha_j} : H^j \longrightarrow H_j$, defined by a metric, give the isomorphisms of cohomologies of elliptic complexes.

We shall call the complex of differential operators (1) the Poincare complex, if there exist isomorphisms A_j from the bundle α_j in the bundle α_{N-j}^t, establishing the isomorphisms of complexes

(1) and (2):

$$\begin{array}{ccccccccc}
0 \to \Gamma(\alpha_0) & \xrightarrow{\nabla_1} & \Gamma(\alpha_1) & \xrightarrow{\nabla_2} & \Gamma(\alpha_2) & \to \dots \to & \Gamma(\alpha_N) \to 0 \\
\downarrow A_0 & & \downarrow A_1 & & \downarrow A_2 & & \downarrow A_N \\
0 \to \Gamma(\alpha_N^t) & \xrightarrow{\nabla_N^t} & \Gamma(\alpha_{N-1}^t) & \xrightarrow{\nabla_{N-1}^t} & \Gamma(\alpha_{N-2}^t) & \to \dots \to & \Gamma(\alpha_0^t) \to 0
\end{array}$$

The following result is a corollary to Theorem 1.

<u>Corollary</u> (Poincare duality).

Under the conditions of the previous theorem for Poincare complex there takes place the isomorphism

$$*_{\alpha_{N-j}}^{-1} \circ A_j : H^j \qquad H^{N-j}$$

<u>Example</u>. Consider the de Rham complex

$$0 \longrightarrow \Lambda^0(M) \longrightarrow \Lambda^1(M) \longrightarrow \Lambda^2(M) \longrightarrow \dots \longrightarrow \Lambda^n(M) \longrightarrow 0$$

Its dual complex has the form:

$$0 \longrightarrow \Lambda^0(M) \xrightarrow{(-1)^n d} \Lambda^1(M) \xrightarrow{(-1)^{n-1} d} \Lambda^2(M) \longrightarrow \dots \longrightarrow \Lambda^n(M) \longrightarrow 0$$

Having taken $A_j = (-1)^{\nu(n,\, j)}$, where $\nu(n, j) = \left[\frac{n}{2}\right] + \left[\frac{n+j}{2}\right]$, we obtain the classical Poincare duality theorem.

3. Dual Spencer Complexes.

In this section we apply the above results to stable Spencer complexes.

Let $E \subset J^k(\alpha)$ be a formally integrable system of differential equations [1,2]; let $E^{(\ell)} \subset J^{k+\ell}(\alpha)$ be its ℓ-th prolongation, $\ell \geqslant 0$.

Denote by $\mathcal{E} \subset \mathcal{J}^k(\alpha)$ ($\mathcal{E}^{(\ell)} \subset \mathcal{J}^{k+\ell}(\alpha)$, respectively) the module of the sections of the bundle $\pi_k : E \longrightarrow M$ ($\pi_{k+\ell} : E^{(\ell)} \longrightarrow M$).

Spencer complex, associated with this system of differential equations, has the form:

$$0 \longrightarrow \mathcal{E}^{(\ell)} \xrightarrow{\mathcal{D}} \mathcal{E}^{(\ell-1)} \underset{F}{\otimes} \Lambda^1(M) \xrightarrow{\mathcal{D}} \dots \longrightarrow \mathcal{E}^{(\ell-n)} \underset{F}{\otimes} \Lambda^n(M) \to 0 \quad (5)$$

The cohomologies of this complex in the term $\mathcal{E}^{(\ell-j)} \underset{F}{\otimes} \Lambda^j(M)$ for sufficiently large values of ℓ are stabilized. Denote by $H^j(E)$ the corresponding stable cohomologies.

We shall start the construction of dual complexes with the absolute Spencer complex:

$$0 \to \Gamma(\alpha) \xrightarrow{j_k} \mathcal{J}^k(\alpha) \xrightarrow{\mathcal{D}} \mathcal{J}^{k-1}(\alpha) \underset{F}{\otimes} \Lambda^1(M) \to \dots \to \mathcal{J}^{k-n}(\alpha) \underset{F}{\otimes} \Lambda^n(M) \to 0 \quad (6)$$

$$(\mathcal{J}^k(\alpha))^t = \mathrm{Hom}\,(J^k(\alpha), \Lambda^n \tau^*) = \mathrm{Diff}_k(\alpha, \Lambda^n \tau^*) =$$

$\mathrm{Diff}_k(\alpha, \mathbb{1}) \underset{F}{\otimes} \Lambda^n(M)$ is a module, dual to the module of the sections $\mathcal{J}^k(\alpha)$.

Therefore

$$(\mathcal{J}^k(\alpha) \underset{F}{\otimes} \Lambda^i(M))^t = \mathrm{Hom}\,(J^k(\alpha) \otimes \Lambda^i \tau^*, \Lambda^n \tau^*) =$$

$$\mathrm{Hom}\,(J^k(\alpha), \mathrm{Hom}(\Lambda^i \tau^*, \Lambda^n \tau^*)) = \mathrm{Diff}_k(\alpha, \Lambda^{n-i} \tau^*)$$

$$= \mathrm{Diff}_k(\alpha, \mathbb{1}) \underset{F}{\otimes} \Lambda^{n-i}(M).$$

Thus the complex, dual to Spencer complex (6) has the form

$$\dots \longrightarrow \mathrm{Diff}_s(\alpha, \mathbb{1}) \underset{F}{\otimes} \Lambda^i(M) \xrightarrow{D^t} \mathrm{Diff}_{s+1}(\alpha, \mathbb{1}) \underset{F}{\otimes} \Lambda^{i+1}(M) \to \dots \quad (7)$$

and operator D^t, dual to Spencer operator D, acts by the following rule:

$$D^t(\nabla \otimes \omega) = (-1)^{n-i}(d \circ \nabla) \wedge \omega + (-1)^{n-i} \zeta_{s,s+1} \nabla \circ d\omega \quad (8)$$

where $\omega \in \Lambda^i(M)$, $\nabla \in \mathrm{Diff}_s(\alpha, \mathbb{1})$,

$\zeta_{s,s+1} : \mathrm{Diff}_s(\alpha, \mathbb{1}) \longrightarrow \mathrm{Diff}_{s+1}(\alpha, \mathbb{1})$ is the natural embedding.

Note that the dual complex (7), as well as complex (6), is acyclic.

Entering upon the construction of the complex, dual to complex (5), we shall describe the module $\mathcal{E}_{(\ell)} = (\mathcal{E}^{(\ell)})^t$. To this end denote by $\mathrm{Ann}\,\mathcal{E}^{(\ell)}$ the annihilator of the system of equations $\mathcal{E}^{(\ell)}$:

$$\mathrm{Ann}\ \mathcal{E}^{(\ell)} = \{\nabla \in \mathrm{Diff}_{k+\ell}(\alpha, \mathbb{1}) \mid \varphi_\nabla(E^{(\ell)}) = 0 \ .$$

Then, $\mathcal{E}_{(\ell)} = \mathrm{Diff}_{k+\ell}(\alpha, \mathbb{1}) \,/\, \mathrm{Ann}\ \mathcal{E}^{(\ell)}$.

Therefore $(\mathcal{E}^{(\ell)} \underset{A}{\otimes} \Lambda^i(M))^t = \mathcal{E}_{(\ell)} \underset{A}{\otimes} \Lambda^{n-i}(M)$

and, consequently, the complex, dual to complex (5), has the form:

$$\ldots \to \mathcal{E}_{(\ell)} \underset{A}{\otimes} \Lambda^i(M) \xrightarrow{D^t} \mathcal{E}_{(\ell+1)} \underset{A}{\otimes} \Lambda^{i+1}(M) \to \ldots \qquad (9)$$

The cohomologies of this complex are stabilized for rather large values of ℓ. In fact, consider the following commutative diagram:

$$\begin{array}{ccc}
 & 0 & & 0 & \\
 & \downarrow & & \downarrow & \\
\ldots \to & g^{\ell+1} \otimes \Lambda^i(M) & \xrightarrow{\ \ \delta\ \ } & g^{\ell} \otimes \Lambda^{i+1}(M) & \to \ldots \\
 & \downarrow & & \downarrow & \\
\ldots \to & \mathcal{E}^{(\ell+1)} \otimes \Lambda^i(M) & \xrightarrow{\ \ \mathcal{D}\ \ } & \mathcal{E}^{(\ell)} \otimes \Lambda^{i+1}(M) & \to \ldots \\
 & \downarrow \pi_{\ell+1,\ell} \otimes 1 & & \downarrow \pi_{\ell,\ell-1} \otimes 1 & \\
\ldots & \mathcal{E}^{(\ell)} \otimes \Lambda^i(M) & \xrightarrow{\ \ \mathcal{D}\ \ } & \mathcal{E}^{(\ell-1)} \otimes \Lambda^{i+1}(M) & \to \ldots \\
 & \downarrow & & \downarrow & \\
 & 0 & & 0 &
\end{array}$$

where by g^s the s-th prolongation of the symbol of the equation E is denoted and by δ – Spencer operator.

Passing on to dual complexes and using formula (8), we obtain the following commutative diagram:

$$\begin{array}{ccc}
 & 0 & & 0 & \\
 & \uparrow & & \uparrow & \\
\ldots \leftarrow & g_{\ell+1} \otimes \Lambda^{n-i}(M) & \xleftarrow{\ \ \delta^t\ \ } & g_{\ell} \otimes \Lambda^{n-i-1}(M) & \leftarrow \ldots \\
 & \uparrow & & \uparrow & \\
\ldots \leftarrow & \mathcal{E}_{(\ell+1)} \otimes \Lambda^{n-i}(M) & \xleftarrow{\ \ \mathcal{D}^t\ \ } & \mathcal{E}_{(\ell)} \otimes \Lambda^{n-i-1}(M) & \leftarrow \ldots \\
 & \uparrow \zeta_{\ell,\ell+1} \otimes 1 & & \uparrow \zeta_{\ell-1,\ell} \otimes 1 & \\
\ldots \leftarrow & \mathcal{E}_{(\ell)} \otimes \Lambda^{n-i}(M) & \xleftarrow{\ \ \mathcal{D}^t\ \ } & \mathcal{E}_{(\ell-1)} \otimes \Lambda^{n-i-1}(M) & \leftarrow \ldots \\
 & \uparrow & & \uparrow & \\
 & 0 & & 0 &
\end{array}$$

where by g_ℓ the modulus, dual to modulus g^ℓ is denoted.

The upper line of the last diagram is a complex, dual to Spencer δ-complex. Therefore it is acyclic for sufficiently high values of ℓ (Poincare δ-lemma). Hence follows the stabilization of cohomologies of Spencer complex (9).

Denote by $H_j(E)$ the corresponding stable cohomologies in the term $\mathcal{E}_{(\ell-j)} \underset{F}{\otimes} \Lambda^j(M)$.

Note that the ellipticity of Spencer complex is equivalent to the ellipticity of the corresponding system of differential equations E (Quillen theorem [1,2]).

The following result is a corollary to Theorem 1.

Theorem 2. Let $E \subset J^k(\alpha)$ be a formally integrable elliptic system of differential equations on a compact oriented manifold M. Then for stable Spencer cohomologies the following isomorphisms take place:

$$H^j(E) \xrightarrow{\sim} H_{n-j}(E), \quad j = 0,1,\dots,n.$$

Examples.

1. Let us consider the differential equation $E \subset J^1(\mathbb{1})$, defined by the operator of the exterior differentiation

$$E = \left\{ [f]^1_x \;\middle|\; \overset{d\,:\,F\,\longrightarrow\,\Lambda^1(M)}{d_x f = 0} \right\}.$$

The equation itself and all its prolongations define a one-dimensional trivial linear bundle over manifold M while the stable Spencer complex for this equation is de Rham complex. The dual Spencer complex is also de Rham complex. Thus theorem 2 is the classical Poincare duality theorem (cf. example in § 2).

The following example is a natural continuation of the previous one.

2. Let α be a vector bundle, provided with a flat connection and let $\nabla : \Gamma(\alpha) \longrightarrow \Gamma(\alpha) \otimes \Lambda^1(M)$ be its covariant differential. Consider the differential equation $E \subset J^1(\alpha)$

$$E = \left\{ [h]^1_x \mid \nabla_x h = 0 \right\}$$

The equation itself and all its prolongations are isomorphic to bundle α, while the stable Spencer complex of this equation is the complex

$$0 \longrightarrow \Gamma(\alpha) \xrightarrow{\nabla} \Gamma(\alpha) \otimes \Lambda^1(M) \longrightarrow \quad \ldots \quad \longrightarrow \Gamma(\alpha) \otimes \Lambda^n(M) \rightarrow 0$$

constructed with respect to a flat connection ∇. Its cohomologies are denoted by $H^i(M,\alpha)$. Dual to this complex is the one, constructed with respect to a flat connection on the conjugate bundle α^* (see example in § 2). Therefore from Theorem 2 it follows that on a compact oriented manifold M the isomorphism $H^i(M,\alpha) \xrightarrow{\sim} H^{n-i}(M,\alpha^*)$ is valid.

3. Let $E \subset J^k(\alpha)$ be a formally integrable system of differential equations of a finite type and let ℓ be such a number that $E^{(\ell+1)} \simeq E^{(\ell)}$. Then Spencer operator

$$\mathcal{E}^{(\ell)} \simeq \mathcal{E}^{(\ell+1)} \xrightarrow{D} \mathcal{E}^{(\ell)} \otimes \Lambda^1(M)$$

defines a flat connection on bundle $E^{(\ell)}$. Thus the stable Spencer cohomologies are isomorphic to the cohomologies considered in the previous example, while the corresponding duality has the above form.

4. Let M be a complex manifold and let α be a holomorphic bundle over M. Denote by $\Lambda^{p,q}(\alpha)$ differential (p,q) forms on manifold M with the values in bundle α and by $\Omega^p(\alpha)$ holomorphic p-forms with the value in bundle α.

The stable Spencer cohomologies of Cauchy-Riemann differential equation, corresponding to the operator $\bar{\partial} : \Lambda^{p,0}(\alpha) \longrightarrow \Lambda^{p,1}(\alpha)$, are Dolbeaut cohomologies $H^j(M,\Omega^p(\alpha))$. It is easy to check that the cohomologies of the dual complex are Dolbeaut cohomologies $H^j(M,\Omega^{n-p}(\alpha^*))$ and Theorem 2 is Serre dual theorem.

Note that the more general duality theorems on the manifold provided with polarization can also be obtained from Theorem 2 by applying it to the corresponding differential equation [5].

4. Dual equations.

In this section we determine the conditions, the fulfilment of which makes it possible to interpret the cohomologies of the dual complex as the cohomologies of a certain Spencer complex.

<u>Theorem 3</u>. (On three isomorphisms). Set three isomorphisms:

$$F_j : \mathcal{E}_{s_0+j} \longrightarrow \tilde{\mathcal{E}}^{N-s_0-j} \quad , \quad j = -1,\ 0,\ 1.$$

for some sufficiently high (stable) values of N and s_o. Let the isomorphisms be such that the diagram

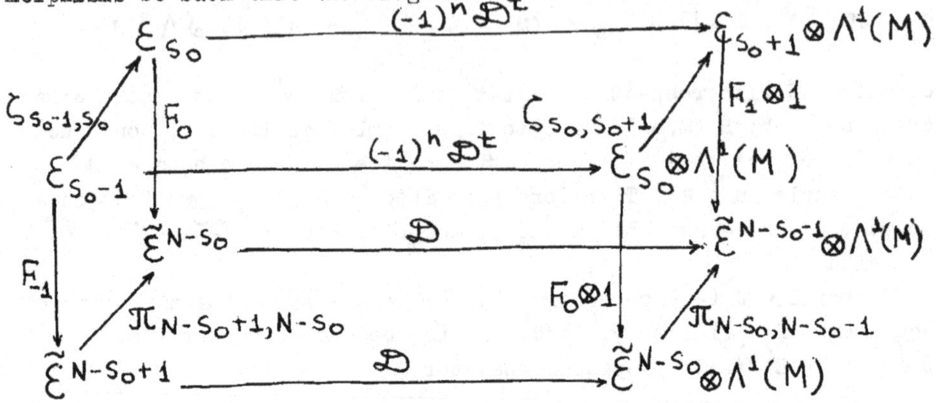

is commutative.

Then if the system of differential equations E is elliptic, the isomorphism of stable Spencer cohomologies takes place

$$H_\ell(E) \xrightarrow{\sim} H^\ell(\tilde{E})$$

and, consequently, with regard to Theorem 2, the isomorphism

$$H^\ell(E) \xrightarrow{\sim} H^{n-\ell}(\tilde{E})$$

To prove Theorem 3 it is sufficient to check that isomorphisms F_{-1}, F_o and F_1 for every value of i define such isomorphisms

$$F_{ji} : \mathcal{E}_{s_o+j} \otimes \Lambda^i(M) \longrightarrow \tilde{\mathcal{E}}^{N-s_o-j} \otimes \Lambda^i(M)$$

$$F_{ji} = F_j \otimes (-1)^{\nu(n,i)} id$$

$$j = -1,0,1, \quad i = 0,1,\dots,n$$

that the diagrams

$$\begin{array}{ccccc}
\mathcal{E}_{s_o-1} \otimes \Lambda^{i-1}(M) & \xrightarrow{D^t} & \mathcal{E}_{s_o} \otimes \Lambda^i(M) & \xrightarrow{D^t} & \mathcal{E}_{s_o+1} \otimes \Lambda^{i+1}(M) \\
\downarrow F_{-1\ i-1} & & \downarrow F_{o\ i} & & \downarrow F_{1\ i+1} \\
\tilde{\mathcal{E}}^{N-s_o+1} \otimes \Lambda^{i-1}(M) & \xrightarrow{D^t} & \tilde{\mathcal{E}}^{N-s_o} \otimes \Lambda^i(M) & \xrightarrow{D^t} & \tilde{\mathcal{E}}^{N-s_o-1} \otimes \Lambda^{i+1}(M)
\end{array} \tag{11}$$

are commutative.

Let us verify the commutativity of these diagrams. Let $\Theta\otimes\omega\in \mathcal{E}_{s_0+j}\otimes\Lambda^t(M)$. Then, using the example from section 2, we obtain the equalities

$$DF_{jt}(\theta\otimes\omega) = (-1)^{\nu(n,t)}DF_j(\theta)\otimes\omega +$$

$$(-1)^{\nu(n,t)}\ \pi_{N-s_0-j,N-s_0-j-1}\, F_j(\theta)\otimes d\omega$$

$$F_{j+1\,t+1}D^t(\theta\otimes\omega) = (-1)^{\nu(n,t)}\big[(F_{j+1}\otimes 1)((-1)^n D^t\theta)\wedge\omega$$

$$+ (-1)^{\nu(n,t)}\ F_{j+1}\,\zeta_{s_0+j,\ s_0+j+1}(\theta)\otimes d\omega$$

for $j = -1,0$.

Therefore for the commutativity of diagram (11) the relations should be satisfied

$$DF_j = (F_{j+1}\otimes 1)(-1)^n D^t$$

$$\pi_{N-s_0-j,\ N-s_0-1}\quad F_j = F_{j+1}\,\zeta_{s_0+j,\ s_0+j+1}$$

for $j = -1,0$.

It remains to note that these relations are contained in the commutative diagram (10).

The conditions of Theorem 3 can be reformulated in another way. To this end note that every isomorphism $F_j\colon \mathcal{E}_{s_0+j}\longrightarrow \tilde{\mathcal{E}}^{N-s_0-j}$ can be identified with a non-degenerate bilinear form

$$F_j:\ \mathcal{E}_{s_0+j}\otimes\tilde{\mathcal{E}}_{N-s_0-j}\longrightarrow F$$

$$(a,b)\longmapsto \langle F_j(a),\ b\rangle,$$

where $a\in\mathcal{E}_{s_0+j}$, $b\in\tilde{\mathcal{E}}_{N-s_0-j}$.

The forms F_j are naturally extended to the forms

$$F_j:(\mathcal{E}_{s_0+j}\otimes\Lambda^\ell(M))\otimes(\tilde{\mathcal{E}}_{N-s_0-j}\otimes\Lambda^t(M))\longrightarrow\Lambda^{\ell+t}(M)$$

$$(a\otimes V,\ b\otimes\omega)\longmapsto\langle F_j(a),b\rangle\ \otimes\ (V\wedge\omega),$$

where $V \in \Lambda^{\ell}(M)$, $\omega \in \Lambda^{t}(M)$.

Holds

Proposition 4. The diagram

$$\begin{array}{ccc} \mathcal{E}_{s_0+j} & \xrightarrow{(-1)^n D^t} & \mathcal{E}_{s_0+j+1} \otimes \Lambda^1(M) \\ \downarrow F_j & & \downarrow F_{j+1} \otimes 1 \\ \tilde{\mathcal{E}}^{N-s_0-j} & \xrightarrow{\quad D \quad} & \tilde{\mathcal{E}}^{N-s_0-j-1} \otimes \Lambda^1(M), \; j = -1, 0 \end{array} \tag{12}$$

is commutative iff the equality

$$F_{j+1}((-1)^n D^t a, b) + (-1)^{n+1} F_j(a, (-1)^n D^t b) = dF_j(a, \zeta_{N-s_0-j-1, N-s_0-j} b)$$

is fulfilled for all $a \in \mathcal{E}_{s_0+j}$, $b \in \tilde{\mathcal{E}}_{N-s_0-j-1}$.

Proof. From the commutativity of diagram (12) we have the identity

$$\langle (F_{j+1} \otimes 1)((-1)^n D^t a), b \rangle = \langle D(F_j a), b \rangle$$

On the other hand, Green's formula for operator D has the form

$$\langle D(\bar{a}), b \otimes V \rangle - \langle \bar{a}, D^t(b \otimes V) \rangle =$$

$$d \left[\langle \bar{a}, \zeta_{N-s_0-j-1,\, N-s_0-j} b \rangle V \right] ,$$

where $V \in \Lambda^{n-1}(M)$, $\bar{a} \in \mathcal{E}^{s_0+j}$.

Using formula(8), let us rewrite Green's formula in the form

$$\langle D(\bar{a}), b \rangle - \langle \bar{a}, D^t(b) \rangle = d \langle \bar{a}, \zeta_{N-s_0-j-1, N-s_0-j} b \rangle$$

Therefore hold the equalities

$$\langle D(F_j a), b \rangle = \langle F_j a, D^t b \rangle +$$

$$d \langle F_j a, \zeta_{N-s_0-j-1, N-s_0-j} b \rangle = (-1)^n F_j(a, (-1)^n D^t b) +$$

$$dF_j(a, \zeta_{N-s_0-j-1,N-s_0-j}\, b)$$

This completes the proof of Proposition 4.

We shall call the system of differential equations autodual, if in the conditions of Theorem 3 the system of differential equations $\tilde{E}$ can be taken as the system of differential equations E.

The autoduality conditions can be reformulated as the condition of the existence of three non-degenerate bilinear forms on equation E, satisfying the conditions of Proposition 4.

Thus, for Lie equation E of a finite type, whose fibres $E^{(\ell)}$ are semi-simple algebras, such bilinear forms can be constructed with the help of Killing's form. In this case theorem 3 leads to the duality theorems obtained in [4].

Note also that for Cauchy-Riemann equation E in the bundle $\Lambda^{p,0}(\alpha)$ it is possible to choose Cauchy-Riemann equation in the bundle $\Lambda^{n-p,0}(\alpha^*)$ as equation E. The corresponding bilinear form is defined by a natural pairing. This gives an alternative proof of Serre's duality theorem.

References

1. Spencer D.C. Overdetermined systems of linear partial differential equations - Bull. of the Amer. Math. Soc., 1969, 75 : 2, p.179-239.

2. Quillen D.G. Formal properties of overdetermined systems of linear partial differential equations, Thesis, Harvard University, Cambridge, Massachusetts, 1964.

3. Zil'berglei̇t L.V., Lychagin V.V. Spencer cohomologies of differential equations. In: Lect. Notes in Math., 1990, vol. 1453, p.121-136.

4. Goldschmidt H. Duality theorems in deformation theory, Transactions of the Amer. Math. Soc., 1985, v.292, N 1, p.1-50.

5. Fischer H.R., Williams F.L. Complex-foliated structures. I. Cohomology of the Dolbeaut - Kostant complexes, Transactions of the Amer. Math. Soc., 1979, v.252, p.163-195.

ON SOME PROBLEMS OF COMPUTATIONAL GEOMETRY AND TOPOLOGY

O.R.Musin
Department of Geography
Moscow State University
119899, Moscow, USSR

Introduction.

The overall practical application of computers brought about the necessity to elaborate algorithms for the solution of a large number of geometric and topological problems. Many of these problems have been considered in geometry before but from somewhat different points of view. An essential difference between the computational and purely mathematical problem-solving lies in the fact that in the first case algorithm is a final result, whereas in the second case it is a proof of a theorem. Naturally, an algorithm should be justified (i.e. the proof of its convergence is required) and estimated from the point of view of its working time and the required memory capacity.

At present it is hard to enumerate all fields of science and technology in which computational geometry is used, but it is possible to point out those of them for which computational geometry serves as a mathematical basis. Among them is computer graphics, a technical discipline aimed at the construction of all sorts of projections and cross-sections of geometric objects, graphs of functions, drawings; the development of programs for computer animation etc. One of the key problems in computer graphics is the problem of removing invisible lines and surfaces. This problem can be solved in many ways involving a number of problems of computational geometry [3,22]. Objects of computer graphics (lines, surfaces, bodies) should be specifically represented in computer memory, i.e. their geometric models should be created. This is also one of the main problems of computer-aided design (CAD) aimed at automation of designers' routine work. A broad class of geometric problems arises upon image processing and pattern recognition. The problem of pattern recognition on an image fed into a computer system's input is, in a sense, an inverse problem of computer graphics [18]. Finally, an active user of computational geometry algorithms is a computer-aided cartography and geoinformatics many of whose problems are similar to computer graphics and image processing problems but which are also specific geometric problems.

At present there exists a certain confusion concerning the term "computational geometry". This word combination has at least two different meanings. The fundamental works [10, 12, 19] treat the discipline investigating the computational complexity of solving geometric problems in the framework of the theory of algorithm analysis. The journal "Discrete and Computational Geometry", almost entirely devoted to these questions, is being published from 1986. In our opinion, this discipline should be called "discrete computational geometry".

In the Soviet literature "computational geometry" is largely associated with the books [11, 21, 22] , devoted to the problems of modelling of curves and surfaces. Evidently, the term "geometric modelling " could be more appropriate here. In the Western literature some other terms can be encountered. In particular, the journal entitled "Computer Aided Geometric Design" is devoted to this discipline. In our opinion, it is natural to consider discrete computational geometry and geometric simulation as independent branches of computational geometry, just as in modern geomentry there exist combinatory and differential geometry, the analogy being sufficiently straightforward.

However, the list of names for computational geometry does not end here. In particular, the authors of [8] suggest using the term "engineering geometry" and at the Department of Mechanics and Mathematics of Moscow University the seminar "Computer Geometry" is being held from 1985. Of course, the name is not that important. More important is the fact that these questions draw much attention: annually dozens of books and hundreds of papers are published and there exist serious applications to practice.

The aim of the present paper is to acquaint the reader with the author's research concerning three problems of computational geometry. A customary phrase "Let us consider manifold M" is quite incomprehensible to a person working with a computer. From his point of view any manifold must be somehow given by a discrete set of initial data and provided with an algorithm allowing one to construct a manifold by these data. There are hundreds (or even thousands) of works dealing with this subject. In topology and mathematical physics the partition of unity is used for the proof of various existence theorems. In §1 it is shown that this is an effective method for the construction of algorithms of computational geometry (cf. [16]).

A qualitative investigation of ordinary differential equations and the search for singular points and separatrices of the corresponding vector field are rather effective methods of the modern mathematical physics. As a rule, such investigation by analytical methods is nontrivial and rather laborious. The methods of computational analysis and computer graphics (a phase portrait is shown on a display's screen) pose the problem of defining singular points and their types with the help of a computer.

A similar problem arises in geoinformatics: automated search for watersheds and talwegs (in topography a watershed is a line, separating the surface run-off of the opposite slopes of hills. Talweg is a line, connecting the lowest points of the bottom of a valley or a ravine) and the isolation of elementary basins by the terrain's model. From the mathematical point of view topography is the function $z = f(x,y)$ and talwegs and watershed are separatrices of the gradient vector field and the boundaries of elementary basins consist of watersheds. For the function $f(x,y)$ (for simplicity let it be a Morse function) the singular points of the gradient vector field may be of two types: index 1 - vertices, hollows; index (-1) - saddle points. Watersheds connect vertices and saddle points and talwegs connect saddle points and hollows.

An interesting review of this problem is given in [6] . Note that the well-known mathematician of the 19th century A.Cayley and the outstanding physicist J.K.Maxwell were much interested in this problem. The latter obtained the algorithm for drawing talwegs and watersheds on topographical maps (the maps with the plotted level curves).

Index is an important characteristic of a singular point. In a two-dimensional case index defines a type of a singular point. §2 deals with a discrete analogue of a singular point index of a vector field and a function and describes a simple algorithm for its calculation.

A number of practical problems in various branches of science and technology leads to the necessity of calculating distances on Riemannian manifolds and the search for shortest paths. These problems are largely manifest in the installation of pipelines or the construction of roads in a broken country.

A discrete analogue of these problems is well known. This is the problem of calculating distances and shortest paths on graphs. The solution is given by Dijkstra algorithm [1, 5] which has a wide application. In the continuous case there are the problems of the search for extremum on infinite dimensional functional spaces. For the Riemannian manifold M with the metric g_{ij} the shortest path among piecewise smooth curves is defined by the condition of the functional's minimum: $\int g_{ij} x_i y_j dt \longrightarrow \min$. Functionals of this type are, as a rule, nonconvex and they have several local minima. Therefore the method of steepest descent (the gradient method) is inapplicable to the determination of the global minimum. For the same reason it is impossible to use the method of "ranging fire" when geodesics are issued from the given point in different directions and among them one finds the geodesic which got into some neighbourhood of another point (the obtained geodesic need not be the shortest).

The analysis of different ways of defining the shortest paths lead to Huyghens principle, which, in our opinion, in some cases may be used to obtain the algorithm for the search for shortest paths. The application of Huyghens principle leads directly to complicated problems of the construction of envelopes and the storage of their parameters in computer's memory. § 3 contains a constructive definition, which in a certain sense is a discrete analogue of Huyghens principle.

In practice the algorithm of the search for shortest paths operates not on a smooth manifold but on its discrete (polyhedral) approximation. Naturally, there arises the problem of evaluating the accuracy of the algorithm operation. The corresponding result is given in 3 (cf. [17]).

Recently there appeared a number of works ([14,20] and some others) treating the algorithms of the search for shortest paths on two-dimensional polyhedrons. We shall deal with this case in detail and analyse the main points of the solution to this problem. Note that the starting point in the analysis is the notion of "cat locus" which goes back to A.Poncare and J.G.K.Whitehead.

The author wishes to express his gratitude to I.K.Babenko, S.N. Serbenyuk, S.P.Tarasov and A.T.Fomenko for valuable comments and helpful discussion of the present paper.

1. Partition of unity and geometric modelling .

The main problem of geometric modelling is the search for methods and the construction of restoration algorithms of the embedding of d-dimensional manifold M in $\mathbb{R}^m$, $F : M \hookrightarrow \mathbb{R}^m$, provided that restriction F on the set of manifolds $M_i \subset M$ is known. This problem is incorrect, since there are many ways of continuing the embedding F from $\{M_i\}$ on M. Therefore a number of methods (least squares method, splines) is based on the variational approach, i.e. F should additionally be the extremum of a functional. As a rule, the variational approach guarantees the uniqueness of the restoration of F. However, in practice they often use the variationally ungrounded methods (Coons portions, Bézier curves and surfaces). Here the main criterion is the method's acceptability for the majority of expert specialists.

The analysis of geometric modelling algorithms allows one to divide them into two groups: constructions by the method of coefficient comparison and constructions by partition of unity. The first method is well known. Here F(x) is represented in the form: $F(x) = \sum_i k_i f_i(x)$, where f_i is a collection of given mappings. Coefficients k_i are usually obtained from the system of linear equations composed in accordance with the problem. For example, if a finite collection of points p_i, $i = 1,2,\ldots,n$ and the collection of values in them Z_i are given on a plane and by these data it is required to construct the function $z = f(p)$, $f(p_i) = z_i$, giving the minimum of the integral $\iint \left(\left(\frac{\partial^2 f}{\partial x^2}\right)^2 + 2\left(\frac{\partial^2 f}{\partial x \partial y}\right)^2 + \left(\frac{\partial^2 f}{\partial y^2}\right)^2 \right) dxdy$, then it turns out [2] that f is represented in the form: $f(p) = \sum_{i=1}^{n} \lambda_i \cdot \|p - p_i\|^2 \ln \|p - p_i\| + Ax + By + C$. Coefficients λ_i, A, B and C are obtained from the system of linear equations $f(p_i) = Z_i$; $i = 1,\ldots,n$; $\sum \lambda_i = 0$; $\sum \lambda_i x_i = 0$; $\sum \lambda_i y_i = 0$. These are the so-called analytical, or D-splines.

The method of coefficient comparison is well known in computational mathematics, whereas the methodsof simulation with the help of partition of unity were not treated as a separate group. The aim of this section is to show that this group of methods is represented in computational geometry as widely as the method of coefficient comparison.

Let M be a compact smooth manifold and let $\{U_\alpha\}$ be a finite covering of M by open sets. The partition of unity, subordinated to the covering $\{U_\alpha\}$ is such a collection of functions $\varphi_\alpha(x) \in C(M)$ that

1) $\varphi_\alpha(x) = 0$ for all α and $x \in M \setminus U_\alpha$,
2) $1 \geqslant \varphi_\alpha(x) \geqslant 0$ for all α and $x \in M$,
3) $\sum_\alpha \varphi_\alpha(x) = 1$ for all $x \in M$.

It is noteworthy that usually [7, part 2, § 8] item 1) is formulated as supp $\varphi_\alpha \subset U_\alpha$. This discrepancy is not essential; all the constructions and the theorem on the existence of partition of unity remain true. On the other hand, our wording makes this item more constructive from the point of view of computational mathematics: when constructing $\{\varphi_\alpha\}$ it is sufficient to check that $\varphi_\alpha = 0$ on the boundary and put φ_α equal to zero outside of U_α.

For the construction of partition of unity two general and very simple ideas are used:

If $\{f_\alpha\}$ is a collection of continuous non-negative functions on M such that $f_\alpha = 0$ outside of U_α and all f_α do not turn zero simultaneously at a point M, then $\varphi_\alpha(x) = \frac{f_\alpha(x)}{\sum_\alpha f_\alpha(x)}$ is partition of unity. In particular, if $r > 0$, then $\varphi_{\alpha,r} = \varphi_\alpha^r / \sum_\alpha \varphi_\alpha^r$ is also partition of unity. Sometimes the latter construction makes it possible to increase the smoothness of partition of unity.

Let $K = M \times N$ be the Cartesion product of two manifolds. If on M and N there are coverings $\{U_\alpha\}$ and $\{V_\beta\}$ and $\{\varphi_\alpha\}$, $\{\psi_\beta\}$ are partitions of unity, corresponding to them, then the collection of functions $\varphi_{\alpha,\beta}(x,y) = \varphi_\alpha(x)\cdot\psi_\beta(y)$, $x \in M$, $y \in N$ is the partition of unity, corresponding to the covering $\{U_\alpha \times V_\beta\}$ of manifold K.

If for every α the local mapping $F_\alpha : \bar{U}_\alpha \to \mathbb{R}^m$ is given, then from them with the help of partition of unity it is possible to "sew" the mapping

$$F(x) = \sum_\alpha \varphi_\alpha(x) F_\alpha(x), \quad x \in M \tag{1.1}$$

Note the simplest properties of this construction:

a) Let $\{\varphi_\alpha\}$ be partition of unity and let all the derivatives of φ_α up to the order r be equal to zero outside of U_α. If $F_\alpha \in C^r(\bar{U}_\alpha)$, then $F \in C^r(M)$.

b) $\min_{\alpha\in\sigma} [F_\alpha(x)]_j \leq [F(x)]_j \leq \max_{\alpha\in\sigma} [F_\alpha(x)]_j$, where σ is the collection of those indices α, for which $x \in U_\alpha$, and $[\ ]_j$ denotes the j-th coordinate $\mathbb{R}^m$.

We shall now present the most wide-spread problems of geometric modelling for $d = 1,2$.

- $d = 1$. p_i ($p_i = F(M_i)$) is an ordered collection of points in $\mathbb{R}^m$. It is required to construct a curve passing through these points or passing near them. In practice they most frequently use B-splines (or another smoothing splines), composite Bezier curves and parametric splines [2,8,23].

For $d = 2$ there exists a great variety of problems. Let us point out some of them.

- M is a simply connected domain on a plane (without loss of generality it is assumed to be a square) and M_i are the points in the nodes of a lattice, covering this square. By the collection $\{M_i, F(M_i)\}$ it is required to construct embedding F. The methods for the construction in this situation can be transferred from a one-dimensional case.

- Let M be the same as above and M_i a collection of segments, constituting the lattice. Here, along with the indicated methods, the method of Coons portions is often used.

- $\{M_i\}$ is an arbitrary collection of points on a plane (irregular net) and M is a convex hull of this set of points. As above, it is required by the collection $\{M_i, F(M_i)\}$ to restore F on M. At present more than a dozen methods for the solution to this problem are known [15], however, everything is not clear here, especially from the point of view of the comparative analysis of constructions F.

Formula (1.1) gives a general scheme of the embedding construction $F : M \hookrightarrow \mathbb{R}^m$ by the collection $\{F_\alpha, \varphi_\alpha, U_\alpha\}$. Let us show that there exist many ways of geometric simulation answering this scheme.

At first let us consider the modelling of curves by the ordered collection of points.

1. Let M be a segment $[0,1]$, $U_o = [0,1)$, $U_n = (0,1]$, $U_i = (0, 1)$, where $o < i < n$. If $\varphi_i(t) = \binom{n}{i}(1-t)^i t^{n-i}$, $t \in M$, then $\{\varphi_i\}$ is partition of unity, subordinated to the covering $\{U_i\}$. Put $F_i(t) = p_i$, then $F(t) = \sum_i \varphi_i F_i$ is called a polynomial curve in the form of Bernstein -

2. $M = [a,b]$, $a = t_o, t_1, \ldots, t_n = b$ is a collection of nodes on M. $U_o = [t_o, t_1)$, $U_n = (t_{n-1}, t_n]$, $U_i = (t_{i-1}, t_{i+1})$, $o < i < n$. Let us give $w_i(t) = \dfrac{t-t_{i+1}}{t_i - t_{i+1}}$, if $t \in [t_i, t_{i+1}]$, $w_i(t) = \dfrac{t-t_{i-1}}{t_i - t_{i-1}}$, if $t \in [t_{i-1}, t_i]$ and $w_i(t) = 0$, if t lies outside of U_i. It is easy to see that $\{w_i\}$ is partition of unity. $F(t) = \sum w_i(t) \cdot p_i$ is a broken line in $\mathbb{R}^m$, connecting points $p_i = F(t_i)$.

The function $w_i(t)$ is non-differentiable at point t_i, however,

$$w_{i,r}(t) = \frac{w_i^r(t)}{\sum_j w_j^r(t)} \in C^{r-1}(M).$$

Thus the partition of unity, subordinated to $\{U_i\}$ of the given smoothness, is obtained.

3. Let M and t_i be the same as above The generalization w_i is well known; these are $B_{i,k}$ B-splines of k-order [2,8]. $B_{i,k}$ is partition of unity, subordinated to the covering $U_i = (t_i, t_{i+k})$.

Construction (1.1), where $F_i(t) = C_i$, gives a spline of the k-th order. It is easy to show that splines of the k-th order can also be obtained with the help of $B_{i,\ell}$, where $\ell < k$ (for example, with the help of $w_i = B_{i-1,2}$), but in this case $F_i(t)$ are polynomials of $k-\ell$ degree.

4. Let us consider the case, when M is a d-dimensional cube and the values of $F : M \longrightarrow \mathbb{R}^m$ in the modes of the lattice, splitting M, are known. Then using the Cartesian product and the constructions described in items 1,2, and 3 it is possible to construct partition of unity and the mapping F.

5. Let M be a rectanlge $[0,n_1] \times [0,n_2]$ and let the net of curves $F(i,v)$, $F(u,j)$ be given on the lines $u = i$, $v = j$, $0 \le i \le n_1$, $0 \le j \le n_2$. The surface, constructed by these curves, is called Coons surface [23], if in a parametric form it is given as

$$\begin{aligned} F(u,v) = {} & (j+1-v)F(u,j) + (v-j)\,F(u,j+1) + (i+1-u)\,F(i,v) + \\ & + (u-i)\,F(i+1,v) - (i+1-u)\,(j+1-v)\,F(i,j) - \\ & - (u-j)\,(j+1-v)\,F(i+1,j) - (i+1-u)\,(v-j)\,F(i,j+1) - \\ & - (u-i)\,(v-j)\,F(i+1,j+1)\,, \end{aligned} \tag{1.2}$$

where $i \le u \le i+1$, $j \le v \le j+1$.

Let us consider the covering of M by the open sets $U_{i,j} = (i-1, i+1) \times (j-1, j+1)$ and the partition of unity $w_{ij} = w_i \cdot w_j$, subordinated to this covering, where w_i are the functions defined in item 2. Let $f_{ij}(u,v) = F(u,j) + F(i,v) - F(i,j)$,

$$\tilde{F}\,(u,v) = \sum_{i,j} w_{ij}(u,v)\; f_{ij}(u,v).$$

Proposition 1.1. $\tilde{F}\,(u,v) = F(u,v)$.

Thus, instead of a bulky formula (1.2) we have a natural construction by scheme (1.1). This remark shows the way of generalizing Coons surfaces on the multidimensional case. If $M = [0,n_1]\cdot\ldots\cdot[0,n_d]$ is a d-dimensional parallelepiped and $U_{i_1,\ldots,i_d}$, $W_{i_1,\ldots i_d}$ are obtained with the help of the Cartesian product from U_i and w_i (item 2), then, depending on initial data, the local maping $f_{i_1,\ldots,i_d}$ and F by formula (1.1) are constructed. For example, if the net of curves $F(i_1,\dots,u_d)\dots,\ F(u_1,\dots,i_d)$ is given, then $f_{i_1,\dots,i_d}(u_1,\dots,u\quad) = F(i_1,\dots,u_d) + \dots + F(u_1,\dots,i_d) - (d-1)$ $F(i_1,\dots,i_d)$.

6. Let $M \subset \mathbb{R}^d$ be a convex hull of the finite set of points $S = \{p_i\}$ and let the values $F_i \in \mathbb{R}^m$ be given at these points. One of the methods for the construction of $F : M \to \mathbb{R}^m$ consists in the preliminary triangulation (simplicial partition) of M so that the set of simpleces vertices coincides with S. Consider $\{U_i\}$, a star covering of M, i.e. U_i consists of all simpleces, containing p_i. Let us construct the partition of unity, subordinated to this covering. If $\Delta = p_i q_1 \dots q_d$ is a simplex from U_i, then any point $p \in \Delta$ can be uniquely represented in the form: $p = t_o p_i + t_1 q_1 + \dots + t_d q_d$, t_j are called barycentric coordinates, $t_j \geqslant 0$, $\sum_j t_j = 1$. Put $\varphi_i(p) = t_o$, then it is evident that $\{\varphi_i\}$ is partition of unity and $F(p) = \sum \varphi_i(p) F_i$ is a piecewise linear embedding of M in $\mathbb{R}^m$.

7. Let us again consider the above problem. It is required to restore F having the collection $\{P_i, F_i\}$. Put $S_i = S \setminus p_i$ and $U_i = M \setminus S_i$, i.e. all the initial points, except for p_i, are deleted out of M. Let $w = 1/\rho(p)$, where ρ is the function, growing with the increase of $\|p\|$, $\rho(o) = 0$, for example, $\rho(p) = \|p\|^r$, $r > 0$.

Then $\varphi_i = \dfrac{w(p-p_i)}{w(p-p_j)}$ is partition of unity, $\varphi_i(p_j) = \delta_{ij}$.

The method of weighted mean interpolation, going back to Gauss, consists in the fact that $F(p) = \dfrac{\sum F_i \cdot \|p-p_i\|^{-r}}{\sum \|p-p_i\|^{-r}}$. If r is an even number, then F(p) is a rational interpolation function with respect to every coordinate, $F(p_i) = F_i$. If the number of initial points grows, there takes place the convergence, however, all partial derivatives F(p) at points p_i are equal to zero. In order to have the convergence of the derivatives, it is possible to act as follows: the least squares method is used to find coefficients of the polynomial mapping of degree k, $f_i(p) = F_i + G_1(p-p_i) + \dots$ by the points of S, $F(p) = \sum \varphi_i f_i$, closest to p_i. Now we have the problem of estimating the order of convergence. Let us fix M and consider an infinite sequence of interpolation nodes $p_1^{\,n},\dots,p_n^{\,n}$. Assume that the limit density of the arrangement of the points $\gamma(x)$ on M nowhere turns zero and it is not equal to infinity, $0 < \gamma(x) < \infty$. Let $\varepsilon_n = \max_i \min_j \|p_i^n - p_j^n\|$ and let $\overline{F}_n$ be an interpolation mapping, constructed by local polynomials of degree k. If $F \in C^{k+1}(M)$ and

$F(p_i) = F_i$, it can be proved that there exists such constant C, independent of n, that $\|F - \overline{F}_n\| < C\,\mathcal{E}_n^{k+1}$. This fact shows the convergence order of the interpolation under consideration.

It is interesting to compare a weighted mean interpolation with Lagrange interpolation. The weighted mean interpolation ($k = 0$) converges and the Lagrange one can diverge by virtue of G.Faber theorem. Here the base Lagrange polynomials satisfy points 1 and 3 of the partition of unity but they do not need to satisfy item 2. This is one more simple explanation of the phenomenon of Runge example (Runge example (1901) considered the function $F(x) = 1/(1+25x^2)$ on segment $[-1, 1]$. By choosing n equidistant points on this segment one may construct an interpolation Lagrange polynomial F_n. It appears that $\|F - F_n\|$ infinitely increases with the increase of n).

2. Discrete analogue of the index of vector field rotations.

In modern geometry the notion of singular points of a vector field or a function is usually defined for smooth vector fields and functions, given on smooth manifolds. In this section we shall transfer these notions onto piecewise linear (PL) case.

Let M be a PL-manifold (possibly with a boundary) with a simplicial partition given on it. If every vertex v_i M corresponds to a real number f_i, we say that a PL-function f is given on M. Assume that a PL-metric is given on M, then by the values of f_i in the vertices of the simplex one may uniquely construct a linear function on the simplex and thus the PL-function gives a piecewise linear function $f : M \to \mathbb{R}$.

We will say that on M, dimM = d, a PL-vector field is given, if each d-dimensional face of F_i is contrasted to the vector $u_i \in \mathbb{R}^d$. Sometimes in applications (as a rule, this is a flat case), discretizing a continuous vector field, they consider it to be given in vertices of M.

Every PL-function, as in a smooth case, can be contrasted to a PL-vector field, which we call a gradient vector field. To this end by the collection of values in vertices (and in a metric) on every d-dimensional simplex we construct a linear function and refer the gradient of this function to the given simplex.

For a PL-vector field the transition from a face to a neighbouring face poses the problem of the agreement of the vectors, corresponding to these faces. Let us clarify this situation for a two-dimensional case. If u_1 and u_2 are vectors on the neighbouring faces, then the transition through a common face gives uncertainty as to whether u_1 should be rotated to u_2 in a clockwise or anticlockwise direction. Such problem does not arise for a PL-vector field obtained by discretization of a continuous vector field, since if a simplicial partition is sufficiently frequent, then u_1 differs little from u_2. In this case the transition should be performed in the direction of a smaller rotation angle.

Now let us give a strict definition of a PL-continuous vector field. Fix an edge of a simplicial partition and consider all non-zero vectors, corresponding to the faces which contain this edge. If these vectors lie in one semispace and if this requirement is satisfied for all edges, then we call this vector field a continuous vector field.

For a two-dimensional case a vector field is not continuous if any neighbouring vectors u_1 and u_2 are collinear and their directions are opposite. A gradient PL-vector field is continuous in our sense, since vectors u_i, corresponding to the faces with a common edge, may be represented in the form: $u_i = a + b_i$, where a is orthogonal to b_i and does not depend on i.

We shall define an index for continuous PL-vector fields. However, this may also be done in the general case, if we additionally give a scheme of vectors' agreement on the neighbouring faces, but in our opinion such a definition is of little practical importance.

Let v be a vertex of a simplicial partition M, P(v) a boundary of a hyperface, dual to v, $\dim P(v) = d - 1$. Note that d-simplices, contained in the star St(v), correspond to vertices of P(v). Thus, a vector of our PL-vector field corresponds to every vertex of P(v).

Let us assume that the field under consideration has "isolated singularities", i.e. all its vectors are non-zero. This is a guarantee of the fact that singularities of the vector field may be found in the vertices of a simplicial partition M.

If it is necessary, let us triangulate all hyperfaces of P(v) and continue a vector field by linearity from vertices onto every simplex. This gives us a piecewise linear image $F : P(v) \longrightarrow \mathbb{R}^d$. Since we consider a continuous PL-vector field with non-zero vectors, for all $x \in P(v)$, $F(x) \neq 0$. The latter allows one to define correctly the mapping $\overline{F} : P(v) \rightarrow S^{d-1}$, $\overline{F}(x) = F(x) / \| F(x) \|$.

<u>Definition 1</u>. We call by the index ind(v) of a PL-vector field

in the vertex v the degree of the mapping $\bar{F} : P(v) \longrightarrow S^{d-1}$.

The notion introduced is transferred literally from the smooth case, therefore, as in the smooth case, the sum of indices of singular points should be a homotopian invariant, equal to Euler characteristic of a manifold.

Theorem 2.1. If M is a closed PL-manifold, then $\sum_{v \in M^0} \text{ind}(v) = \chi(M)$, where M^0 is a set of vertices of a simplicial partition M.

The proof of the theorem is analogous to that of the smooth case [13, §6], therefore there is no need to give it here and we restrict ourselves to short remarks.

The main construction in the proof is as follows: M is embedded into an Euclidean space E of a high dimension; a tubular neighbourhood $N(M) \subset E$ is taken and a vector field is continued from M onto N(M) so that on the boundary of the tube it is directed "outwards". It is easy to do the same in the PL-case.

A continuous, or rather piecewise linear, vector field on M is constructed as follows: a vector field is assumed to be equal to zero in vertices of M and on P(v) it is the same as in the definition; it is continued inside the tube by linearity, connecting v with the vertices and splitting a d-face, dual to v, into simpleces. Note that v and the vertices of P(v) are vertices of such a triangulation. The vector field, continued "along the normal" from M onto N(M), has singular points only in the vertices of M and it is easy to show that the vector field's index on N(M) is equal to the vector field's index on M at these points.

The definition, given above, is constructive from the point of view of computational geometry. The coefficients of the piecewise linear mapping $F : P(v) \longrightarrow \mathbb{R}^d$ are found explicitly. In order to calculate ind(v) it is necessary to choose a vector $r \in \mathbb{R}^d$, $\|r\| = 1$. For almost all vectors r the points of $F^{-1}(r)$ lie inside the hyperfaces of P(v). The search for those hyperfaces of P(v) which cover r is reduced to the solution of the system of linear equations and verification of linear inequalities. Having defined whether the mapping on these faces is oriented or not and having attributed the sign +1 or −1 to these faces, respectively, one may find ind(v), which is equal to the sum of the signs. The computer's realization of this algorithm is easy and proceeds in a standard way.

This algorithm may be used for the calculation of singular points' indices of differentiable vector fields. For this purpose let us triangulate the neighbourhood of a point v and attribute a vector, equal to the value of the field in its centre, to every simplex of a higher dimension (here another variants are also possible). The mapping $F : P(v) \longrightarrow S^{d-1}$ is a piecewise linear approximation of the corresponding differentiable mapping. Hence follows the fact that if the triangulation is sufficiently frequent, then the index of point v in a PL-sense coincides with the desired one.

Let us dwell on this in more detail and indicate a more concrete partition for the calculation algorithm of a singular point's index of a differentiable vector field. If a singular point $v \in M$, $\dim M = d$, then choose a d-dimensional simplex Δ, containing v and not containing any other singular points. Connecting v with the vertices of Δ we obtain the partition of Δ into simplexes. In order to construct more dense partitions of Δ it is possible to carry out a barycentric subpartition of the boundary of Δ and to connect v with new vertices.

It is interesting to consider the question of when the partition should be stopped and one may consider the index correctly defined.

We suggest using the method, quite acceptable from the practical point of view. At first several sequential partitions of Δ without calculating the index are carried out. The number of preliminary subpartitions is an input parameter, determined by an expert evaluation (i.e. informally). Then by constructing new partitions of Δ the indices (in a PL-sense) are ontained and the calculations stop if the indices obtained by the next and previous subpartition coincide. If we fix an initial number of steps, then whatever it is, one may always give an example of a singularity, for which the suggested algorithm works unsatisfactorily. In computational mathematics such problem arises very often, for example, in the search for all the roots of equation $f(x) = 0$ on a segment by numerical methods. Here a collection of points on a segment, which help to find solutions, is such an input parameter. Moreover, if a priori we know nothing about the behaviour of the function $y = f(x)$, then it is impossible to define this collection. Perhaps some additional investigations of the index's relation in a differentiable and PL-cases will help somehow to solve this problem.

Now let us consider singular points of PL-functions. Let a real number f_i be attributed to each vertex v_i of a simplician partition of a manifold M. Consider a vertex v_0, the function of which is equal to f_0. If in the neighbouring vertices, i.e. the vertices, connected by the edges with v_0, the function's values differ from f_0, then vertex v_0 can be either an isolated singularity or it can be non-singular.

Let $M(f_0) \subset M$ be a polyhedron, consisting of simplexes in the vertices of which the function assumes the value f_0 and let V_0 be a connected component $M(f_0)$ containing v_0. Denote by $St(V_0)$ the union of all stars of the vertices contined in V_0. The set of simplexes of $St(V_0)$, in which the values of the function are strictly less than $St(V_0)\setminus V_0$ form a subpolyhedron in $St(V_0)$. Denote this subpolyhedron by V_-. V_- may be empty. Now everything is ready for the definition of a PL-index of the set V_0.

Definition 2. $\operatorname{ind}(V_0) = \chi(V_0) - \chi(V_-)$.

Let in the vertices, neighbouring v_0, the values differ from f_0, i.e. V_0 consists of one point v_0, then $\operatorname{ind}(v_0) = 1 - \chi(V_-)$. In this case one may use definition 1 for $\operatorname{ind}(v_0)$ of the gradient vector field of function f.

Theorem 2.2. If v_0 is an isolated singularity of a PL-function f on M, then by definition 2 the index's value coincides with the value of the index of the gradient vector field of function f by definition 1.

Proof. We shall prove that the index of the gradient vector field is equal to $1 - \chi(V_-)$.

At first let us consider the case when $d = \dim M$ is even. Let us sew $St(v_0)$ with its doublet along the boundary P and denote the obtained manifold by W. Note that $St(v_0)$ is homeomorphic to a d-dimensional ball and W is homeomorphic to a d-dimensional sphere S^d. A new vertex v_0' is added into the set of vertices $W^0 \subset W$ as compared with $St(v_0)^0$. We attribute the value f_0 to this vertex and denote by τ a gradient vector field of a PL-function on W. Then $\operatorname{ind}_\tau(v_0) = \operatorname{ind}_\tau(v_0')$ (the subscipt shows which vector field defines ind).

Let us consider the restriction of function f on P^0 and denote by η the corresponding vector field on P. The set of vertices P^0 consists of V_-^0 and V_+^0. (The value of f in the vertices V_+^0 is greater than f_0). Directly from definition 1 follow the equalities: $\operatorname{ind}_\tau(v) = \operatorname{ind}_\eta(v)$ if $v \in V_-^0$ and $\operatorname{ind}_\tau(v) = -\operatorname{ind}_\eta(v)$ if $v \in V_+^0$.

If the restriction of function f on P^0 is continued by linearity onto simplexes and the submanifold, at the points of which the function is less than f_0, is denoted by $\overline{V}_+ \subset P$ and its closed complement is denoted by $\overline{V}_-$, then $P = \overline{V}_+ \cup \overline{V}_-$. The manifold P has an odd dimension and hence $\chi(\overline{V}_+) = \chi(\overline{V}_-)$. Since V_+ is a retract of V_+, and V_- is a retract of V_-, then $\chi(\overline{V}_+) = \chi(V_+)$ and $\chi(\overline{V}_-) = \chi(V_-)$.

Vector field η is directed "outwards" at the boundary of $\overline{V}_-$ and "inwards" at the boundary of V_+, consequently, $\sum_{v\in V_-} \mathrm{ind}_\eta\, v = \chi(V_-)$, $\sum_{v\in V_+} \mathrm{ind}\ (v) = -\chi(V_+)$.

From theorem 2.1 it follows that $\sum_{v\in \bar{W}^c} \mathrm{ind}_\tau\, v = \chi(W)$. Using the above equalities, we obtain:

$$\sum_{v\in W^o} \mathrm{ind}_\tau\, v = \mathrm{ind}_\tau\, v_o + \mathrm{ind}_\tau\, v_o' + \sum_{v\in V_+^o} \mathrm{ind}_\tau\, v + \sum_{v\in V_-^c} \mathrm{ind}_\tau\, v =$$

$$= 2\ \mathrm{ind}_\tau\, v_o + \sum_{v\in V_-^c} \mathrm{ind}_\eta\, v - \sum_{v\in V_+^c} \mathrm{ind}_\eta\, v = 2\ \mathrm{ind}_\tau\, v_o + \chi(V_+) + \chi(V_-) =$$

$= 2\ \mathrm{ind}_\tau\, v_o + 2\,\chi(V_-)$. Since $X(W) = X(S^d) = 2$, we have $2\ \mathrm{ind}_\tau v_o + 2\,X(V_-) = 2$. From the latter equality follows the proof of the theorem for an even case.

We reduce the proof for an odd d to the even case. Consider the suspension $\Sigma\, St(v_o)$, which we denote by W. W is a piecewise linear manifold, homeomorphic to d+1-disk. Two vertices are added into W as compared with $St(v_o)^o$. Let us attribute the numbers greater than f_o to these verices and denote by η a gradient vector field on W and by τ on $St(v_o)$. Note that $\mathrm{ind}_\tau (v_o) = \mathrm{ind}_\eta (v_o)$. On the other hand, since d+1 is even, it follows that $\mathrm{ind}_\eta (v_o) = 1 - \chi(W_-)$, where W_- denotes a subset W, on whose simplexes the function assumes the values less than f_o . By the construction of the function W_- coincides with V_-, consequently, $\mathrm{ind}_\tau (v_o) = 1 - \chi(V_-)$. Q.E.D.

From the computational point of view the algorithm for the determination of index by definition 2 is quite simple. All simplexes from $St(V_o)$ are looked over and k_i-number of simplexes of dimension $i = 0,1,\dots,d-1$, in whose veritices the function assumes the values less than f_o , are defined. $\mathrm{ind}(V_o) = \chi(V_o) - \sum_{i=0}^{d-1}(-1)^i k_i$. Note that for the calculation of k_i only comparison operations are used.

In definition 2 the specific character of a PL-manifold is nowhere required. This definition can be used in the case when M is an arbitrary simplicial complex. The fact that in this general case the sum of indices of singular points is a homotopic invariant speaks for the correctness of the definition introduced.

Denote by $\{V(i)\}$ a set of the connected subcomplexes of M in whose veritices the function is constant, $M^o = \bigcup V(i)^o$, and by $V_-(i)$ the corresponding subcomplexes, as in definition 2. If the values of the function in any two neighbouring veritices are different, then the collection $\{V(i)\}$ is a set of all vertices of M^o.

<u>Theorem 2.3</u>. $\sum_i \text{ind}\,(V(i)) = \chi(M)$.

<u>Proof</u>. This is a purely combinatorial topological fact and the proof is carried out by elementary reasonings in the framework of this discipline. Among the simplexes of complex M isolate those ones on which the function is constant, i.e. the ones belonging to $\bigcup_i V(i)$. Denote by N a set of the remaining simplexes. $\chi(M) = \sum_k (-1)^k |M^k| = \sum_{i,k} (-1)^k |V(i)^k| + \sum_k (-1)^k |N^k|$, where $|P^k|$ denotes a number of simplexes of dimension k in P. By the definition $\sum_i \text{ind}\, V(i) = \sum_i [\chi(V(i)) - \chi(V\_(i))]$. Making the corresponding reductions, we have to prove the equality

$$\sum_{k=1}^{d} |N^k| = \sum_i \sum_{k=1}^{d} (-1)^{k-1} |V\_(i)^{k-1}| .$$

Note that each simplex from $V\_(i)^{k-1}$ can be uniquely compared with a simplex from N^k, hence $|N^k| = \sum_i |V\_(i)^{k-1}|$ and, consequently, the desired equality is true.

In conclusion note that the notion of the index of a function's singularity can be also generalized for CW-complexes. However, here not only the vertices (cells of dimension 0) can be singular, but also the cells of higher dimensions. Let us consider an example. Let M be a square divided into n n square cells, in whose nodes 0 and 1 are arranged in a staggered order. Any vertex of M is either a local minimum (if 0 is attributed to it) or a local maximum. Therefore the index of any inner vertex is 1 and that of a boundary vertex is either 0 or 1. Thus, the sum of indices is geater than $(n-1)^2$. On the other hand, $\chi(M) = 1$ and we see that the sum of vertices' indices is never equal to the Euler characteristic. The fact is that the cells whose index in this case is equal to -1 are also singular. The sum of indices of the vertices and cells is equal to $\chi(M)$.

For a two-dimensional polyhedron the notion of index of a face Γ can be introduced. Let k be a number of local minima in the vertices of face Γ, i.e. a number of vertices, the function's values in which are less than in two neighbouring ones. Let $\text{ind}(\Gamma) = 1 - k$. It appears (this is an easy exercise) that the sum of all vertices and faces is equal to the Euler characteristic of the polyhedron.

3. Distances and shortest paths.

The aim of this section is to give constructive definitions of distances and shortest paths on metric spaces, to evaluate the accuracy of the determination of distances on a surface by a polyhedral approximation and to consider the behaviour of the distance function on two-dimensional polyhedrons.

3.1. Let K be a metric arcwise connected compactum with metric (Further we shall consider only arcwise connected compacta and subcompacta and for brevity we shall call them simply compacta and subcompacta). For any two points p and q from K there exists the shortest path, probably not unique, connecting these points 9, Ch.2 . Denote by $D_K(p, q)$ the length of this path. D_K is a metric on K, which, generally speaking, can differ from . (For example, when K is a Euclidean plane, in which a circle is punctured). If V is a subcompactum of K, then metric D_V need not necessarily coincide with the restriction of metric D_K on V, in other words, the shortest path in K between the points p,q V can fall outside of the limits of V. Moreover, it cannot be longer than the shortest path in V and therefore $D_V(p,q) \gg D_K(p,q)$.

The distance function $R(q) = D_K(p,q)$ determines the shortest path between p and q . The main idea of the construction of function R on K lies in the fact that K is divided into subcompacta F_i; at point p the "wave" V is emitted , which moves step-wise (discretely) with each step enclosing a bew boundary subcompactum F_n, on which by the values of R on V and D_G , R is defined on the boundary. One can observe that the boundary G, a "wave's front", consisting of the subcompacta, adjoining V, moves uniformly, not allowing a break in any direction. For this purpose a boundary subcompactum F_n, the closest to p, is chosen.

Let us pass on to the algorithm for the determination of function R on K. First consider the connection D_V and D_K, where V is a subcompactum of K and ∂V is a boundary of V. The proposition given below will be used to determine function R on the "wav's" boundary.

<u>Proposition 3.1</u>. If p is an inner point of V, $a = \inf_{r \in \partial V} D_V(p,r)$, $q \in \partial V$, $\mathcal{E} = D_V(p,q) - a$, $O_{\mathcal{E}}(q)$ is a closed $\mathcal{E}$ - neighbourhood in metric D_K of point q and $U(q) = V \cup O_{\mathcal{E}}(q)$, then $D_K(p,q) = D_{U(q)}(p,q)$. In particular, if $\mathcal{E} = 0$, i.e. q is the nearest to p point of the boundary in metric D_V , then $D_K(p,q) = D_V(p,q)$.

Proof. First let us consider the case $\mathcal{E} = 0$. Assume that between points p and q in K there is a path shorter than in V, then it intersects the boundary ∂V at point r, such that $D_V(p, r) < D_V(p, q)$. The latter inequality contradicts the condition that point q is the nearest to p.

Now let $\mathcal{E} > 0$. The shortest path, connecting p and q in K can intersect ∂V at point r, removed from p in metric D_V at a distance $\tilde{a} \geqslant a$. Therefore $D_K(r,q) \leq \mathcal{E}$, the shortest path, connecting r and q, lies in $O_\mathcal{E}(q)$ and thus $D_K(p,q) = D_{U(p)}(p,q)$.

Assume that K is decomposed into subcompacta F_i, $K = \bigcup_i F_i$. We shall call such a subpartition correct, if metric D_{F_i} coincides with restriction D_K on F_i. In the most interesting cases for applications (manifolds, polyhedrons) K can always be decomposed correctly. For example, such is the case of the simplicial partition of a polyhedron with the Euclidean metric on simplexes.

Let us fix a point $p \in K$ and for a correct partition $K = \bigcup_{i=1}^N F_i$ give the definition of the function $D_K(p, *)$ on K based on proposition 3.1.

Basic construction. Let $p \in F_1$. At the first step assume that $V = F_1$ and a set G consists of the points of subcompacta F_j which have a non-empty intersection with F_1. Put $R(q) = D_{F_1}(p,q)$ for all $q \in V$. If $q \in G$, then we set $R(q) = D_{V \cup G}(p,q)$.

Let at n-step there exist the sets $V = \bigcup F_i$, $G = \bigcup F_{jm}$ and let the functions R on V and $\tilde{R}$ on G be known; moreover, $\tilde{R}(q) = D_U(p,q)$, where $U = V \cup G$, $q \in G$.

Let us find such a subcompactum F_{jm}, constituting G (let it be F_n), which is the closest one to point p with respect to metric D_U, i.e. $a_n = \inf_{q \in F_n} \tilde{R}(q) = \inf_{q \in G} \tilde{R}(q)$. Then follow two stages: first we define R on F_n, exclude F_n from G and include it in V, second, we include in G all those subcompacta F_j, not belonging to U, which have a non-empty intersection with F_n and redefine function $\tilde{R}$ on G.

Let point $q \in F_n$, $\mathcal{E} = \tilde{R}(q) - a_n$ and let $O_\mathcal{E}(q)$ be a closed - neighbourhood of point q in K. Put $R(q) = D_{U(q)}(p,q)$, where $U(q) = V \cup O_\mathcal{E}(q)$, i.e. $R(q) = \inf_{r \in O_\varepsilon(q) \cap V} R(r) + D_{O_\mathcal{E}(q)}(r,q)$. From the proposition it follows that $R(q) = D_K(p,q)$.

Now let us consider all F_j neighbouring F_n and not entering $U = V \cup G$. If there are none, then pass on to the next step. Denote by Q the union of these subcompacta. Put $\tilde{R}(q) = \inf_{r \in Q \cap U} (D(r,q) + \tilde{R}(r))$, $q \in Q$. It is easy to see that $\tilde{R}(q) = D_{U \cup Q}(p,q)$. Let us include Q into G, i.e. $G = G \cup Q$, and pass on to n+1 step. The process will be continued until all F_j are included in V, i.e. coincides with K.

As it was noted at n-step for q F_n, $R(q) = D_K(p,q)$. Then R on F_n is not reconstructed. Hence follows

Theorem 3.1. The function R(q) on K, constructed above, coincides with $D_K(p,q)$.

Now let us show how to find the shortest path $\gamma(p,q)$ on K, if $R(\cdot)$ is known. Note that point z lies on the shortest path $\gamma(p,q)$ (on one of the shortest paths if they are several) iff the following equality is satisfied:

$$D_K(p, q) = D_K(p, z) + D_K(z, q) \tag{3.1}$$

In this case K is decomposed into subcompacta F_i correctly, i.e. the metric $\rho = D_{F_i}$ coincides with D_K on F_i. Put $q_1 = q$ and consider all subcompacta $F_{1\ell}$, $\ell = 1,\dots,m$, containing q_1. For points z, lying on the boundary of F_1, we seek the solution to equation $R(z) + (z,q_1) = R(q_1)$. If point q_2, distinct from q_1, is the solution to this equation, then by (3.1) it lies on the shortest path. Now consider the sets F_{2j}, containing q_2, and find analogously point q_3 and so on. Continue this process until we come to p. It is evident that if the diameters of F_i are separated from zero, then the process will sometime stop, since at every step $R(q_i)$ becomes less than $R(q_{i-1})$. It may turn out that at some step there are several solutions to the equation. In this case all the solutions should be exhausted and all the shortest paths constructed. Thus, for the shortest path the collection q_i of the points lying on it is determined. If it is additionally assumed that there exists a way of seeking the shortest paths on F_j (note that q_{i-1} and q_i belong to the same F_j), then this collection defines completely.

Remark. It is easy to see that the above method for determining the distance is quite close to a simple and natural Dijkstra algorithm [5] for calculating distances on a graph. This similarity allowed the authors of [14] to call their algorithm, which is in fact a realization of the above construction on two-dimensional polyhedrons, "continuous Dijkstra". However, the Dijkstra algorithm itself for the same reason can be called "discrete Huyghens", since the main idea of the algorithm is contained in the Huyghens principle.

3.2. For the numerical realization of the above scheme on a computer it is necessary to isolate a finite collection of supporting points

$q_i \in K$, for which the values of $R(q_i)$ will be calculated. Besides, it is required to have an interpolation algorithm of $R(q)$ at any point $q \in K$ by the collection $\{q_i, R(q_i)\}$. For the general case K the finite collection need not exist, however, if K is a polyhedron, then such a collection always exists.

For applications it is interesting to consider the case when K is a differentiable surface which, as it was mentioned in § 1, is usually given by the embedding $F : M^d \hookrightarrow \mathbb{R}^m$. The metric on K is induced by the embedding. For an arbitrary differentiable surface it is impossible to indicate the finite collection of supporting points for the calculation of function R and therefore the surface should be approximated by the polyhedron. The simplest method for this is as follows: $\mathbb{R}^m$ is decomposed by the cubic ε-lattice and polyhedron P consists of those small cubes of the lattice which contain the points of K. In another, also widely known method, on K a finite set of points p_i is chosen and polyhedron P is a simplician approximation of vertices p_i.

In the algorithm the shortest path γ is sought not on K but on its polyhedral approximation P. The lengths of the shortest paths $\gamma(p,q)$ on K and the paths $\gamma_P(p,q)$ on P can differ. It is natural to pose the question of the convergence of γ_P to γ while P approaches K. The author did not find the answer to this question in the literature on the numerical analysis. The following result provides such an answer.

Theorem 3.2. If N is a tubular neighbourhood with radius ε of twice differentiable compact manifold $K \subset \mathbb{R}^m$, then there exist ε_0 and constant c, independent of ε, such that for $\varepsilon \leq \varepsilon_0$ $|D_K(p,q) - D_N(p,q)| \leq c \cdot \varepsilon \cdot D_K(p,q)$, where $D_K(p,q)$ and $D_N(p,q)$ are the lengths of the shortest paths $\gamma_K(p,q)$ and $\gamma_N(p,q)$ on K and N respectively.

Proof. For sufficiently small $\varepsilon \leq \varepsilon_0$, N is a normal fibration over K, i.e. there exists a mapping $pr : N \to K$, such that for any point $p \in K$, $pr^{-1}(p)$ is a m-d - dimensional ball with radius ε, normal to K at point p. Let $\gamma = pr(\gamma_N)$ and let τ be a natural parameter on γ (i.e. $\|\dot{\gamma}\| = \|\frac{d\gamma}{d\tau}\| = 1$). Then we may write $\gamma_N(\tau) = \gamma(\tau) + \alpha(\tau)$, where $\alpha(\tau) \in pr^{-1}(\gamma(\tau))$, $\|\alpha(\tau)\| \leq \varepsilon$. Since $\dot{\gamma}(\tau)$ is a vector tangent to K, then $\dot{\gamma}(\tau) \cdot \alpha(\tau) \equiv 0$, whence $\dot{\gamma}(\tau) \cdot \dot{\alpha}(\tau) = -\ddot{\gamma}(\tau) \cdot \alpha(\tau)$. Denote by ℓ the length of path γ, then

$$0 \leq \ell - D_N(p,q) = \int_0^{\ell} (\|\dot{\gamma}(\tau)\| - \|\dot{\gamma}_N(\tau)\|) \, d\tau =$$

$$= \int_0^{\ell} \frac{\dot{\gamma}^2 - \dot{\gamma}_N^2}{1 + \|\dot{\gamma}_N\|}\, d\tau = \int_0^{\ell} \frac{-\dot{\alpha}^2 - 2\dot{\alpha}\dot{\gamma}}{1 + \|\dot{\gamma}_N\|}\, d\tau \leqslant \int_0^{\ell} \frac{-2\dot{\alpha}\dot{\gamma}}{1 + \|\dot{\gamma}_N\|}\, d\tau =$$

$$= \int_0^{\ell} \frac{2\ddot{\gamma}\alpha}{1 + \|\dot{\gamma}_N\|}\, d\tau \leqslant 2 \int_0^{\ell} \|\alpha \cdot \ddot{\gamma}\|\, d\tau.$$

Write $\alpha(\tau) = \|\alpha(\tau)\| \cdot e(\tau)$, where $e(\tau)$ is a single vector of the normal, therefore $k(\tau) = e(\tau) \cdot \ddot{\gamma}(\tau)$ is a curvature of the normal cross-section at point $p_{\tau} = \gamma(\tau) \in K$. k is a continuous function, dependent on singular vectors of a tangent and normal fibration over K, i.e. this is the function on a compactum, dependent only on the surface and its embedding in $\mathbb{R}^m$. Denoting by $\varkappa_K$ the greatest value of the function, we obtain:

$$\begin{aligned} \ell - D_N(p,q) &\leq 2 \int_0^{\ell} \|\alpha(\tau)\| \cdot \| e(\tau) \cdot \ddot{\gamma}(\tau) \| \, d\tau \leqslant \\ &\leqslant 2 \cdot \varepsilon \int_0^{\ell} |k(\tau)| \cdot d\tau \leqslant 2 \varkappa_K \cdot \varepsilon \cdot \ell . \end{aligned} \tag{3.2}$$

Since $\gamma(p,q)$ is a path on surface K, its length is not less than $D_K(p,q)$, whence $D_K(p,q) \geqslant D_N(p,q)$. If we put $C = 2\varkappa_K + \dfrac{4\varkappa_K^2 \cdot \varepsilon_0}{D_K(p,q)}$, then by (3.2) $D_K(p,q) - D_N(p,q) \leqslant C \cdot \varepsilon \cdot D_K(p,q)$. This completes the proof of the theorem.

The estimate of the constant in the theorem is efficient. For example, for a domain in Euclidean space $\varkappa_K = 0$, therefore $D_K = D_N$. If K is a sphere in $\mathbb{R}^m$ with radius ρ, then $\varkappa_K = \frac{1}{\rho}$. The order of convergence cannot be improved. In the case when K is a plane curve, the difference in the lengths of the curve and the shortest path in ε-neighbourhood of the curve has order ε.

If P is simplicial approximation of surface K and the maximal diameter of the simplex is equal to ε, then in the uniform metric the deviation of P from K has order ε^2. From the theorem it follows that the difference of the shortest paths on P and K also has order ε^2.

3.3. To realize the construction given in item 1 it is necessary to study the behaviour of the distance function and to determine the set of supporting points.

The distance function R on the Riemannian manifold M is closely connected with the notion of the set of partition. The point $q \in M$

belongs to the set of partition C(p), iff the geodesic γ, started at point p, either cannot be continued beyond q or, if continued, stops to be the shortest, i.e. for point q', lying on γ beyond q, there exists a shorter path between p and q than γ. Note one important property: $M \setminus C(p)$ is homeomorphic to d-dimensional ball, d = dim M [4; §5].

It is interesting to note that though this notion goes back to A.Poincare and J.G.K.Whitehead little is known about it in the general case. Only in dimension two did it become possible to obtain the local construction of C(p); Meyers (1935) proved that C(p) is a one-dimensional polyhedron, i.e. a graph.

For piece-wise manifolds no investigations of the set of partition were carried out, but, undoubtedly, here the situation is considerabby simplified as compared with the smooth case. We shall give the basic results for dimension 2, since in practice this case is the most frequent one. In [14,20] the exact algorithms of the search for the shortest path on two-dimensional polyhedrons are obtained. In dimension three and beyond such algorithms are not yet worked out. The analysis of the structure of a set of partition shows the difficulties which had to be overcome in the construction of the algorithm. It is noteworthy that the authors of the above papers were not familiar with the notion of C(p), they did not analyse it specially, but in the process of algorithm construction they had to study its various properties.

Theorem 3.3. If p is a point of a polyhedron M, then C(p) is a graph whose vertices of degree 1 are vertices of a positive curvature of the polyhedron. Any edge of C(p) consists of a finite number of rectilinear and hyperbolic (in an face's coordinates) sections.

Recall that the curvature of the polyhedron's vertex is equal to 2π minus the sum of the angles by the vertex. A number of the edges entering this vertex is called a degree of the graph's vertex.

Proof. The verification of the basic statements of the theorem is not difficult and it is done in the framework of the elementary geometry. We shall not dwell on the proof in detail but we shall formulate the fundamentals.

Any section of the shortest path (of the geodesic) on the boundary is rectilinear. If the geodesic does not go through the vertices of the non-zero curvature, then on the development it is a straight line. The behaviour of the geodesic entering the vertex of the non-zero curvature is different depending on positiveness or negativeness of the curvature. If v is a vertex of a positive curvature, then the geodesic cannot be continued beyond v, since it stops to be the shortest. In the case of a negative curvature there exists the whole angle of extensions, i.e. the extension may be a ray, lying in the angle, whose value is equal to the value of the curvature.

If the point $q \in C(p)$ is not a vertex of a positive curvature, then there exist at least two different shortest paths $\gamma_1(p,q)$ and $\gamma_2(p,q)$ between p and q. Let us make a development along γ_1 and γ_2, i.e. starting from point q we shall lay on the surface the faces through which goes γ_i. At first assume that γ_1 and γ_2 do not go through the vertices of a negative curvature, then on a plane development the point p is represented by two images and the point q is equidistant from them. The geometric place for the points equidistant from two points of a plane is a straight line. Hence the neighbourhood of q in $C(p)$ is a rectilinear segment. If apart from γ_1 and γ_2 there are other shortest paths between p and q, then q is a vertex of the graph $C(p)$, whose degree is equal to the number of short paths between these points. If γ_1 and γ_2 go through the vertices of a negative curvature, then the development from q along γ_i is made only up to such vertex V_i. Let $c = \| V_2 - q \| - \| V_1 - q \|$. The geometric place of points q', satisfying the equation $\| V_2 - q' \| - \| V_1 - q' \| = c$, is a straight line if $c = 0$ and a hyperbola, if $c \neq 0$. Hence the neighbourhood of q in $C(p)$ is either a section of a straight line or a section of a hyperbola. The above arguments prove the theorem.

In conclusion without going into details let us describe the behaviour of the distance function R on a polyhedron M. Cut M along $C(p)$, then $U = M \setminus C(p)$ is homeomorphic to a disk. If on M there are no vertices of a negative curvature, then U is a plane polygon and the function $R(q)$ is equal to $\| p - q \|$ on U.

In the case when on M there are vertices of a negative curvature, every such vertex v_i generates a domain $A_i \subset U$ which is an angle of the geodesic extension $\gamma(p, v_i)$. Inside A_i there also may be vertices of a negative curvature, generating A_j. If point $q \in A_i$ but it does not lie within the inner A_j, then $R(q) = R(v_i) + \| q - v_i \|$.

References

1. Aho A.V., Hopcroft J.E., Ullman J.D. The design and analysis of computer algorithms, Addison-Wesley, Reading, Massachusetts, 1974.

2. Vasilenko V.A. Splines. Theory, algorithms, programs. Novosibirsk, 1983.

3. Giloi W. Interactive computer graphics - data structures, algorithms, languages, Prentice-Hall, 1978.
4. Gromoll D., Klingenberg W., Meyer W. Riemannsche Geometrie im Grossen. Springer-Verlag, 1968.

5. Dijkstra E.W. A note on two problems in connection with graphs / Numer. Math. - 1959 - N 1, p.269-271.

6. Douglas D. Experiments to locate ridges and channels to create a new type of digital elevation model / Cartographica - 1986 - V.23, N 4, p.279-303.

7. Dubrovin B.A., Novikov S.P., Fomenko A.T. Modern Geometry. Moscow, 1979.

8. Zav'yalov Yu.S., Leus V.A., Skorospelov V.A. Splines in engineering geometry. Moscow, 1985.

9. Kolmogorov A.N., Fomin S.V. Elements of function theory and functional analysis. Moscow, 1976.

10. Korneenko N.M. Efficient algorithms of combinatory computational geometry / Kibernetika i vychislitel'naya tekhnika. Vyp.4. Moscow, 1988, p.179-203.

11. Kornishin M.S., Paimushin V.N., Snegirev V.F. Computational geometry in problems of hull mechanics. Moscow, 1989.

12. Lee D.T., Preparata F.P. Computational geometry - A survey. IEEE Transactions on Computers, 1984, V.C-33, No.12, p.1072-1101.

13. Milnor J.W. Topology from the differentiable viewpoint. Charlottesville, Virginia, 1965.

14. Mitchell J., Mount D., Papadimitriou C. The discrete geodesic problem / SIAM J. Comput. - 1987 - V.15, N 1, p.193-215.

15. Musin O.R., Serbenyuk S.N. Digital models of "relief" of continuous and discrete geofields / Geographic data bank for thematic mapmaking. Moscow, 1987, p.156-170.

16. Musin O.R. Partition of unity and interpolation of functions / VII All-Union seminar "Theoretical fundamental and construction of numerical algorithms for the solution of problems of mathematical physics". Abstracts. Kemerovo, 1988, p.86-87.

17. Musin O.R. Shortest paths on piece-wise linear manifolds / IX All-Union geometric conference. Abstracts. Kishinev, 1988, p.221-222.

18 Pavlidis T. Algorithms for graphics and image processing. Computer Science Press, 1982.

19. Preparata F.P., Shamos M.I. Computational geometry. An itroduction. Springer-Verlag, 1985.

20. Sharir M., Schorr A. On shortest paths in polyhedral spaces / SIAM J. Comput. - 1986 - V.15, N 1, p.193-215.

21. Starodetko E.A. Elements of computational geometry. Minsk, 1986.

22. Foley J.D., van Dam A. Fundamental of interactive computer graphics, Addison-Wesley, Reading, Massachusetts, 1982.

23. Faux I.D., Pratt M.J. Computational geometry for design and manufacture / Ellis Horwood Ltd., 1979.

INTRODUCTION
TO MASLOV'S OPERATIONAL METHOD
(NON-COMMUTATIVE ANALYSIS
AND
DIFFERENTIAL EQUATIONS)

V.E.Nazaikinskiĭ, B.Yu.Sternin
and V.E.Shatalov
Department of Applied
Mathematics
Moscow Institute
of Electronic Engineering
Bolshoy Vuzovsky per.3
109028, Moscow, USSR

1. General non-commutative analysis

The object of study in Maslov's operational method [1 , 2] is the analysis of functions of several linear operators $A_1, \ldots, A_n$. Consider first of all the question of how we may define such functions.

We start from the well-known approach to the definition of functions of a single operator, which goes as follows. We need to learn how one may substitute an operator A into a function $f(x)$ instead of the numeric variable x, thus obtaining the operator $f(A)$. For naturalness reasons, the desired result of the substitution is clear for some particular functions $f(x)$; namely, if $f(x) = x$, then it should be $f(A) = A$; if $f(x) = 1/(\lambda - x)$, then we should obtain the resolvent $R_\lambda(A)$; if $f(x) = e^{ikx}$, then $f(A)$ should be equal to the element $\exp(ikA)$ of the one-parametric group, generated by the operator iA; if $f(x) = \theta(\lambda - x)$ and A is a self-adjoint operator, then $f(A)$ should be equal to the projector $E_\lambda(A)$, i.e. to the projector on the space generated by such eigenvectors that the corresponding eigenvalues do not exceed λ.

Using series or integral representations (Taylor series, Cauchy integral, Fourier integral etc.) we are able to construct general functions from those mentioned above, and this immediately leads to the following definitions:

$$f(A) = \int_{-\infty}^{\infty} f(\lambda)\, dE_\lambda(A); \tag{1}$$

$$f(A) = \frac{1}{2\pi i} \oint_{\partial C} f(\lambda) R_\lambda(A)\, d\lambda; \tag{2}$$

$$f(A) = \frac{1}{\sqrt{2\pi}} \int_{-\infty}^{\infty} \tilde{f}(k) e^{ikA}\, dk; \tag{3}$$

the list may be easily continued.

Of course, any of the definitions (1 - 3) is valid for its own function and operator classes; e.g. the function $f(x)$ in (2) should be analytic in some domain C containing the spectrum of the operator A. However, if more than one definition may be applied, all they give the same result. Note also that for fixed A the mapping $f \longmapsto f(A)$ is a homomorphism of the ring of functions into the ring of operators.

Now let us try to make things clear for the case of several operators. We immediately meet a series of difficulties. Really, for example, let

$$f(x, y) = xy. \tag{4}$$

The attempt to define the function $f(A, B)$ of two operators A and B fails if their commutator

$$[A, B] = AB - BA \tag{5}$$

is not equal to zero since one cannot decide which of the two variants

$$\begin{aligned} f(A, B) &= AB, \\ f(A, B) &= BA \end{aligned} \tag{6}$$

to prefer.

A simple solution was invented by R. Feynman [5]. He proposed to supply operators with indices, or numbers, prescribing the order in which the operators act (Feynman ordering of operators). These numbers are placed over the operators. Thus in the expression $f(A, B)u$ the operator A acts first on the element u of a linear space so that for the function (6)

$$f(\overset{1}{A},\overset{2}{B}) = BA. \tag{7}$$

On the opposite,

$$f(\overset{2}{A},\overset{1}{B}) = AB. \tag{8}$$

Note that in this notation

$$\overset{1}{A}\overset{2}{B} = BA = \overset{2}{B}\overset{1}{A}. \tag{9}$$

Now the definitions (1-3) may be generalized for a function of n operators A_1, ..., A_n with a chosen order of their action, say $\overset{1}{A_1},\ldots,\overset{n}{A_n}$:

$$f(\overset{1}{A_1},\ldots,\overset{n}{A_n}) = \int\ldots\int f(\lambda_1,\ldots,\lambda_n)\,dE_{\lambda_n}(A_n)\ldots dE_{\lambda_1}(A_1), \tag{10}$$

or

$$f(\overset{1}{A_1},\ldots,\overset{n}{A_n}) = \frac{1}{(2\pi i)^n}\oint\ldots\oint f(\lambda_1,\ldots,\lambda_n)R_{\lambda_n}(A_n)\ldots R_{\lambda_1}(A_1)\,d\lambda, \tag{11}$$

or

$$f(\overset{1}{A_1},\ldots,\overset{n}{A_n}) = \frac{1}{(2\pi)^{n/2}}\int\ldots\int \tilde{f}(k_1,\ldots,k_n)e^{ik_nA_n}\ldots e^{ik_1A_1}\,dk. \tag{12}$$

Like the case of functions of a single operator, these definitions do not contradict one another when more than one definition may be used. However, the mapping $f \longmapsto f(\overset{1}{A_1},\ldots,\overset{n}{A_n})$ is not a homomorphism of rings unless the operators $A_1,\ldots,A_n$ are pairwise commutative, and the study of its algebraic structure is a complicated stand-alone problem. (When commutativity holds the situation is quite analogous to one-dimensional case.)

As R. Feynman has mentioned, the presence of numbers over operators often allows to treat the latter as if they were commutative; the numbers automatically govern the order of their action; e.g. the following expansion is valid:

$$e^A e^B = e^{\overset{2}{A}+\overset{1}{B}} = \sum_{n=0}^{\infty}\frac{(\overset{2}{A}+\overset{1}{B})^n}{n!}. \tag{13}$$

Of course, this doesn't mean that non-commutativity plays no role once the numbers over operators are introduced. In fact the necessity arises to develop a general non-commutative functional analysis comprising a collection of rules and formulae which supply the researcher with necessary technics to deal with functions of noncommuting operators.

We present now some examples.

Consider the notion of differential. We define the differential of an operator function f(A) in a usual way:

$$< df(A), H > \overset{\text{def}}{=} \lim_{t\to 0} \frac{f(A+tH)-f(A)}{t}. \tag{14}$$

A simple example shows, however, that the equality

$$< df(A), H > = f'(A)H \tag{15}$$

is not generally true; in other words, to calculate df(A) one cannot merely calculate the derivative f'(x) and then substitute the operator A into the result. Really, let $f(x) = x^2$, then

$$\frac{df}{dx}(A) = 2A,$$

on the other hand,

$$f(A+tH)-f(A) = (A+tH)^2 - A^2 = (A+tH)(A+tH) - A^2 =$$

$$= A^2 + t(AH+HA) + t^2H^2 - A^2 = t(AH+HA) + t^2H^2 =$$

$$= t(AH+HA) + O(t^2), \tag{16}$$

so that if the commutator $[H,A] = HA-AH$ does not vanish then the relation (15) doesn't hold.

Thus we have a problem of explicit calculation of df(A).

Here is the solution of this problem. Let $\mathcal{A}$ be an algebra, and let $D : \mathcal{A} \to \mathcal{A}$ be a differentiation of $\mathcal{A}$. Then the following formula holds:

$$Df(A) = \overset{2}{DA}\, \frac{\delta f}{\delta x}(\overset{1}{A}, \overset{3}{A}), \tag{17}$$

where

$$\frac{\delta f}{\delta x}(x,y) \overset{\text{def}}{=} \frac{f(x)-f(y)}{x-y} \tag{18}$$

is a difference derivative of the function f.

Let us give a sketch of proof of (17). It suffices to consider the case $f(x) = e^{itx}$ (the general case follows then via Fourier in integral). We have

$$\frac{\delta e^{itx}}{\delta x}(x,y) = i\int_0^t e^{i((t-\tau)x+\tau y)}\,d\tau .$$

It is clear that

$$D(e^{iAt})\big|_{t=0} = D(1) = 0.$$

Next,

$$\frac{d}{dt}\left(D(e^{iAt})\right) = D\left(\frac{d}{dt}e^{iAt}\right) = i\left(A\,D(e^{iAt}) + D(A)\cdot e^{iAt}\right),$$

so that

$$D(e^{iAt}) = i\int_0^t e^{iA(t-\tau)}\,D(A)\,e^{iA\tau}\,d\tau = \overset{2}{\overline{DA}}\;\frac{\delta e^{itx}}{\delta x}(\overset{1}{A},\overset{3}{A}),$$

q.e.d.

Note that most simply this formula is proved for polynomials. Using Leibniz rule we have

$$D(A^n) = D(A\cdot\ldots\cdot A) = \sum_{k=0}^{n-1} A^k D(A) A^{n-k-1} = \overset{2}{\overline{DA}}\;\frac{\delta(x^n)}{\delta x}(\overset{1}{A},\overset{3}{A}),$$

since

$$\frac{\delta(x^n)}{\delta x}(x,y) = \frac{x^n-y^n}{x-y} = \sum_{k=0}^{n-1} x^k y^{n-k-1}.$$

The formula (17) is one of the baseline formulae of non-commutative analysis. In particular, the theorem below follows from (17):

<u>Theorem</u> (Yu.L. Daletskii, S.G. Krein, [7]). The equality

$$\frac{d}{dt}f(A(t)) = \overset{2}{\overline{A'(t)}}\;\frac{\delta f}{\delta x}(\overset{1}{A(t)},\overset{3}{A(t)})$$

holds.

Here the algebra $\mathcal{A}$ consists of families of operators, depending on the parameter t, $D = \frac{d}{dt}$.

We obtain another corollary of (17), taking $D = \mathrm{ad}_B$ where ad_B is the operator of commutation with some fixed $B \in \mathcal{A}$, $\mathrm{ad}_B(X) = [B,X]$. Namely, we have

$$\mathrm{ad}_B(f(A)) = \overset{2}{\overline{\mathrm{ad}_B(A)}}\;\frac{\delta f}{\delta x}(\overset{1}{A},\overset{3}{A}).$$

Another example arises when we consider superposition of functions. Here even the problem of its proper definition and adequate notations is not trivial. In fact, the expression $f(g(A,B))$ is ambiguous: it may be considered either as the result of substitution of the operators A,B into the function $h(x,y) = f(g(x,y))$ or as the result of substitution of the operator $C = g(A,B)$ into the function $f(x)$ ("true" superposition for functions of operators). If A and B do not commute and f is not a linear function, these two interpretations lead, as a rule, to different results. Let $f(x) = x^2$, $g(x,y) = x+y$. Then

$$h(x,y) = (x+y)^2, \quad h(\overset{1}{A},\overset{2}{B}) = (\overset{1}{A} + \overset{2}{B})^2 = A^2 + 2BA + B^2;$$

on the other hand,

$$C = \overset{1}{A} + \overset{2}{B} = A + B,$$

$$f(C) = (A + B)^2 = (A+B)(A+B) = A^2 + AB + BA + B^2 \neq h(\overset{1}{A},\overset{2}{B}).$$

In order to avoid this ambiguity we adopt the convention that the expression $f(g(\overset{1}{A},\overset{2}{B}))$ always has the former interpretation, while the "true" superposition is denoted by $f(\llbracket g(\overset{1}{A},\overset{2}{B}) \rrbracket)$. That is,

$$f(\llbracket g(\overset{1}{A},\overset{2}{B}) \rrbracket) \overset{\text{def}}{=} f(C), \qquad \text{where } C = g(\overset{1}{A},\overset{2}{B}). \tag{19}$$

(Thus the "autonomous brackets" $\llbracket\ \rrbracket$ [1] define the order of calculations in operator expressions - first the expression in these brackets is evaluated and then the resulting operator is used in further evaluations.)

It is natural to pose the problem of explicit evaluation of "true" superposition, namely of its representation via functions of operators, which don't contain autonomous brackets.

2. Special non-commutative analysis

Consider now the application of general non-commutative analysis to construction of functional calculus for a given fixed tuple of operators $\overset{1}{A}_1, \ldots, \overset{n}{A}_n$. Surely there is no need to prove its importance, since particular examples of such calculus are the well-known algebra of classical pseudo-differential operators (which are functions of the differentiation operators $-i\frac{\partial}{\partial x^j}$ and multiplication operators x_j, $j = 1, \ldots, n$) and also the calculus of Fourier-Maslov integral operators (see, e.g., [6]), which form the module over the algebra of pseudo-differential operators.

The special non-commutative analysis studies, as its main object, algebraic operations (in particular, multiplication) in the set

$$\mathcal{A} = \{ f(\overset{1}{A}_1, \ldots, \overset{n}{A}_n)\}_{f\in\Sigma}$$

of functions of the operators $\overset{1}{A}_1,\ldots,\overset{n}{A}_n$ for a certain class Σ of symbols - functions $f(x_1,\ldots,x_n)$. The requirement that the set A be an algebra, i.e. that the product of the operators $f(\overset{1}{A}_1,\ldots,\overset{n}{A}_n)$ and $g(\overset{1}{A}_1,\ldots,\overset{n}{A}_n)$ could be represented as a function of $\overset{1}{A}_1,\ldots,\overset{n}{A}_n$,

$$[\![f(\overset{1}{A}_1,\ldots,\overset{n}{A}_n)]\!] \cdot [\![g(\overset{1}{A}_1,\ldots,\overset{n}{A}_n)]\!] = h(\overset{1}{A}_1,\ldots,\overset{n}{A}_n), \qquad (20)$$

imposes rigid restrictions on the operators $\overset{1}{A}_1,\ldots,\overset{n}{A}_n$.

In particular, it is clear that (20) implies, that any commutator $[A_j, A_k]$ may be represented as a function of $\overset{1}{A}_1,\ldots,\overset{n}{A}_n$, i.e the operators $\overset{1}{A}_1,\ldots,\overset{n}{A}_n$ form a so-called Poisson algebra [4],

$$[A_j, A_k] = \varphi_{jk}(\overset{1}{A}_1,\ldots,\overset{n}{A}_n). \qquad (21)$$

Here we restrict ourselves to the case when this Poisson algebra is in fact a nilpotent Lie algebra, i.e. the functions $\varphi_{jk}(x)$ are linear,

$$\varphi_{jk}(x) = \sum_{i=1}^{n} c_{jki}\, x_i , \qquad (22)$$

and all the commutators of order N are equal to 0 for N large enough,

$$[A_{j1}, [\ldots [A_{jN-1}, A_{jN}] \ldots]] = 0.$$

(This case was first considered in [1] and covers a number of applications. On general case see, e.g., [3], [4] and papers cited therein.)

Thus let the operators $\overset{1}{A}_1,\ldots,\overset{n}{A}_n$ be fixed and suppose that they realize a representation of a nilpotent Lie algebra $\mathcal{G}$, i.e.

$$[A_j, A_k] = A_jA_k - A_kA_j = -i\sum_{i=1}^{n} c^{i}_{jk}A_i, \quad j,k = 1, \ldots, n,$$

c^i_{jk} are structure constants of $\mathcal{G}$ in some basis $\{a_j\}$ (the factor -i is introduced for the sake of convenience, so that the structure constants be real when the A_j's are self-adjoint).

As we have told already, we are interested of the algebraic structure of the set $M = \{f(\overset{1}{A_1},\ldots,\overset{n}{A_n})\}$, in particular of existence of products and inverse elements in M. First of all, let us specialize the definition of $f(\overset{1}{A_1},\ldots,\overset{n}{A_n})$ in our particular case. We shall assume, that the operators $\overset{1}{A_1},\ldots,\overset{n}{A_n}$ are self-adjoint and consequently (under certain auxiliary assumptions) they realize the derived representation T_* of an unitary representation $T : G \to U(H)$ of the connected simply connected Lie group G, corresponding to the Lie algebra $\mathcal{G}$, in a Hilbert space H:

$$A_j = -iT_*(a_j).$$

The operators A_j are called the generators of the representation T in this situation.

Consider the coordinates of second genus on the group G. These are the coordinates $(x_1,\ldots,x_n)$, introduced via the mapping

$$\exp_2 : \mathcal{G} \to G$$
$$x = \sum_{i=1}^{n} x_i a_i \mapsto \exp_2(x) \overset{\text{def}}{=} \exp(x_n a_n)\ldots\exp(x_1 a_1)$$

where $\exp : \mathcal{G} \to G$ is the usual exponential mapping.

The mapping $\exp_2$ is a diffeomorphism of a neighbourhood of 0 in the Lie algebra onto a neighbourhood of unity in the Lie group; in our case (nilpotent Lie algebra) it is a global diffeomorphism of onto G (for special choice of the basis $(a_1, \ldots, a_n)$).

Using the introduced coordinates, we may write

$$f(\overset{1}{A_1},\ldots,\overset{n}{A_n}) = \frac{1}{(2\pi)^{n/2}} \int_{\mathbb{R}^n} \tilde{f}(x_1,\ldots,x_n) e^{ix_nA_n}\ldots e^{ix_1A_1} dx_1\ldots dx_n =$$

$$= \frac{1}{(2\pi)^{n/2}} \int_{\mathbb{R}^n} \tilde{f}(x_1,\ldots,x_n)\, T(\exp(x_na_n))\ldots T(\exp(x_1a_1))\, dx_1\ldots dx_n =$$

$$= \frac{1}{(2\pi)^{n/2}} \int_{\mathbb{R}^n} \tilde{f}(x_1,\ldots,x_n)\, T(\exp_2(x_1a_1+\ldots+x_na_n))\, dx_1\ldots dx_n =$$

(thus we see, that the integration is in fact over G)

$$= \frac{1}{(2\pi)^{n/2}} \int_G \tilde{f}(g) T(g) d\mu(g).$$

Here $d\mu(g)$ is the Haar measure on G, the Jacobian equals 1 due to nilpotency, $\tilde{f}(g)$ is a "group" Fourier transform of f, $\tilde{f}(\exp_2(x)) = \tilde{f}(x)$.

In what follows we write f(A) instead of $f(\overset{1}{A}_1, \ldots, \overset{n}{A}_n)$ to save space.

Now we are able to calculate the product of two elements of M, say, $f_1(A)$ and $f_2(A)$, via integrals over the group:

$$f_1(A) f_2(A) = \frac{1}{(2\pi)^n} \int_{G\times G} \tilde{f}_1(g) \tilde{f}_2(h) T(g) T(h) d\mu(g) d\mu(h) =$$

$$= \frac{1}{(2\pi)^n} \int_{G\times G} \tilde{f}_1(g) \tilde{f}_2(h)\, T(gh) d\mu(g) d\mu(h) =$$

(change of variables $gh = k$, $d\mu(g^{-1}k) = d\mu(k)$)

$$= \frac{1}{(2\pi)^n} \int_{G\times G} \tilde{f}_1(g) \tilde{f}_2(g^{-1}k) T(k) d\mu(g) d\mu(k) =$$

$$= \frac{1}{(2\pi)^{n/2}} \int_G \tilde{f}(k) T(k) d\mu(k) = f(A),$$

where

$$\tilde{f}(k) = (\tilde{f}_1 * \tilde{f}_2)(k) = \frac{1}{(2\pi)^{n/2}} \int_G \tilde{f}_1(g) \tilde{f}_2(g^{-1}k) d\mu(g)$$

is the convolution of f_1 and f_2 with respect to Haar measure.

Let $\mathcal{L}$ be the left regular representation of the group g,

$$\mathcal{L}(g)\, h(k) = h(g^{-1}k)$$

for any function h on the group G. Using the representation $\mathcal{L}$ the function $\tilde{f}$ may be put into the form

$$\tilde{f} = \left[\frac{1}{(2\pi)^{n/2}} \int_G \tilde{f}_1(g) \mathcal{L}(g) d\mu(g) \right] \tilde{f}_2 = f_1(L)(\tilde{f}_2),$$

where $L = L_1, \ldots, L_n$ are the generators of the representation $\mathcal{L}$ (L_j is simply the right-invariant vector field on G satisfying the condition $L_j|_e = -i \frac{\partial}{\partial x_j}$).

Consequently,

$$f = [\mathcal{F}_{-1} \circ f_1(L) \circ \mathcal{F}](f_2) = f_1(\ell)(f_2),$$

where $\mathcal{F}$ is the group Fourier transformation,

$$\ell_j = \mathcal{F}_{-1} \circ L_j \circ \mathcal{F}$$

are some pseudo-differential operators (in our case in a special basis these operators will be differential ones).

Thus we have shown that the product $f_1(A)f_2(A)$ belongs to the set M and its symbol - such a function f that $f_1(A)f_2(A) = f(A)$ - is given by

$$f = f_1(l)(f_2),$$

where $l = (l_1, \ldots, l_n)$ are the operators of left regular representation, acting in the space of symbols.

More accurate considerations show that M is in fact only a module over the algebra $M_0 \subset M$ of the operators $f(A_1, \ldots, A_n)$ with classical symbols. This module is an analogue of the module of Fourier-Maslov integral operators over pseudodifferential operators mentioned above.

How to calculate the inverse element ? Suppose that we wish to invert the operator f(A). It is clear from above that it suffices to solve the equation

$$f(l)g = 1,$$

where g is an (unknown) symbol of the inverse operator, or the equivalent to it pseudo-differential equation

$$f(L)g = \delta_e \tag{23}$$

on the group G (here δ_e is the Dirac δ -function at the point $e \in G$).

As a rule, one cannot obtain a precise solution of this equation. Fortunately, for applications it is often sufficient to obtain asymptotic solutions. In the case of asymptotics with respect to powers of the operators $A_1, \ldots, A_k$ the equation (23) reduces the problem of construction of asymptotically inverse operator to the problem of smooth asymptotics for solutions of (23). (More precisely, we should speak of partial smoothness, defined by directions of vector fields $L_1, \ldots, L_k$). Note that the phase space for our problem turns out to be the cotangent space T^*G, while the Hamiltonian function is given by left shifts of the principal part of the symbol f:

$$H_0 = L^*(f_0)$$

$L : T^*G \longrightarrow T^*_eG$ being the projection induced by left shifts.

This paper is an account of the lecture delivered by the authors at the seminaire "Differential-geometrical and computer-algebraic methods of investigations of non-linear problems (Rachov, autumn of 1989).

References

1. V.P. Maslov, Operational methods, Nauka, Moscow, 1973.
2. V.P. Maslov, Asymptotic methods of solution of pseudodifferential equations, Nauka, Moscow, 1987.
3. V.P. Maslov and V.E. Nazaikinskiĭ, Asymptotics of operator and pseudo-differential equations, Consultants Bureau, New York, 1988.
4. M.V. Karasev and V.P. Maslov, Global asymptotic operators of regular representation, Dokl. Acad. Nauk SSSR 257 (1) (1981), 33-38.
5. R.P. Feynman, An operator calculus having applications in quantum electrodynamics, Phys. Rev. 84 (2) (1951), 108-128.
6. V.E. Nazaikinskiĭ, V.G. Oshmjan, B.Yu.Sternin and V.E. Shatalov, Fourier integral operators and canonical operator, Uspekhi Mat. Nauk 36 (2) (1981), 81-140.
7. Yu.L. Daletskiĭ and S.G. Kreĭn, A formula for differentiating with respect to parameter of functions of Hermitian operators, Dokl. Acad. Nauk SSSR 76 (1) (1951), 13-66.

THE PROBLEM OF REALIZATION OF HOMOLOGY CLASSES FROM POINCARE UP TO THE PRESENT

Yu.B.Rudyak
Department of Mathematics
Moscow Institute of Civil Engineers
Yaroslavskoe shosse 26
129337, Moscow, USSR

Introduction.

This paper is mainly a survey. The choice of the theme is governed not only by the scientific interests of the author, but also by his desire to introduce a reader to a sufficiently large field of ideas and methods of modern algebraic topology by considering a concrete problem.

The problem of the realization of homology classes by the manifolds is usually associated with Steenrod, but in fact it must be associated with Poincare. Namely, Poincare has developed the highly dimensional integration theory and hence has laid the basis for the algebraic topology. In the process of this development he at first considered chains in a manifold as submanifolds, but later he was forced to adopt a more abstract (and in fact modern) treatment of the chains as simplicial decompositions. The question of the relation of chains to submanifolds remained in the background in view of the correct construction of homology theory, but with the development of topology this question was highlighted again, and Steenrod formulated it explicitly in [1] as a problem of the realization of homology classes by manifolds.

Thom [2] proved the realizability of every homology class mod 2 and constructed non-realizable integer homology classes. For this purpose Thom (by invertion of one idea of Pontryagin) developed the machinery which made a good connection between homotopy and geometric topology and, no doubt, made a big progress in topology. There appeared the chances of wide applications of homotopy methods to geometric problems and vice versa. Later Thom's constructions were carefully investigated and generalized. There arose new directions, ideas and methods which essentially created the image of contemporary algebraic topology.

In this paper we give a contemporary statement of the results on the realizability. The main attention is devoted to the conceptual but not computational aspects. The bibliography allows a reader to restore all the missing or indicated arguments, and in this case the reader will be able to go to the modern level of one of the branches of algebraic topology.

Throughout the paper SX denotes the suspension over a space X (or the reduced suspension for pointed X), X denotes the disjoint union of X and the point pt. Further, θ^n denotes a trivial n-dimensional vector bundle over the given (clear from the context) base. The product of vector bundles ξ and η (over different bases) is denoted by $\xi \times \eta$, and the Whitney sum of ξ and η over the same base is denoted by $\xi \oplus \eta$. The unit closed interval $[0,1]$ is denoted by I. Finally, $p_i : X_1 \times \dots \times X_n \longrightarrow X_i$ denotes the projection on the i-th factor.

1. Statement of a problem.

Let M^n be a closed connected topological manifold. It is well-known that $H_n(M^n; Z/2) = Z/2$; let $[M]$ be a non-trivial element of this group. Every (continuous) map $f : M \to X$ gives the element $z = f_*[M] \in H_*(X; Z/2)$, where $f_* : H_*(M; Z/2) \to H_*(X; Z/2)$, and they say that z is realized by the map $f : M \to X$ or by the singular manifold (M,f). It seems that the term "singular manifold" comes from the "ancient" times when there was not such a big difference between the map $f : M \to X$ and the image $f(M) \subset X$ with $f(M)$ being a tumbled manifold with some singularities etc.; but now the term "singular manifold" denotes simply a map $f: M \to X$.

Similar arguments work also for the groups $H_*(X)$ of integer homology. Let x be an arbitrary but fixed point of M^n, and let U_x be a closed neighbourhood of x which is homeomorphic to disk D^n. Consider the map $\varepsilon : M^n \to U_x/\partial U_x$ which collapses the complement of the interior U_x in M to the point. Since $U_x/\partial U_x$ is homeomorphic to S^n, there arises a homomorphism $\varepsilon_* : H_n(M^n) \longrightarrow H_n(U_x/\partial U_x) \xrightarrow{\approx} H_n(S^n) = Z$. Let $s \in H_n(S^n)$ be a generator.

<u>1.1.Definition.</u> The element $[M] \in H_n(M)$ is called the orientation of M if $\varepsilon_*[M] = \pm s$.

It can be proved that the orientability of M does not depend on a choice of x and U_x. Geometrically, an orientation of a manifold is a family of the compatible orientations of charts. Further, there is a global criterion of orientability: a closed connected M is orientable iff $H_n(M) = Z$; otherwise $H_n(M) = 0$. So, every closed connected M has just two orientations - free generators of $H_n(M)$.

Now, let X be a topological space and $f : M^n \to X$ be a (continuous) map of a closed connected oriented M. There arises the element $f_*[M] \in H_n(X)$, where $f : H_n(M) \to H_n(X)$. We say that a homology class $z \in H_n(X)$ can be realized by a singular manifold if it has a form $z = f_*[M]$ for some singular manifold (M,f).

It seems natural to ask whether every homology class can be realized by a singular manifold. The answer is "no", and a deeper problem is to describe all the realizable homology classes. This is the problem of the realization of homology classes.

This problem can be modified as follows. First, it is possible to speak about the realization not by topological but smooth or piecewise linear manifolds. Second, not just manifolds but some manifolds-like polyhedra can be used as models for the realization of homology classes.

2. Bordisms.

From this point and to the end of §4 we assume that every manifold is smooth, i.e. we consider the realization of homology classes by singular manifolds (M,f) with a smooth M.

2.1. Definition. Two singular manifolds $f : M^n \to X$ and $g : N^n \to X$ with closed M,N are bordant if there exists a singular manifold $F : W^{n+1} \to X$ such that ∂W is a disjoint union $M \cup N$ and $F|M = f$, $F|N = g$. Here F is called the bordism between f and g.

So, we have the relation (usually called the bordism relation) on a set of all singular manifolds $M^n \to X$ with closed manifolds M^n. It is evident that the bordism relation is the equivalence relation. The corresponding (quotient) set of bordism classes is denoted by $\mathfrak{N}_n(X)$ and is called nonoriented bordisms of X. The bordism class of (M,f) is denoted by [M,f].

$\mathfrak{N}_n(X)$ admits a natural commutative group structure. Put $(M,f) + (N,g) = (f \cup g : M \cup N \to X)$ for $f : M \to X$ and $g : N \to X$; here $\cup$ is a disjoint union. It is evident that $\mathfrak{N}_n(X)$ becomes a commutative semigroup with respect to +. Further, the map $\varepsilon : S^n \to * \in X$ to point is the neutral element of this semigroup. Indeed, (M,f) and $(M \cup S^n, f \cup \varepsilon)$ are bordant, the bordism is $(M \times I \cup D^{n+1}, f \cup \delta)$ with $F(m,t) = f(m)$, $m \in M$, $t \in I$ and $\delta(x) = *$, $x \in D^{n+1}$. Finally, just the same F gives a bordism between $(M \cup M, f \cup f)$ and (S^n, ε), so every element of $\mathfrak{N}_n(X)$ has the order 2, so $\mathfrak{N}_n(X)$ is a commutative group.

There is the Steenrod - Thom map

(2.2) $$\bar{\mu} : \mathfrak{N}_n(X) \longrightarrow H_n(X; Z/2)$$

with $\bar{\mu}[M,f] = f_*[M]$, where $[M] \in H_n(M; Z/2) = Z/2$ is the generator; $\bar{\mu}$ is well-defined. Indeed, let (W,F) be a bordism between (M,f) and $(N;g)$ and $i : M \to W$, $j : N \to W$ be the obvious inclusions. Then $i_*[M] = j_*[N]$ and $f_*[M] = F_*i_*[M] = F_*j_*[N] = g_*[N]$. It is easy to see that $\bar{\mu}$ is a homomorphism. Now it is obvious that the problem of realization of mod 2 classes is just the description of the image of $\bar{\mu}$.

Similarly one can define the group $\Omega_n(X)$ of oriented bordisms of X.Here M,N and W must be oriented and W must be oriented so that $\partial W = M \cup (-N)$; here ∂W is the oriented boundary of W and $(-N)$ is N with the opposite orientation. Further, $-[M,f] = [(-M),f]$. Note that $\Omega_n(X)$ can consist of the elements of arbitrary order. As above, the Steenrod - Thom homomorphism

(2.3) $$\mu : \Omega_n(X) \longrightarrow H_n(X)$$

is defined, and the problem of the realization of integer homology classes is the description of im μ.

The next step to attack the realizability problem is the homotopy interpretation of bordism groups. This is one of the main results of Thom [2].

Let ξ be an n-dimensional vector bundle over CW-space X. Let us choose the Riemannian metric on ξ (i.e. the continuous family of inner products in every fibre). This is possible because the set of positive defined quadratic forms is convex and hence contractible, so the bundle associated with ξ, whose fibre consists of these quadratic forms, has a cross-section. Let $D(\xi)$, resp. $S(\xi)$ be a subbundle of ξ with unit disks, resp unit spheres, as fibres.

<u>2.4.Definition.</u> The quotient space $D(\xi)/S(\xi)$ is called the Thom space $T\xi$ of bundle ξ.

<u>2.5.Proposition.</u> (i) The Thom space of a trivial bundle θ^n over X is homeomorphic to S^nX^+.

(ii) Let ξ and η be vector bundles over X and Y, respectively. Then $T(\xi \times \eta)$ is homeomorphic to $T(\xi) \times T(\eta)$.

The proof is simple and can be found e.g. in [3]. ∎

Now let M^n be a (smooth closed) manifold which is smoothly embedded in R^{N+n}, $N \gg n$.(Such an embedding exists in view of the well-known Whitney theorem,see e.g.[12]).Let F_x be a normal N-plane to M at $x \in M$, and ε_x be a maximum of such r_x that the open disk in F_x of radius r at x

does not intersect F_y for all $y \neq x$. It can be proved by the standard manner that $\varepsilon_x > 0$ and ε_x is a continuous function of x. The compactness of M implies that $\varepsilon = \min \varepsilon_x > 0$. Choose some $\delta < \varepsilon$, $\delta > 0$.

2.6. Definition. The set $U = \{z \in R^{N+n} \mid \rho(z,M) < \delta\}$ is called a tubular neighbourhood of M^n in R^{N+n}.

This construction shows that U is fibered over M with open δ-disks as fibres. Let $\nu = \nu^N$ be a normal bundle of M^n in R^{N+n}.

2.7. Lemma. U is a subbundle of ν.

Proof. Consider the set $V = \{(x,y) \mid y \in F_x\}$ in $M \times R^N$. Put $p = p_1 | V : V \to M$ where $p_1 : M \times R^{N+n} \to M$ is a projection. It is easy to see that p is a projection in a vector bundle ξ over M. Of course, $\xi \oplus \tau_M = \theta^{N+n}$, so ξ is equivalent to ν. Now the lemma follows from the fibrewise inclusion $U \subset V$. ∎

2.8. Corollary. The quotient space $R^{N+n}/R^{N+n} \setminus U$ is a Thom space $T\nu$ of normal bundle ν. ∎

Now, R^m can be considered as S^m without one point, so we have the map

$$(2.9) \qquad c : S^{N+n} \longrightarrow S^{N+n}/S^{N+n} \setminus U = T$$

This map plays a very important role in the topology of manifolds. We shall call it a natural collapse.

Now we recall the classification of vector bundles, see e.g. [3].

2.10. Theorem. There is a space BO_n and an n-dimensional vector bundle γ_n over BO_n such that

(i) Every n-dimensional vector bundle ξ over X is equivalent to the bundle $f^*\overline{\gamma}_n$ for some $f : X \to BO_n$.

(ii) Maps $f, g : X \to BO_n$ are homotopic iff $f^*\overline{\gamma}_n$ and $g^*\overline{\gamma}_n$ are equivalent.

The space BO_n is called the classifying space and the bundle $\overline{\gamma}_n$ is called the universal n-dimensional vector bundle.

Now we define oriented vector bundles. Let ξ be any n-dimensional vector bundle over a connected CW-space X, and let R^n_x be a fibre over some $x \in X$. The inclusion of R^n_x to the total space of ξ induces an inclusion $j_x : S^n \to T\xi$ (because $D^n_x / S^{n-1}_x = S^n$). Let $s \in H^n(S^n) = Z$ be a generator.

2.11. Definition. The element $u \in H^n(T\xi)$ is called the orientation of ξ iff $j^*_x(u) = \pm s$.

Since for every $x, y \in X$ inclusions $j_x, j_y : S^n \to T\xi$ are homotopic, the orientability of a bundle does not depend on a choice of x. Geometrically, orientation of a bundle is a family of compatible orientations of fibres. There is also a global criterion of orientability: $H^n(T\xi) = Z$ for orientable ξ and $H^n(T\xi) = 0$ for non-orientable ξ. This is easily proved if X is considered as CW-space with just one 0-cell and hence $T\xi$ as CW-space with just one n-cell.

We recommend the reader to compare definition 2.11 with definition 1.1. Later we shall show that a manifold is orientable iff its normal bundle is.

Two oriented vector bundles are called equivalent if there is the equivalence of bundles which transfers one orientation to another. As in 2.10, there exist the space BSO_n and the universal n-dimensional oriented vector bundle γ_n over BSO_n which classify n-dimensional oriented vector bundles, see [3], [4].

The Thom spaces $T\bar{\gamma}_n$ and $T\gamma_n$ denote MO_n and MSO_n resp.

2.12. Theorem. (The homotopy interpretation of bordisms).

There are natural with respect to X isomorphisms (where $N \gg n$)

$$\mathfrak{N}_n(X) = \pi_{N+n}(X^+ \wedge MO_N)$$

$$\Omega_n(X) = \pi_{N+n}(X^+ \wedge MSO_N)$$

Proof. We prove only the first isomorphism, the second can be proved in the similar way with the comment that a normal bundle of an orientable manifold is orientable, see 2.17. We shall construct mutually inverse maps $\mathfrak{N}_n(X) \longrightarrow \pi_{N+n}(X^+ \wedge MO_N)$ and $\pi_{N+n}(X^+ \wedge MO_N) \longrightarrow \mathfrak{N}_n(X)$. Note that the projection $p_2 : X \times BO_N \longrightarrow BO_N$ gives the bundle $\xi = p^*\bar{\gamma}_n$ over $X \times BO_N$, and it is easy to see that $T\xi = X^+ \wedge MO_N$. Let $a \in \mathfrak{N}_n(X)$ be presented by a singular manifold $f: M^n \longrightarrow X$, and let ν be a normal bundle of M^n in R^{N+n}. By 2.10 ν can be determined by a map $t : M^n \longrightarrow BO_N$, so there arises $h : M^n \longrightarrow X \times BO_N$, such that $h(m) = (f(m), t(m))$, and $h^*(\xi) = \nu$. The composition

$$S^{N+n} \xrightarrow{\ c\ } T\nu \xrightarrow{\ Th\ } T\xi = X^+ \wedge MO_N$$

where c is the collapse (2.9) gives an element of $\pi_{N+n}(X^+ \ MO_N)$.

Now let $g : L^n \longrightarrow X$ be a singular manifold which is bordant to (M,f), and let $F : W^{n+1} \longrightarrow X$ be a bordism between (M,f) and (L,g). We can assume that W^{n+1} is embedded in $S^{N+n} \times I$ so that $W \cap (S^{N+n} \times 0) = M$, $W \ (S^{N+n} \times 1) = L$ and W is normal to $S^{N+n} \cap \{0,1\}$. As above, one can construct the collapse $c: S^{N+n} \times I \longrightarrow T\nu_W$ and make a composition $S^{N+n} \times I \xrightarrow{c} T\nu_W \longrightarrow T\xi = X^+ \wedge MO_N$. So, (M,f) and (L,g) give homotopic maps $S^{N+n} \longrightarrow X^+ \wedge MO_N$. So, we constructed a well-defined map

$\mathfrak{N}_n(X) \to \pi_{N+n}(X^+ \wedge MO_N)$.

Now we shall construct the map $\pi_{N+n}(X^+ \wedge MO_N) \to \mathfrak{N}_n(X)$. For this we must recall the transversality. Let E be a total space of a vector bundle ξ over CW-space Z. We can treat Z as a subspace of E via zero section of ξ. Let M be a (smooth) manifold. Let E be embedded in some Y as an open subset.

2.13. Definition. A map $f: M \to Y$ is called transversal to Z if $f^{-1}(Z)$ is a submanifold in M and the normal bundle of the inclusion $f^{-1}(Z) \subset M$ is equivalent to $f^*\xi$.

2.14. Lemma. Every map $g: M \to Y$ is homotopic to a map $f: M \to Y$ which is transversal to Z. Moreover, if for some closed subset L and open $U \supset L$ the map $g|_U: U \to Y$ is transversal to Z, then f can be chosen such that $f|L = g|L$.

Proof can be found in [2], see also [12]. Note that in [2,12] the lemma is proved for manifolds Z, but it does not matter.

Thus, let an element of $\pi_{N+n}(X^+ \wedge MO_N)$ be given by a map $f': S^{N+n} \to X^+ \wedge MO_N$. By 2.14 there exists $f: S^{N+n} \to X^+ \wedge MO_N$ homotopic to f' and transversal to $X \times BO_N$. Then $f^{-1}(X \times BO_N)$ is a manifold of codimension N in S^{N+n}, i.e. some manifold M^n. The map $M^n \xrightarrow{f} X \times BO_N \xrightarrow{p_1} X$ gives a singular manifold in X.

Now let $g': S^{N+n} \to X^+ \wedge MO_N$ be homotopic to f', and let g be a transversal approximation of g' as above. Then f and g are homotopic, and by 2.14 the homotopy F between f and g can be chosen transversal to $X \times BO_N$. So, we have a map

$$F: S^{N+n} \times I \to X^+ \wedge MO_N$$

and $F^{-1}(X \times BO_n)$ gives a bordism between $f^{-1}(X \times BO_N)$ and $g^{-1}(X \times BO_N)$.

So, we constructed a well-defined map $\pi_{N+n}(X^+ \quad MO_N) \to \mathfrak{N}_n(X)$. It is easy to prove that the maps $\pi_{N+n}(X^+ \quad MO_N) \rightleftarrows \mathfrak{N}_n(X)$ are mutually inverse, Q.E.D. ∎

Now we shall prove that a manifold is orientable iff its normal bundle is. Let us recall that finite CW-spaces X,Y are called N-dual (à la Alexander-Spanier-Whitehead) if X is a deformation retract of $S^{N+n} \setminus Y$ with respect to some cellular embedding $Y \subset S^{N+n}$. The N-duality is a symmetric relation, and in this case we have the isomorphism

$$\tilde{H}^i(X) = \tilde{H}_{N-i}(Y) \tag{2.15}$$

where $\tilde{H}$ is a reduced (co)homology, see [4]. (NOte that the homotopy type of Y is not uniquely determined by the homotopy type of X, but the so-called stable homotopy type of Y (in fact, the homotopy type of

S^kY, $k \gg \dim Y$)is.)

2.16. Theorem. (Milnor, Spanier, see [3], [4]). For every closed manifold M the space $T\nu^N$ is (N+n)-dual to M^+. ∎

2.17. Theorem. A closed manifold M is orientable if its normal bundle is.

Proof. By (2.15) and 2.16 we have

$$H_n(M) = \tilde{H}_n(M^+) = \tilde{H}^N(T\nu^N).$$

Thus $H_n(M) = Z$ iff $H^N(T\nu^N) = Z$. ∎

3. A modern approach: spectra.

The reader noted that we deal permanently with the inequality $N \gg n$, i.e. with the so-called stable situation. This situation has a convenient formalization: its main tool is the conception of a spectrum. Of course, we could continue the discussion on the "$N \gg n$"- level, but we prefer to acquaint the reader with some present viewpoints. There are different categories of spectra proposed by different authors. We consider this one not so convenient technically but explicit conceptually. For deeper investigations it is preferable to use the categories introduced by Adams, see e.g. [4], and May [13].

Throughout this section we suppose that all the spaces and maps are pointed.

3.1. Definition. The spectrum E is a sequence $\{E_n, s_n\}_{n=0}^{\infty}$ of CW-spaces E_n and cellular maps $s_n : E_n \to SE_{n+1}$. (Sometimes we write $E = \{E_n\}$ if the maps s_n are obvious). The map $f : E \to F$ (with $F = \{F_n, t_n\}$) of spectra is a family $f_n : E_n \to F_n$ with $t_{n+1} \circ Sf_n = f_{n+1} \circ s_n$.

3.2. Examples. 1. Every space X generates the spectrum $\Sigma^{\infty} X =$ $= \{X, SX, S^2X, \dots, S^nX, \dots\}$ with $s_n = 1_{S^nX}$. In particular, there is the spectrum $S = \Sigma^{\infty} S^0 = \{S^n\}$ of spheres.

2. The Eilenberg-MacLane spectrum $H(\pi)$ of abelian group π. Here $H(\pi)_n$ is the Eilenberg-MacLane space $K(\pi, n)$ and the map s_n is adjoint to the homotopy equivalence $w_n : K(\pi, n) \to \Omega K(\pi, n+1)$.

3. The bundle $\overline{\gamma}_n \oplus \theta^1$ over BO_n is classified by a map $BO_n \to BO_{n+1}$. So there arises the map $s_n : T(\overline{\gamma}_n \oplus \theta^1) \to T\overline{\gamma}_{n+1}$ of Thom spaces. But $T(\overline{\gamma}_n \oplus \theta^1) = ST\overline{\gamma}_n$, so $s_n : SMO_n \to MO_{n+1}$. Thus, we have the Thom spectrum $MO = \{MO_n, s_n\}$. Similarly, we have the Thom spectrum $MSO =$ $= \{MSO_n, s_n\}$.

4. For every space X and every spectrum E we have the spectrum

$X\wedge E = \{X\wedge E_n, 1_X\wedge s_n\}$.

Now we shall connect spectra and extraordinary (co)homology theories.

3.3. Definition. An extraordinary homology theory is the family $\{E_n, \partial_n\}$ where E_n is a functor from CW-pairs (X,A) to abelian groups and $\partial_n : E_n(X,A) \to E_{n-1}(A,\emptyset) = E_{n-1}(A)$ is the morphism of functors. Here $\{E_n, \partial_n\}$ must satisfy all the Steenrod-Eilenberg axioms excluding the dimension axiom, see [4], chapter 7, or [5], chapter V, § 5. The definition of extraordinary cohomology theories is similar.

Note that now one says simply "(co)homology theory" instead of "extraordinary (co)homology theory". For (H, ∂) there is the term "classical (co)homology theory".

Let [X,Y] be the set of homotopy classes of maps $X \to Y$ (recall that spaces, maps, suspensions, etc. are pointed). For CW-space X and spectrum E let us consider the sequences

(3.5) $$[X,E_n] \longrightarrow [SX,SE_n] \xrightarrow{(s_n)_*} [SX, E_{n+1}] \longrightarrow \dots$$

(3.6) $$[S^{n+k}, X\wedge E_k] \to [S^{n+k+1}, X\wedge SE_k] \xrightarrow{(s_k)_*} [S^{n+k+1}, X\wedge E_{k+1}] \longrightarrow \dots$$

Since $[S^2X, Y]$ is a natural abelian group, these sequences are the sequences of abelian groups and homomorphisms. The(direct)limit of (3.5) is denoted by $\widetilde{E}^n(X)$, and that of (3.6) is denoted by $\widetilde{E}_n(X)$. The group $\widetilde{E}_n(S^0) = \widetilde{E}^{-n}(S^0)$ is called the coefficient group (n-dimensional) of E. It is a direct limit of the sequence

$$\dots \to \pi_{n+N}(E_N) \longrightarrow \pi_{n+N+1}(SE_N) \longrightarrow \pi_{n+N+1}(E_{N+1}) \to \dots$$

In particular, $\widetilde{E}_n(X) = \pi_n(X\wedge E)$.

The definitions of $\widetilde{E}$ imply

3.7. Lemma. The maps $s_n: E_n \to E_{n+1}$ induce the isomorphisms

$$\widetilde{E}_n(X) \xrightarrow{\approx} \widetilde{E}_{n+1}(SX); \quad \widetilde{E}^n(X) \xrightarrow{\approx} \widetilde{E}^{n+1}(SX). \blacksquare$$

Let us consider the CW-pair (X,A) and the Puppe sequence (see [4])

$$A \hookrightarrow X \to X\cup CA \to SA \to SX \to \dots$$

Applying the functor $[\ , E_n]$ to it, we obtain the exact sequence

$$\dots \to [SX, E_n] \to [SA, E_n] \to [X\cup CA, E_n] \to [X, E_n] \to [A, E_n].$$

Since the direct limit preserves an exactness, we have the exact sequence

$$\ldots \longrightarrow \tilde{E}^n(SX) \longrightarrow \tilde{E}^n(SA) \longrightarrow \tilde{E}^n(X \cup CA) \longrightarrow \tilde{E}^n(X) \longrightarrow \tilde{E}^n(A) \ ,$$

i.e. (in view of 3.7) the exact sequence

$$(3.8) \quad \ldots \to \tilde{E}^{n-1}(X) \longrightarrow \tilde{E}^{n-1}(A) \xrightarrow{\tilde{\delta}^n} \tilde{E}^n(X \cup CA) \longrightarrow \tilde{E}^n(X) \longrightarrow \ldots$$

A similar sequence can be considered for E_n.

Put

$$E_n(X,A) = \tilde{E}_n(X \cup CA), \text{ so } E_n(X) = \tilde{E}_n(X^+),$$

$$E^n(X,A) = \tilde{E}^n(X \cup CA), \text{ so } E^n(X) = \tilde{E}^n(X^+),$$

$$\delta^k : E^k(A) = E^k(A^+) \xrightarrow{\tilde{\delta}^k} E^k(X^+ \cup CA^+) \to E^k((X \cup CA)^+) = E^k(X,A)$$

and ∂_k similarly.

<u>3.9.Theorem.</u> $\{E_n, \partial_n\}$ $((E^n, \delta^n)$, respecively) is the extraordinary homology (and cohomology) theory.

<u>Proof</u> (for cohomology). The validity of the exactness axiom follows from the exactness of (3.8). The validity of the homotopy axiom is evident. The excision property follows from the equivalence $X \cup CA \sim X/A$, see [4]. Indeed, from this relation and homotopy invariance

$$E^n(X,A) = E(X \quad CA) = E^n(X/A) = E^n(X \ U \ / \ A \ U) = E^n(X \ U, \ A \ U).$$

Thus, every spectrum produces the (extraordinary) homology theory and cohomology theory. Vice versa, every cohomology theory can be produced as above from a spectrum, cf. [4], ch.9. $E_n(X)$ is isomorphic to ker $(E_n(X) \quad E_n(pt))$ and is called the reduced (extraordinary) homology group. Note, see [4], that the Alexander-Spanier-Whitehead duality is also valid in this case, i.e.

$$\tilde{E}_i(X) = \tilde{E}^{n-i}(Y)$$

for n-dual X,Y.

Note that classical cohomology $H^*(X;\pi)$ can be given as above by the spectrum $H(\pi)$, so the duality implies

$$H_n(X;\pi) = \pi_n(X \wedge H(\pi)) = \lim_N \pi_{n+N}(X \wedge K(\pi, N)).$$

Of course, this can be proved by checking the Steenrod-Eilenberg axioms for the group at the right side.

For any spectra E, F the group $F^i(E)$ - the inverse limit of the sequence

$$\ldots \leftarrow \tilde{F}^{i+n}(E_n) \overset{\approx}{\leftarrow} \tilde{F}^{i+n+1}(SE_n) \longleftarrow \tilde{F}^{i+n+1}(E_{n+1}) \longleftarrow \ldots$$

is defined.

In particular, for any abelian groups $\pi; \tau$ there is the group $H^i(H(\pi);\tau)$. An element $\varphi \in H^i(H(\pi);\tau)$ is the family $\{\varphi_n\}$ with $\varphi_n \in H^{i+n}(K(\pi,n);\tau)$.

For every CW-space X and every $x \in H^n(X;\pi)$ one can define $\varphi(x) \in H^{i+n}(X;\tau)$ by the composition

$$X \xrightarrow{x} K(\pi, n) \xrightarrow{\varphi_n} K(\tau, i+n)$$

It is easy to see that for any $f : Y \longrightarrow X$ and any $x \in H^n(X;\pi)$ we have $f^*\varphi(x) = \varphi(f^*(x))$, i.e. φ induces the natural transformation

$$\varphi: H^*(X;\pi) \longrightarrow H^{*+i}(X;\tau)$$

of cohomology theories. The elements of $(H^i(H(\pi);\tau)$ are called stable cohomology operations of dimension i. It is a powerful computing tool, see [6], [7]; we shall use it in §4. Note that the groups $H^*(H(\pi);\tau)$ are computed for all finitely generated abelian groups, see [7].

Now let us return to bordisms. In the above notations 2.12 trasforms to

(3.10) $\quad \mathfrak{N}_n(X) = MO_n(X); \quad \Omega_n(X) = MSO_n(X).$

In fact it is possible to introduce geometrically the relative bordism groups $\mathfrak{N}_n(X,A)$ and the homomorphisms $\mathfrak{N}_n(X,A) \rightarrow \mathfrak{N}_{n-1}(A)$ and to show that there arises the extraordinary homology theory. Similarly for Ω, see [8]. Further, it is possible to define the cobordism theories $MO^*(X)$ and $MSO^*(X)$.

Let ξ be an n-dimensional vector bundle over a connected CW-space X. Let $j : S^n \rightarrow T\xi$ be the inclusion from definition 2.11. Let

$\bar{s} \in H^n(S^n; Z/2)$ and $s \in H^n(S^n)$ be the generators.

3.11. Proposition- definition. (i) There is the element $\bar{u} \in H^n(T\xi; Z/2)$ with $j^*(\bar{u}) = \bar{s}$. Besides, $H^n(T\xi; Z/2) = Z/2 = \{\bar{u}\}$. (ii) If ξ is orientable, there is the element $u \in H^n(T\xi)$ with $j^*u = s$. Besides, $H^n(T\xi) = Z = \{u\}$. The elements $\bar{u}$ and u are called Thom classes (of ξ).

Proof. Without loss of generality we may suppose that X has only one 0-cell. The cellular decomposition of X produces the cellular decomposition of $T\xi$ so that $T\xi$ has just one n-cell and has no cells of less than n positive dimensions. The closure of the n-cell is the image $j(S^n)$ in $T\xi$. This cell gives an n-cycle z. It is easy to see (from the cellular decomposition of X) that if mz is a boundary, then m is even. At all events, $z \bmod 2$ is not a boundary mod2, so $H_n(T\xi; Z/2) \neq 0$. But $H_n(T\xi; Z/2)$ is cyclic, so $H_n(T\xi; Z/2) = Z/2$ and $H^n(T\xi; Z/2) = Z/2$. Of course, for the generator $\bar{u}$ of this group we have $j^*\bar{u} = \bar{s}$.

If ξ is orientable, we can take the orientation u (or $-u$) as a Thom class. Since $j^*: H^n(T\xi) \to H^n(S^n)$ is epic, $j^*u = s$. ∎

The family $\bar{u}_n \in H^n(MO_n; Z/2)$, $1 \leq n < \infty$ of Thom classes gives a family of maps $\bar{u}_n : MO_n \to K(Z/2, n)$ and hence (it is easy to see) the map $\bar{u} : MO \to HZ/2$ of spectra. Similarly one can construct the map $u : MSO \to HZ$. These maps induce the homomorphisms $\bar{u}_* : MO_n(X) \to H_n(X; Z/2)$; $u_* : MSO_n(X) \to H_n(X)$ for every X and every n, and it can be proved (on the basis of (2.9)) that they coincide with (2.2) and (2.3), respectively.

Thus, the investigations of the Steenrod-Thom maps are reduced to the investigations of $\bar{u} : MO \to HZ/2$ and $u : MSO \to HZ$. This is a very important conceptual step: the geometric problem is reduced to the investigation of the universal objects, namely, Thom spectra, and hence to a problem of the stable homotopy theory.

Of course, conceptions are not sufficient for the solution. The necessary computing information is based on the following facts, see e.g. [2],[3],[4].

3.12. Theorem (Thom isomorphism). For every vector n-dimensional bundle ξ over X there is an isomorphism $\varphi_{Z/2} : H^k(X; Z/2) \to \tilde{H}^{n+k}(T\xi; Z/2)$. Further, if ξ is orientable, then for every abelian group G there is an isomorphism $\varphi_G : H^k(X; G) \to \tilde{H}^{n+k}(T\xi; G)$. In particular,

$$H^k(BO_n; Z/2) = \tilde{H}^{n+k}(MO_n; Z/2); \quad H^k(BSO_n) = \tilde{H}^{n+k}(MSO_n). \ ∎$$

The cohomology groups of BO_n and BSO_n are well-known. We need only the following extract information.

3.13. Proposition. For every odd prime p we have $H^{N+2k+1}(MSO_N; Z/p) = 0$.

4. The realizability of mod 2 homology classes. The non-realizability of integer homology classes.

First, consider the non-oriented case. Suppose that there exists a cross-section of $\bar{u}$, i.e. the map $\sigma : HZ/2 \to MO$ such that $\bar{u}\sigma = 1_{HZ/2}$. Then every composition

$$H_n(X; Z/2) \xrightarrow{\sigma_*} MO_n(X) \xrightarrow{\bar{u}_*} H_n(X; Z/2)$$

is an identity map and hence $\bar{u}_*$ is epic, i.e. every mod 2 homology class can be realized.

In fact such σ does not exist, but there is a good approximation of σ, and now we shall describe it. Consider the spectrum $E = \{E_n, s_n\}$ where E_n is a (2n)-skeleton of $K(Z/2,n)$ and s_n is the restriction of the corresponding cellular map $SK(Z/2,n) \to K(Z/2, n+1)$, see example 2 of 3.2. The inclusions $E_n \to K(Z/2, n)$ form a map $e : E \to HZ/2$.

4.1. Lemma. The homomorphism

$$e_* : E_i(X) \to H_i(X; Z/2)$$

is an isomorphism for every X and every i.

Proof. A map e_* has the form $e_* : \pi_i(X \quad E) \to \pi_i(X \quad HZ/2)$. Let K^r be an r-skeleton of $K(Z/2, N)$ for some $N \gg i$. In this case all the maps in the sequence

$$\pi_{i+N}(X \wedge E_N) = \pi_{i+N}(X \wedge K^{2N}) \to \pi_{i+N}(X \quad K^{2N+1}) \to \dots \to \pi_{i+N}(X \wedge K(Z/2,N))$$

are isomorphisms in view of the Freudenthal Suspension Theorem, see [4]. Further, $\pi_{i+N}(X \wedge E_N) \to \pi_{i+N+1}(X \wedge E_{N+1})$ are also isomorphisms by the same reasons, so e_* is an isomorphism. ∎

Thom computed the group $H^*(MO_n;Z/2)$ and proved (using the cohomology operations $H^*(X;Z/2) \to H^*(X;Z/z)$, i.e. elements of $H^*(HZ/2; Z/2)$) that the homomorphism $\bar{u}_n^* : H^i(K(Z/2,n);Z/2) \to H^i(MO_n; Z/2)$ is monic for $i<2n$. On this basis he (if we formulate the present version of his result) constructed the map $\tau : E \to MO$ such that the composition $E \xrightarrow{\tau} MO \xrightarrow{\bar{u}} HZ/2$ coincides with e. The arguments in the beginning of this section imply now

4.2. Theorem. The map $\bar{u}_*: MO_n(X) \to H_n(X; Z/2)$ is epic for every X and every n. So, every mod 2 homology class can be realized by a singular smooth manifold.

The discussion above shows that it is a good idea to have a category of spectra in which e is an equivalence and hence τ can be treated as a map (morphism) $HZ/2 \to MO$. In fact such a category exists, see [4].

Now consider the oriented case. The map $MSO_k(X) \to H_k(X)$ is epic iff the dual map $MSO^{n-k}(Y) \to H^{n-k}(Y)$, where Y is n-dual to X, is epic. Thus, we replace the problem of the realization of homology classes by the problem of the realization of cohomology classes, i.e. the problem of the description of the image of $MSO^*(Y) \to H^*(Y)$ but for finite Y.

Let $\varphi \in H^d(HZ;Z/p)$, $p > 2$ be any odd-dimensional (i.e. $d = 1$ mod2) stable cohomology operation. If there exists the class $x \in H^n(X)$ with $\varphi(x) \neq 0$, then x cannot be realized. Indeed, x is given by a map $x: X \to K(Z, n)$ or, equivalently, a map $x: S^N X \to K(Z,N+n)$. If x can be realized, then for some N this map has a form $S^N X \xrightarrow{t} MO_{N+n} \xrightarrow{u_{N+n}} K(Z, N+n)$. Let $\iota_s \in H^s(K(Z,s))$ be a fundamental class. Then $0 \neq \varphi(x) = \varphi(x^* \iota_{N+n}) = x^* \varphi(\iota_{N+n}) = t^* u^* \varphi(\iota_{N+n}) = t^* \varphi(u_{N+n})$. But $\varphi(u_{N+n}) = 0$ in view of 3.13.

So, for the construction of the non-realizable homology class z it is sufficient to find a pair (φ, x) with $\varphi(x) \neq 0$ and $\dim\varphi = 1$ mod2 and to put $z = h(x)$, where $h: H^n(X) \to H_{N-n}(Y)$ is a N-duality isomorphism. It can be proved that $H^{N+5}(K(Z,N); Z/3) = Z/3$, for $N > 6$ see [7], so we can choose φ as a 5-dimensional operation which generates this group and x as a fundamental class $\iota_N \in H^N(K(Z,N);Z/3)$ restricted on (N+6)-skeleton. Deeper investigations in the same spirit enable us to prove

4.3. Theorem. (see [2]). Every integer homology class of dimension ≤ 6 can be realized by a smooth singular manifold. There is a non-realizable homology class in $H_7(K(Z/3 + Z/3, 1))$. ∎

It can be proved that this class cannot be realized even by a topological manifold.

Thus, there exist non-realizable classes. Nevertheless, we have

4.4. Theorem. For every $z \in H_*(X)$ there exists a natural N such that Nz can be realized.

Proof. It is sufficient to prove that

$$MSO_n(X) \otimes Q \xrightarrow{u_*} H_n(X) \otimes Q = H_*(X; Q)$$

is epic (here Q is a field of rationals). The spectrum $S = S^n$ of spheres produces the extraordinary homology theory $\Pi_n(X) = S_n(X) = \lim \pi_{n+N}(S^N X)$. It is the well-known result of Serre that $\Pi_n(pt)$ is finite for $n > 0$, so $\Pi_n(X) \otimes Q$ satisfies all Steenrod-Eilenberg axioms (including the dimension axiom). Hence $\Pi_n(X) \otimes Q = H_n(X) \otimes Q$. Moreover, the generator ι_n of $\pi_n(K(Z,n)) = Z$ gives a map $\iota_n : S^n \to K(Z,n)$, and the family $\{\iota_n\}$ (compatible with respect to suspensions) induces the map of spectra $\iota : S \to HZ$ and hence the natural transformation $\iota_* : \Pi_*(X) \to H_*(X)$. It is easy to check that $\iota_* : \Pi_*(X) \otimes Q \to H_*(X) \otimes Q$ is an isomorphism for X = pt and hence it is an isomorphism for every X in view of general theorems of stable homotopy theory, see e.g. [4]. theorem 7.55.

The inclusions $j : S^n \to MSO_n$, see definition 2.11, give a map $j : S \to MSO$ of spectra and hence the map $j_* : \Pi_*(X) \to MSO_*(X)$ of homology theories. It is easy to see that the composition

$$\Pi_*(X) \xrightarrow{j_*} MSO_*(X) \xrightarrow{u_*} H_*(X)$$

coincides with ι_* above. Hence the composition

$$\Pi_*(X) \otimes Q \xrightarrow{j_*} MSO_*(X) \otimes Q \xrightarrow{u_*} H_*(X) \otimes Q$$

is an isomorphism and hence $u_* : MSO_*(X) \otimes Q \to H_*(X) \otimes Q$ is epic. ∎

A much deeper investigation of MSO shows that the number N in 4.4 can be taken odd.

Thom computed also the bordism group $\mathfrak{N}_*(pt) = \pi_*(MO)$. The total group $\mathfrak{N}_*(pt)$ is a graded ring with respect to the product of manifolds, and it is a polynomial ring

$$\mathfrak{N}_*(pt) = Z/2\,[x_i \mid \dim x_i = i,\ i \text{ is natural. } i \neq 2^s - 1]$$

Later different authors computed different bordism groups. A good survey of these results in contained in [9].

5. New models. Manifolds with Sullivan singularities.

Sullivan introduced manifolds with a special kind of singularities in 1966 for his investigations of Hauptvermutung for manifolds. It occurs that these gadgets are useful for some other fields and, in particular, for the problem of the realization of homology classes. Now we recall the background, see [10].

Let P be a closed manifold and CP be a cone over P. Consider a manifold M and an isomorphism (diffeomorphism, homeomorphism, etc.) $\varphi : \partial M \xrightarrow{\cong} P \times A$. The polyhedron

$$V = M \cup_{\varphi} (CP \times A)$$

where ∂M is identified with the subspace $P \times A$ of $CP \times A$ by φ, is called the closed manifold with a (Sullivan) singularity of the type P.

It is easy to see that for an orientable M and the orientation preserving φ we have $H_n(V) = Z$, $n = \dim M$. So it is possible to talk about the realization of integer homology classes by manifolds with such singularities.

One can define manifolds with boundaries having singularities of the type P and construct the bordism groups $MO_*^P(X)$, $MSO_*^P(X)$, etc. Note that every closed manifold M can be considered as a closed manifold with singularity of the type P (with $A = \emptyset$), so we have the ignoring map $r : MO_*(X) \longrightarrow MO_*^P(X)$, etc. Further, the introduction of a singularity can be iterated. So, we have bordism groups $M_*^{\{P_1,\dots,P_n\}}(X)$ and $M_*^{\{P_1,\dots,P_n,\dots\}}(X) = \lim_{n\to\infty} M_*^{\{P_1,\dots,P_n\}}(X)$, where $M = MO$, MSO etc. and limit is considered with respect to the sequence of ignoring maps

$$\dots \longrightarrow M_*^{\{P_1,\dots,P_n\}}(X) \longrightarrow M_*^{\{P_1,\dots,P_n,P_{n+1}\}}(X) \longrightarrow \dots$$

Moreover, one can define the relative bordism groups $M_*^{\{P_1,\dots\}}(X,A)$ and show that there arises an extraordinary homology theory, see [10].

5.1. Theorem For every X the sequence

$$\dots \longrightarrow M_n(X) \xrightarrow{\times [P]} M_{n+d}(X) \xrightarrow{r} M_{n+d}^P(X) \xrightarrow{\partial_*} M_{n-1}(X) \longrightarrow \dots$$

is exact. Here $d = \dim P$, symbol $\times [P]$ denotes the multiplication by bordism class of P, i.e. a singular manifold $f : M \longrightarrow X$ goes to $M \times P \xrightarrow{p_1} M \xrightarrow{f} X$ and $\partial_* : M_{n+d}^P(X) \longrightarrow M_{n-1}(X)$ assigns to $f : M \cup_{\varphi} CP \times A \longrightarrow X$ the map $f|A \times \{*\} : A \longrightarrow X$, where $*$ is the vertex of CP.

See the proof in [10]. ∎

This exact sequence is an analog (and in fact a generalization) of the Bokstein exact sequence

$$\dots \longrightarrow H_n(X) \xrightarrow{\times m} H_n(X) \longrightarrow H_n(X;Z/m) \longrightarrow H_{n-1}(X) \longrightarrow \dots$$

where m is an integer.

Now we shall show the applications of manifolds with singularities to the problem of the realization of homology classes. Consider the space BU_n which classifies complex n-dimensional vector bundles. There arise the Thom space MU_n and the Thom spectrum MU, the latter produces bordism theory of (smooth) manifolds with complex structure in a stable normal bundle. It can be proved that $\pi_*(MU) = MU_*(pt) = Z[x_i \mid \dim x_i = 2i,\ i = 1,2,\dots]$, (Novikov and Milnor, see [9]).

Let $\{P_1,\dots,P_n,\dots\}$ be manifolds which give the polynomial generators $x_1,\dots, x_n,\dots$. By 5.1 there is the exact sequence

$$\xrightarrow{\partial_*} MU_*(pt) \xrightarrow{\times x_1} MU_*(pt) \longrightarrow M_*^{P_1}(pt) \xrightarrow{\partial_*} \dots$$

Since $\times x_1$ is monic (because a polynomial ring has no zero divisors), $\partial_* = 0$ and hence

$$MU_*^{P_1}(pt) = MU_*(pt)/(x_1) = Z[x_i \mid i > 1].$$

Iterating these arguments, we have

$$MU_*^{\{P_1,\dots,P_n\}}(pt) = Z[x_i \mid i > n] \quad \text{and hence}$$

$MU_*^{\{P_1,\dots,P_n,\dots\}} = Z$. Now the Steenrod-Eilenberg theorem and the fact that $MU_*^{\{P_1,\dots,P_n\dots\}}$ is an extraordinary homology theory imply that $MU_*^{\{P_1,\dots,P_n,\dots\}}(X) = H_*(X)$, and in fact

$$\mu\colon MU_*^{\{P_1,\dots,P_n,\dots\}}(X) \to H_*(X) \quad \text{is an isomorphism.}$$

Thus, every class z $H_*(X)$ can be realized by an almost complex (and hence smooth) manifold with finite number of Sullivan singularities. However, in this case we need to consider manifolds with arbitrary finite number of singularities. It occurs that the use of piecewise linear manifolds instead of smooth ones simplifies the situation.

Let $MSPL_*(X)$ be the bordism theory of piecewise linear manifolds. Let p be an odd prime and CP^{p-1} be a complex projective space of the complex dimension $p-1$. Let $Z_{(p)}$ be a subring of Q which consists just of the fractions m/n such that $(m,n) = 1$ and $(n,p) = 1$.

<u>5.2. Theorem.</u> For every X the homomorphism

$$\mu : MSPL_*^{CP^{p-1}}(X) \otimes Z_{(p)} \to H_*(X) \otimes Z_{(p)}$$

is epic.

See the proof in [11].

5.3. Corollary. Every homology class $z \in H_*(X)$ can be realized by a map $f : V \to X$, where V is a disjoint union of polyhedra of the form

$$M \cup_\varphi C(CP^{p-1}) \times A$$

Here M is a piecewise linear manifold with $\partial M = CP^{p-1} \times A$ and p runs all odd primes.

See the proof in [11].∎

References

1. Eilenberg S. Problems in Topology. - Ann. Math., 1949, 50, p.246-260.
2. Thom R. Quelques propriétés globales des variétés différentiables.- Comm. Math. Helv., 1954, 28, p.17-86.
3. Husemoller D. Fibre bundles. - New York, McGraw-Hill, 1966.
4. Switzer R. Algebraic topology - homotopy and homology. - Berlin-Heidelberg - New York, Springer-Verlag, 1975.
5. Borisovich Yu.G. et al. Introduction to topology. Moscow, "Mir", 1985.
6. Mosher R., Tangora M. Cohomology operations - New York, Harper and Row, 1968.
7. Cartan H. Séminaire H.Cartan 1954-1955 and 1959-1960, Paris.
8. Conner P., Floyd E. Differentiable periodic maps. - Berlin - Heidelberg - New York, Springer-Verlag, 1964.
9. Stong R. Notes on cobordism theory. - Princeton, Princeton Univ. Press, 1968.
10. Baas N. On the bordism theory of manifolds with singularities. - Math. Scand., 1973, 33, p.279-302.
11. Rudyak Yu.B. On realization of homological classes by PL-manifolds with singularities. - Matematicheskie zametki, 1987, vol.41, N 5, p.741-749 (in Russian, there exists an English translation).
12. Dubrovin B., Fomenko A., Novikov S. Modern geometry - Methods and applications. Part II. The geometry and topology of manifolds, Berlin - Heidelberg - New York - Tokyo, Springer-Verlag, 1985.
13. May J.P. E_∞-ring spaces and E_∞-ring spectra. Lect. Notes in Math., 577. Berlin - Heidelberg - New York, Springer-Verlag, 1977.

ORIENTED DEGREE OF FREDHOLM MAPS OF NON-NEGATIVE INDEX AND ITS APPLICATION TO GLOBAL BIFURCATION OF SOLUTIONS

V.G.Zvyagin and N.M.Ratiner
Department of Mathematics
Voronezh State University
394693, Voronezh, USSR

Introduction

Elworthy and Tromba [I] have presented the theory of oriented degree for Fredholm maps of non-negative index. However, a number of limitations (the most significant one being the presence of smooth functions with bounded non-empty support on a model space) imposed on manifolds and maps in the construction of the degree, narrow its application considerably. In this paper we use the method of finite-dimensional reduction to construct the oriented degree for Fredholm maps of nonnegative index. This method being actually the global version of Liapunoff-Schmidt's redaction for branching equations has been used in [2] for the determination of oriented degree for Fredholm maps of zero index, then in [3] for the determination of a degree for Fredholm maps of zero index. In [4] , it has been used in the construction of oriented degree for completely continuous perturbations of Fredholm maps of non-negative index, and in [5] for obtaining a version of oriented degree of completely continuous perturbations of Fredholm maps of non-negative index. However, the degree obtained in [5] allows homotopies only in a group of Fredholm maps preserving a fixed Fredholm structure over a manifold, which limits its application considerably.

This paper proposes the theory of degree of completely continuous perturbations of proper Fredholm maps of non-negative index allowing homotopies in the whole class of maps, studies the properties of this degree, calculates the degree of a given Fredholm map of index one and gives an application to the problem of global bifurcation of a non-linear elliptic boundary value problem.

In conclusion, one aspect is to be noted: irrespective of a Fredholm maps index, C^I-smoothness is sufficient for the degree determination, which is essential for a number of applications.

The results of Sections 2 to 4 are announced in [7].

I. Preliminarily Notions and Information

For reader's convenience, this Section describes the necessary information on Fredholm structures on Banach manifolds as well as the analogue of Pontryagin framed bordisms used for infinite-dimensional case, introduced and studied in [I].

Let E be a real Banach space, and X be C^r - smooth, $r \geqslant I$, Banach manifold with model space $E \times R^q$, where R^q is q-dimensional Euclidean space.

The Fredholm structure (Φ -structure, for short) X_Φ on X is a maximal atlas $\{(U_i, \varphi_i)\}$ with the following property: for any two chart $\varphi_i : U_i \longrightarrow E \times R^q$ and $\varphi_j : U_j \longrightarrow E \times R^q$ of this atlas, for which $U_i \cap U_j \neq \emptyset$, the Frechet derivative $D(\varphi_i \circ \varphi_j^{-I})y$ for any $y \in \varphi_i(U_i \cap U_j)$ belongs to the group $GLc(E \times R^q)$, where $GLc(E \times R^q)$ is the group of linear continuous operators of space $E \times R^q$ of the form $I + k$, where I is an identical operator and k is completely continous one. Various Fredholm structures can exist on a single Banach manifold.

Group $GLc\ (E \times R^q)$ has two components of linear connection. As usual, let $GLc^+ (E \times R^q)$ be a component containing an identical operator, and $GLc^- (E \times R^q)$ be the second component.

The Fredholm structure X_Φ on X is called oriented if there exists a (maximal) sub-atlas O_X ($O_X \in X_\Phi$) such that for any intersecting chart (U_i, φ_i), (U_j, φ_j) the following condition is satisfied:

$$D(\varphi_i \circ \varphi_j^{-I}) (\varphi_j(x)) \in GLc^+(E \times R^q), \quad x \in U_i \cap U_j .$$

Atlas O_X is called an orientation. If orientation O_X is fixed on manifold X, the manifold is called oriented.

Both oriented and non-oriented Fredholm structures may exist on a single Banach manifold.

Note that if X_Φ is any oriented structure on X, then the natural Φ -structure $X_\Phi \times [0,I]$ on $X \times [0,I]$ is the oriented structure as well.

Vector bundle $\pi : B \longrightarrow X$ with the fiber $E \times R^q$ is called GLc-bundle provided it is equipped with the maximal collection of trivillizations$\{(U_i, \tau_i)\}$ such that for all i,j satisfying $U_i \cap U_j \neq \emptyset$ the map $\tau_j \circ \tau_i^{-I}| (U_i \cap U_j) \times E \times R^q : (U_i \cap U_j) \times (E \times R^q) \to (U_i \cap U_j) \times E \times R^q$ is of the form $(x, v) \longrightarrow (x, v + a(x)v)$, where $a(x)$ is a completely continuous operator for all $x \in U_i \cap U_j$. If $a(x)$ is a local finite-dimensional operator, then π is called a layer bundle.

It should be noted that the tangent bundle TX_Φ to a manifold X with a Fredholm structure X_Φ is a GLc-bundle.

Let π_I, π_2 be two GLc-bundles. Isomorphism $\tau : \pi_I \to \pi_2$ of these bundles is a GLc-isomorphism if in a local presentation it has the form $(x,v) \longrightarrow (x, v, + b(x)v)$, where $b(x)$ is a linear completely continuous operator for all x 's.

Let now X be a Banach manifold, on which an oriented Fredholm structure X is fixed.

<u>Definition I.</u> A Glc-framed q-dimensional (singular) sub-manifold of structure X_Φ is the tripet (M, i, τ), where M is a q-dimensional closed manifold, $i : M \longrightarrow X$ - is a continuous map and

$$\tau : TM \oplus (M \times E) \longrightarrow i^* (TX_\Phi) \tag{I}$$

is a GLc- isomorphism of vector bundles (here $i^*(TX_\Phi)$ is an inverse image of bundle TX_Φ).

GLc-isomorphism (I) is called a GLc-framing of a pair (M, i) in structure X_Φ.

<u>Remark I.</u> If TX_Φ is an oriented bundle and $M \times E$ is trivial, then M is an oriented manifold.

A set of GLc-framed q-dimensional sub-manifolds of structure X_Φ is denoted by $Sq(X_\Phi)$.

<u>Definition 2.</u> Two triplets (M_0, i_0, τ_0) and (M_I, i_I, τ_I) from $Sq(X_\Phi)$ are called GLc-framed bordant in the structure X_Φ if there exists a $(q + I)$-dimensional compact manifold Y with the boundary $\partial Y = M_0 \cup M_I$, a continuous map $i : Y \longrightarrow X \times [0,I]$ such that $i|_{M_r} : M_r \longrightarrow X \times \{r\}$, $i|_{M_r} = i_r$, $r=0,I$, and a GLc-isomorphism of vector bundles

$$\tau : TY \oplus (Y \times E) \longrightarrow i^* (T(X_\Phi \times [0,I])) \qquad (2)$$

coinciding with τ_0 and τ_1 on $TM_0 \oplus (M_0 \times E)$ and $TM_I \longrightarrow (M_I \times E)$ respectively.

Remark 2. Formula (2) deals with the bundle $T(X_\Phi \times [0,I])$ tangent to the oriented Φ -structure on $X \times [0,I]$ defined via structure X_Φ in a natural manner. However, other Φ -structures can exist on $X \times [0,I]$ coinciding with X over $X \times \{0\}$ and $X \times \{I\}$.

Following [I], let us consider a set of classes of bordant GLc-framed triplets in structure X_Φ, denoting it by $F_q(X_\Phi)$, while a class of a triplet (M, i, τ) in $F_q(X_\Phi)$ by $[M,i,\tau]$.

The set $F_q(X_\Phi)$ has a natural group structure. However, the use of this group as a set of values for degree of proper Fredholm maps restricts admissible homotopies considerably (in this case, the homotopies must belong to the class of Fredholm maps preserving the given Φ -structure X_Φ). In order to avoid this restriction, let us introduce, following [I], another object $\widetilde{F}_q(b)$ in which the degree of Fredholm maps and their completely continuous perturbations will take on their values.

Definition 3. Two structures X_{Φ_0} and X_{Φ_1} on X are said to be concordant if there exists a Fredholm structure C on $X \times [0,I]$, such that for some $0 < \varepsilon < \frac{1}{2}$ on $X \times [0,\varepsilon)$ and on $X \times (I-\varepsilon, I]$ it coincides with $X_{\Phi_0} \times [0,\varepsilon)$ and $X_{\Phi_1} \times (I-\varepsilon, I]$ respectively.

Concordance is an equivalence relation on a class of all Fredholm structures on manifold X. We denote a set of classes of concordant structures on a Banach manifold X by ConX.

Let us consider $S_q(b)$ for a class $b \in \mathrm{Con}X$, i.e. the union of all sets $S_q(X_\Phi)$ for oriented Φ -structures $X_\Phi \in b$. Let us introduce a bordism relation between GLc-framed submanifolds in class b (and not in X_Φ structure, as in definition 3).

Definition 4. Let X_{Φ_0} and X_{Φ_1} be two oriented structures on X in class b. Triplets $(M_0, i_0, \tau_0) \in S_q(X_{\Phi_0})$ and $(M_I, i_I, \tau_I) \in S_q(X_{\Phi_1})$ are called GLc-framed bordant in class $b \in \mathrm{Con}\, X$, if there exists a structure C providing concordance of X_{Φ_0} and X_{Φ_1}, $(q+I)$-dimensional compact manifold Y with $\partial Y = M_0 \cup M_1$, continuous map $i : Y \longrightarrow M \times [0,I]$ with $i|_{M_r} : M_r \longrightarrow X \times \{r\}$, $i|_{M_r} = i_r$, $r = 0, I$, and GLc-isomorphism of vector bundles

$$\tau : TY \oplus (Y \times E) \longrightarrow i^{*} (TC),$$

coinciding with τ_0 and τ_1, on $TM_0 \oplus (M_0 \times E)$ and $TM_I \oplus (M_I \times E)$ respectively. (Here TC is the tangent bundle to structure C).

Bordant relation of GLc-framed submanifolds in class b is an equivalence relation on the set $S_q(b)$. The quotient set $S_q(b)$ over equivalence relation is denoted by $\widetilde{F}_q(b)$.

For the most important cases $\widetilde{F}_q(b)$ is calculated ([I]). So, if X is (q+I)-connected, then all Fredholm structures on X are oriented, and $\widetilde{F}_q(b)$ is in one-to-one correspondence with the set obtained from a qth-stable homotopy group of spheres via the identification of v and -v.

2. Determination of Oriented Degree

This section deals with the determination of a degree for Fredholm maps of index $q \geqslant 0$ and their completely continuous perturbations which take values in the set $\widetilde{F}_q(b)$ for some class b defined by the Fredholm map. This approach appears to be constructive and allows the calculation of the degree in a number of cases.

Let X be a submanifold of topological space $\hat{X}$, on which a Banach manifold structure is given. Denote by $\bar{X}$ the closure of X in $\hat{X}$, and denote the boundary of manifold X by ∂X.

Let $f : \bar{X} \longrightarrow E$ be a proper map with values in a Banach space E. (The map $f : \bar{X} \longrightarrow E$ is called proper if $f^{-I}(K)$ for any compact $K \in E$ is a compact in $\bar{X}$). We shall assume that the restriction $f = f|_X : X \longrightarrow E$ is a C^I-smooth Fredholm map of index $q \geqslant 0$ (or ΦqC^I-map, for short). It is known ([I]) that every ΦqC^I-map $f : X \longrightarrow E$ induces Φ-structure $X_\Phi = \{X, f\}_\Phi$ on a Banach manifold X with the model space $E \times R^q$ such that for every chart $(U, \varphi) \in X_\Phi$, $U \subset X$, $\varphi : U \longrightarrow E \times R^q$ and any point $x \in U$, the following is true:

$$D(f \circ \varphi^{-I}) (\varphi (x)) = P + a$$

where $P : E \times R^q \longrightarrow E$ is the projection, α is a completely continuous operator. We shall assume that the Φ-structure X_ϕ is oriented.

Let us denote by $b \in \mathrm{Con}X$ the class of concordant structures containing the structure X_ϕ. First let $k : \overline{X} \longrightarrow E$ be a continuous bounded finite-dimensional map. Further on, for the map $f + k : \overline{X} \longrightarrow E$ and a point $y \in E \backslash (f+k)(\partial X)$ we will construct an element $\tilde{d}(f+k, X, y) \in \widetilde{F}_q(b)$ which will be called the degree of the map $f+k$ with respect to the point y .

So, let $F^m \subseteq F$ be a finite-dimensional space such that the image $k(\overline{X})$ is contained in F^m . The map $f+k$ is proper as well, therefore $Q = (f+k)^{-I}(y)$ is a compact set, and, by the condition $y \in E \backslash (f+k)(\partial X)$, it belongs to X.

Let us assume, that $Q \neq \emptyset$. According to the principle of finite-dimensional reduction (Ref. [6], p.I7), there exists a finite-dimensional space $F^n \subset E$ for which the map f is transversal to F^n on some open neighbourhood Ω_Q of the set Q . Without loss of generality we may assume that $y \in F^n$ and $F^m \subseteq F^n$. Then $M^{n+q} = f^{-I}(F^n) \cap \Omega_Q$ is a $(n+q)$-dimensional C^I-smooth sub-manifold in X, while $Q \in M^{n+q}$.

Let us fix orientation O_x of structure X_ϕ on X. .

Then O_x will induce the oriented structure on manifold M^{n+q} as well. Let N^{n+q} be a neighbourd of Q in M^{n+q} with C^I-smooth boundary. We call the restriction of the map $f+k$ on N^{n+q} (i.e. the map $h = (f+k)\big|_{N^{n+q}} : N^{n+q} \longrightarrow F^n$) the reduced map.

The map h is continuous. Consider $\delta = \min\limits_{x \in N^{n+q}} \|(f+k)(x) - -y\| = 0$. Next, we choose C^I-smooth map $h'_\varepsilon : N^{n+q} \longrightarrow F^n$ so that $\|h'_\varepsilon(x) - (f+k)(x)\| \le \varepsilon < \delta/2, x \in N^{n+q}$.

Further, we approximate the C^I-smooth map h'_ε by a map h_ε which is transversal to the point $y \in F^n$ and

$$\|h_\varepsilon(x) - h'_\varepsilon(x)\| \le \varepsilon < \delta/2 .$$

This can be done by the introduction of a more smooth structure on the manifold and by the use of the Thom transversality theorem.

Consider $W^q = h_\varepsilon^{-I}(y)$. Then W^q is a closed C^I-smooth sub-manifold in $N^{n+q} \subset X$. Let $i : W^q \longrightarrow X$ be an identical embedding. We shall define GLc-framing of pair (W^q, i) in structure X_ϕ as follows: first, we construct a finite-dimensional framing of W^q in N^{n+q} , then "enrich" it with GLc-framing of the manifold N^{n+q} in structure X_ϕ .

Now let us introduce a new notation for convenience.

Let $f : X \longrightarrow E$ be C^r -map ($r \geqslant I$), and we denote by $D^*f : TX \longrightarrow X \times E$ the C^{r-I}-morphism of vertor bundles mapping $v \in T_xX$ into $(x, Df_x(v))$.

Framing W^q in N^{n+q} is given in a standard way: since $W^q = h_\varepsilon^{-I}(y)$ and y is a regular value, the differential $Dh_\varepsilon(x)$ maps T_xW^q into zero, and it maps normal space isomorphly on F^n Therefore the trivilization of normal bundle $\nu_N W^q$ for submanifold W^q in N^{n+q} may be defined as follows:

$$\xi_n = Dh_\varepsilon \Big|_{\nu_N W^q} : \nu_N W^q \longrightarrow W^q \times E^n .$$

Framing $\tau_n : TW^q \oplus (W^q \times F^n) \longrightarrow TW^q \oplus \nu_N W^q \approx TN^{n+q} \mid W^q$ is defined by $\tau_n = I_{TW^q} \oplus \xi_n^{-I}$, where $I_{TW^q} : TW^q \to TW^q$ is an identical map.

Let us denote by $F^{\infty-n}$ a closed subspace in E the complement to F^n , and by $p^{\infty-n} : E \longrightarrow F^{\infty-n}$ - the projection corresponding to the split $E = F^n \times F^{\infty-n}$.

Next, due to the assumption that $f|_{\Omega_Q}$ is transversal to F^n , we shall define the trivilization of a the normal bundle $\nu_x N^{n+q}$ for N^{n+q} in structure X_ϕ , using the superposition

$$\nu_x N^{n+q} \xrightarrow{D^*f} N^{n+q} \times E \xrightarrow{I_{N^{n+q}} \times p^{\infty-n}} N^{n+q} \times F^{\infty-n}$$

Let us introduce the notation

$$\xi_{\infty-n} = (I_{N^{n+q}} \times p^{\infty-n}) \circ D^*f \Big|_{\nu_x N^{n+q}} .$$

Define the following framing

$$\tau_{\infty-n} : TN^{n+q} \oplus (N^{n+q} \times F^{\infty-n}) \longrightarrow TN^{n+q} \oplus \nu_x N^{n+q} \approx TX_\phi \big|_{N^{n+q}}$$

by $\tau_{\infty-n} = I_{TN^{n+q}} \oplus \xi_{\infty-n}$.

Finally, the framing of the pair (W^q , i) in structure X_ϕ is defined as follows:

$$\tau : TW^q \oplus (W^q \times E) = TW^q \oplus (W^q \times F^n) \oplus (W^q \times F^{\infty-n}) \longrightarrow$$

$$\tau_n \oplus I_{W^q \times F^{\infty-n}} \longrightarrow TN^{n+q} \oplus (W^q \times F^{\infty-n}) \xrightarrow{\tau_{\infty-n}} TX_\Phi\big|_{W^q} \approx i^* (TX_\Phi) .$$

Definition 5. If $(f + k)^{-I} (y) \neq \emptyset$, then we say that the class of GLc-framed bordisms of the triplet (W^q, i, τ) in $\widetilde{F}_q$ (b) constructed before is the degree $d(f + k, X, y)$ of map $f + k$ with respect to the point y . If $(f + k)^{-I} (y) = \emptyset$ we say that $d (f + k, X, y,) = 0 \in \widetilde{F}_q$ (b).

Remark 3. The above construction of degree $d (f + k, X,y)$ uses a finite-dimensional sub-space F^n , for which $f|_{\Omega_Q} \pitchfork F^n$, a neighbourhood N^{n+q} of compact Q in $f^{-I}(F^n)$ and an approximation $h_\varepsilon : N^{n+q} \longrightarrow F^n$ and a split $E=F^n \times F^{\infty-n}$. The proof of the independence of degree $\widetilde{d}$ (f+k, X,y) from this arbitrary choice is actually a standard one and is based on the following lemma.

Lemma I. Let $h_r:\bar{N}^{n+q} \longrightarrow F^n$, $r = 0,I$, be two C^I -smooth maps for which $y \in F^n \setminus h_r(\partial N^{n+q})$ is a regular value. Let $H : N^{n+q} \times [0,I] \longrightarrow F^n$ be a continuous homotopy connecting h_0 and h_I such that $H(x,t) \neq y$ for $x \in N^{n+q}, t \in [0,I]$. Then the triplets $(W^q_0, i_0, \tau_0), (W^q_I, i_I, \tau_I)$ (where $W_r^q = h_r^{-I}(y)$, $i_r: W^q_r \longrightarrow N^{n+q} \subset X$ are identical embeddings and τ_r are GLc-framings constructed with the use of Dh_r and Df , $r = 0,I$) will be GLc-framed bordant.

Remark 4. Let us assume that $(f+k)^{-I}(y) \neq \emptyset$. The class of triplet (W^q, i, τ) in the group $Fq(X_\Phi)$ is also called an oriented degree. We will denote this degree with symbol d(f+k, x, y). If $(f+k)^{-I}(y) = \emptyset$, then d(f+k, X,y) = 0 . The degree d(f+k, x, y) and its properties have been studied in [5] . Its advantage is the ability to take on values in a group. The disadvantage (as is stated in the introduction) is the fact that the homotopies in the theory of the degree d(f+k, x,y) are allowable only in the class of Fredholm maps preserving a fixed Fredholm structure, while in applications it is neccessary to consider the homotopies which change the given Fredholm structure.

3. Properties of the Oriented Degree

The oriented degree d(f+k, X, y) defined in the preceding section features usual properties. Now we shall formulate them.

Property I. Degree $\tilde{d}(f + k, X, y)$ is constant for all y from a single component of the set $E \setminus (f + k)(\partial X)$.

Property 2. If $\tilde{d}(f+k, X, y_0) \neq 0$, then equation $f(x) + k(x) = y$ is solvable for all y from the same component of the set $E \setminus (f+k)(\partial X)$, as y_0.

Let us consider the property of a homotopical invariance of degree $\tilde{d}(f + k, X, y)$.

Property 3. Let $F : \bar{X} \times [0,I] \longrightarrow E$ be a proper map, continuously differentiable on $X \times [0,I]$, and let $K:\bar{X} \times [0,I] \longrightarrow E$ be a continuous bounded finite-dimensional map. Let X be connected and for every $t \in [0,I]$, the map $f_t : X \longrightarrow E, f_t(x,t)=F(x,t)$ be a $\Phi q C^I$-map. Then for every $y \in E \setminus (F+K)(\partial X \times [0,I])$, the following equality holds:

$$\tilde{d}(f_0 + k_0, X,y) = \tilde{d}(f_I + k_I, X,y)$$

where $k_t(x) = K(x,t)$.

Proof. It is evident, that following the above assumptions, degree $d(f_t + k_t, X, y)$ is defined for every $t \in [0,I]$. Without loss of generality, we may suppose that $F(x,t)= f_0(x)$ for $0 \leq t < \varepsilon$ and $F(x,t) = f_1(x)$ for $I-\varepsilon < t \leq I$, where $0 < \varepsilon < \frac{1}{2}$. Further, $F : X \times [0,I] \longrightarrow E$ is a Fredholm map of index $(q + I)$. Considering the induced structure $C = \{X \times [0,I], F\}_\Phi$ on $X \times [0,I]$, we can see that it coincides with $X_{\Phi_0} = \{X,f_0\}_\Phi$ on $X \times \{0\}$ and with $X_{\Phi_1} = \{X, f_1\}_\Phi$ on $X \times \{I\}$. The structure C provides concordance of structures X_{Φ_0} and X_{Φ_1}, so they belong to the same class $b \in \mathrm{Con}\, X$.

Using the construction given in Section 2 for the determination of degree $\tilde{d}(f+k, X, y)$ for map $F+K : X \times [0,I] \longrightarrow E$, we will choose a split $E=F^n \times F^{\infty-n}, \dim F^n < \infty$, such that $F|_{\Omega_Q} \pitchfork F^n$, where Ω_Q is an open neighbourhood of the compact $Q = (F + K)^{-I}(y)$. The sub-space F^n can be chosen sufficiently large, and the neighbourhood Ω_Q can be chosen such that $f_i|_{\Omega^i_Q} \pitchfork F^n$, where $\Omega^i_Q = \Omega_Q \cap (X \times \{i\})$. Besides, we can suppose that $y \in F^n$ and $K(\bar{X} \times [0,I]) \subseteq F^n$. Then a smooth $(n+q+I)$-dimensional manifold $M^{n+q+I}=F^{-I}(F^n)$ has the boundary

$$\partial M^{n+q+I} = (M^{n+q+I} \cap (X \times \{0\})) \cup (M^{n+q+I} \cap (X \times \{I\})).$$

Let us denote by M_i, $i = 0,I$, the component of boundary ∂M^{n+q+I} lying in $X \times \{i\}$. It is evident that $M_i = f_i^{-I}(F^n) \cap \Omega^i_Q$.

Now let us consider the restriction $H = (F + K)|_{M^{n+q+I}}$: $M^{n+q+I} \longrightarrow F^n$.

It should be noted that $H|_{M^i} = (f_i + k_i)|_{M^i}$ the finite-dimensional map, was used for the determination of degree $\tilde{d}(f_i + k_i, X, y)$, $i = 0,I$.

Let N be a closed sub-set in M^{n+q+I} satisfying the condition $Q \subset N \subset \bar{N} \subset M^{n+q+I}$, and let $\delta > 0$ such that $\|H(x,t) - y\| \geqslant \delta$ for $(x,t) \in \partial N$.

Let us choose a C^I-smooth map $H_\varepsilon : \bar{N} \longrightarrow F^n$, for which $\|H_\varepsilon(x,t) - H(x,t)\| \leqslant \varepsilon < \delta/2$, $(x,t) \in \bar{N}$, and y is a regular value. Note that H_ε can be chosen so that y is a regular value for $h_{\varepsilon,i} = H_\varepsilon|_{N_i}$, where $N_i = N \cap \Omega^i_Q$, $i = 0,I$.

Then $W^{q+I} = H_\varepsilon^{-I}(y)$ is a $(q+I)$-demensional manifold with the boundary $\partial W^{q+I} = h^{-I}_{\varepsilon,0}(y) \cup h^{-I}_{\varepsilon,I}(y) = W^q_0 \cup W^q_I$, where $W^q_i = h^{-I}_{\varepsilon,i}(y)$, $i = 0,I$. Next, we shall recall that the dergree $d(f_i+k_i, X,y)$, $i=0,I$, is defined by the class of equivalence of the triplet (W^q_i, j, τ_i) , where $j : W^q_i \longrightarrow X \times [0,I]$ is an identical embedding and the GLc-framing

$$\tau_i : TW^q_i \oplus (W^q_i \times F^n) \oplus (W^q_i \times F^{\infty-n}) \longrightarrow TW^q_i \oplus \nu_{N_i} W^q_i \oplus \nu_X N_i \approx TX_\Phi|_{W_q}$$

is defined by the formula $\tau_i = I_{TW^q_i} \oplus (D^* h_{\varepsilon,i}|_{\nu_{N_i} W^q_i})^{-I} \oplus (D^* f_i|_{\nu_X N_i})^{-I}$
(here the formula for τ_i corresponds to the split of the bundles in which τ_i acts).

Manifold W^{q+I} together with the identical embedding in $X \times [0,I]$ and GLc-framing:

$$\tau : TW^{q+I} \oplus (W^{q+I} \times F^n) \oplus (W^{q+I} \times F^{\infty-n}) \longrightarrow TW^{q+I} \oplus \nu_N W^{q+I} \oplus \nu_C N \approx TC|_{W^{q+I}}$$

$$\tau = I_{TW^{q+I}} \oplus (D^* H_\varepsilon|_{\nu_N W^{q+I}})^{-I} \oplus (D^*(p^{\infty-n} \circ F)|_{\nu_C N})^{-I},$$

where $C = \{X \times [0,I], F\}_\Phi$ and $\nu_C N$ is the normal bundle for N in structure C on $X \times [0,I]$, defines the equivalence of these triplets in $\tilde{F}_q(b)$, where b is the class of concordanced structures containing $X_{\Phi_0} = \{X, f_0\}_\Phi$ and $X_{\Phi_1} = \{X, f_I\}_\Phi$. So, $d(f_0 + k_0, X, y) = d(f_I + k_I, X, y)$. The property is proved.

Remark 5. So, Property 3 says that degree $\tilde{d}(f+k, X,y)$ remains invariant even if the induced Fredholm structure changes during the homotopy.

4. Oriented Degree for Completely Continuous Perturbations of Fredholm Maps

Let now X be a bounded sub-set in a metric space, on which, as before, a Banach manifold structure is given. As usual, let $\bar{X}$ be the closure and ∂X the boundary of the manifold X.

Let E be a Banach space and $f:\bar{X} \longrightarrow E$ a proper map, whose restriction $f|_X$ is $\Phi_q C^I$ -map, $q \geqslant 0$. We denote by $k:\bar{X} \longrightarrow E$ a completely continuous map. In this Section we shall determine the degree $\tilde{d}(f + k, X,y)$ for the map $f + k : \bar{X} \longrightarrow E$ with respect to a point $y \in E \setminus (f + k)(\partial X)$.

We will approximate the completely continuous map $k:\bar{X} \longrightarrow E$ through a continuous finite-dimensional map $k_\varepsilon : \bar{X} \longrightarrow E$ (for instance, using the Schauder's projector) so that

$$\|k_\varepsilon(x) - k(x)\| < \delta/2, \ x \in X, \text{ where } \delta = \min_{x \in \partial X} \|(f+k)(x) - y\| \quad (3)$$

Definition 6. Determine the degree for $f + k$ by $\tilde{d}(f+k,X,y) = \tilde{d}(f + k_\varepsilon, X,y)$.

In order to make this definition correct, it is neccessary to show its independence from the choice of the finite-dimensional approximation k_ε.

Let k_{ε_I} and k_{ε_2} be two finite-dimensional, continuous, bounded maps satisfying (3). Let us consider the following homotopy:

$$\Phi(x,t) = f(x) + t\, k_{\varepsilon_I}(x) + (I-t)k_{\varepsilon_2}(x), \ x \in X, \ t \in [0,I].$$

Because of (3), we have $\Phi(x,t) \neq y$ for $x \in \partial X$, $t \in [0,I]$, therefore, due to Property 3 of the preceding section:

$$\tilde{d}(f + k_{\varepsilon_I}, X,y) = \tilde{d}(f + k_{\varepsilon_2}, X,y),$$

which means the correctness of Definition 6.

Remark 6. Of course, degree $d(f+k, X,y)$ is defined similarly, where $f:\overline{X} \longrightarrow E$ is a proper map, whose restriction $f|_X$ is a $\Phi_q C^I$ -map, $q \geqslant 0$, and $k:\overline{X} \longrightarrow E$ is completely continuous. Here $d(f + k_\varepsilon, X, y)$ is the degree according to Remark 4.

Remark 7. For degree $\tilde{d}(f + k, X, y)$ introduced in Definition 6 properties I to 3 of Section 3 are true.

Note the additivity of degree $d(f + k, X, y)$, which will be used in the following section.

Let now G^i, $i = I, \ldots, m$ be a family of open sub-sets in X, providing $G^i \cap G^j = \emptyset$, if $j \neq i$. Let us denote the restriction of Fredholm structure X_Φ on G^i by G^i_Φ. Then every triplet (M^q, i, τ) from $S_q(G^i_\Phi)$ will be at the same time the triplet from $Sq(X_\Phi)$. Therefore, the below embedding is well defined:

$$\chi_j : F_q(G^j_\Phi) \longrightarrow (F_q(X_\Phi)).$$

Property 4. Let $y \in E \setminus (f+k)(\overline{X} \setminus \bigcup_{j=I}^{m} G^j)$. The following equality holds

$$d(f + k, X, y) = \sum_{j=I}^{m} \chi_j \; d(f + k, G^j, y)$$

The proof of this property follows directly from the fact that $F_q(X_\Phi)$ is a group and the addition in it is induced by a disjoint union of manifolds.

Remark 8. For degree $\tilde{d}(f+k, X, y)$, property 4 is not valid because $\tilde{F}_q(X_\Phi)$ is not a group.

5. Calculation of Degres d(f+k,x,y) for One Special Class of Maps

It is said that the Banach space E satisfies the property of finite-dimensional approximation if for any compact set M in E and any $\varepsilon > 0$ there exists a linear, continuous finite-dimensional operator T , which provides $\|Tx - x\| \leqslant \varepsilon$ for all $x \in M$.

Let F , E be real Banach spaces, and let E satisfy the property of finite-dimensional approximation. Let us assume that for linear Fredholm operator $A:F \longrightarrow E$ of index I there exist the following splits: $F = F_I \times F_2 \times R^I$ and $E = A(F_I) \times F_2$ where $A(F_2 \times R^I) \subset 0 \times F_2$, $\operatorname{Ker} A \subset F_2 \times R^I$, $\dim F_2 < \infty$, so

$A|F_I : F_I \to A(F_I)$ appears to be a linear isomprphism. Let B be a ball in Banach spase F with the centre in zero point and let $k:\bar{B} \to E$ be a completely continuous map. We denote $B^2 = B \cap (F_2 \times R^I)$. Let $k(\bar{B}^2) \subset F^2$.

Theorem I. Suppose that $(A + k)x \neq 0$ for $x \in \partial B^2$ and map $A+k : \bar{B}^2 \to F^2$ is homotopic to a finite-dimensional suspension over a m-multiple Hopf map (where m is an integer) with the use of a non-zero homotopy on ∂B . Then $d(A + k, B, 0) = m \pmod 2$.

I. Before proving theorem I, let us introduce some notions. We shall call a sub-manifold M^q in Euclidean space R^{n+q} a framed one if there is a trivilization of the normal bundle $\nu_{n+q} M^q$ to M^q in R_{n+q} :

$$\xi : \nu_{n+q} M^q \longrightarrow M^q \times R^{n+q}$$

A pair (M^q , ξ) is a framed sub-manifold in R^{n+q} . The relation of a framed bordism is introduced in a natural way. It is easy to see that this definition is equivalent to the one given by Pontryagin [II].

We will denote a group composed of classes of bordant framed q-dimensional sub-manifolds in R^{n+q} similarly to [II] , by Π^q_{n+q}.

A framed sub-manifold (M^q, $\Sigma\xi$) in R^{n+q+I} is called the suspension over a framed sub-manifold (M^q , ξ) in R^{n+q} , where the isomorphism of bundles $\Sigma\xi$ is defined by

$$\Sigma\xi : \nu_{n+q+I} M^q = \nu_{n+q} M^q \oplus (M^q \times R^I) \xrightarrow{\xi \oplus I_{M^q} \times R^I} M^q \times R^{n+q+I}$$

A suspension of finite multiplicity (of multiplicity k for the integer k) is defined as k-multiple iteration of operator Σ :

$$\Sigma^k \xi = \underbrace{\Sigma \ldots \Sigma}_{k} \xi \quad , \quad \Sigma^k (M, \xi) = (M, \Sigma^k \xi) .$$

It is known ([II]) that there exists the isomorphism

$$\pi_{n+q}(S^n) \longrightarrow \Pi^q_{n+q} \tag{4}$$

from (n+q) homototopy group of S^n in Π^q_{n+q} . It is to be noted that in the presence of isomorphism (4), the Freudenthal suspension Σf over map $f : S^{n+q} \to S^n$ is transferred to the suspension over the respective framed manifold.

For the further convenience we shall use the words "framing of a finite-dimensional manifold M^q " to denote not the trivialization of normal bundle ξ, but the isomorphism of bundles

$$\tau = I_{TM^q} \oplus \xi^{-I} : TM^q \oplus (M^q \times E) \longrightarrow TM^q \oplus \nu_{E^m \times R^q} M^q = M^q \times E^n \times R^q \qquad (5)$$

Further, isomorphisms of type $\tau : TM^q \oplus (M^q \times E^n) \longrightarrow M^q \times E^n \times R^q$ (not necessarily of form (5)) will be considered. In order to differentiate them, we shall call isomorphisms (5) the correct framing, and isomorphism τ of an arbitrary type - simply the framing.

The concept of suspension is generalized easily for GLc-framed sub-manifolds in structure X_Φ in the case when the tangent bundle TX_Φ is trivial: $TX_\Phi \approx X \times E \times R^q$. In fact we shall say that the framed τ of a singular sub-manifold (M, i) is an infinite-dimensional suspension over the finite-dimensional framing if there exists the split $E = E^n \times E^{\infty - n}$, $\dim E^n = n < \infty$, such that $i(M^q) \subset E^n \times R^q$, and GLc-framing has the following form $\tau : TM^q \oplus (M^q \times E^n) \oplus (M^q \times E^{\infty-n}) \xrightarrow{\tau^n \oplus I_{M^q \times E^{\infty-n}}} T(E^n \times R^q)|_{i(M^q)} \oplus (M^q \times E^{\infty-n}) \approx M^q \times E \times R^q$.

It is also necessary to note that

$$\Pi_3^I \approx \pi_3(S^2) \approx Z, \quad \Pi_4^I \approx \pi_4(S^3) \approx Z_2, \ldots, \Pi_{n+I}^I \approx \pi_{n+I}(S^n) \approx Z_2,$$

The integer m in group Z corresponds to m-multiple Hopf map, which, if we consider S^3 as a unity sphere in a complex two-dimensional space $\mathbb{C}^2$, and suppose S^2 to be identical to $R^3 \setminus \{0\}$, can be written as follows

$$(\lambda, \omega) \longrightarrow (\bar{\lambda}^m \omega, \ |\lambda|^2 - |\omega|^2) \qquad (6)$$

The map $\Sigma^k \chi_m$, $k \geq I$, under the isomorphism $\pi_{k+3}(S^{k+2}) \approx Z_2$ corresponds to zero in group Z_2, if m is even, and to unit, if m is odd (Ref. [II]).

It should be notes also that in the next Section we will consider a Hopf map χ_m as the map from $R^4 \approx \mathbb{C}^2$ into R^3, given by the same formula (6). For any ball T with the centre at zero point we have $\chi_m(x) \neq 0$ for $x \in \partial T$. The suspension over this map will be as follows

$$\Sigma \chi_m : T \times R^I \longrightarrow R^3 \times R^I$$

$$\Sigma\chi(x,t) = (\chi_m(x),t)$$

2. Proof of Theorem I. To calculate $d(A+k, B, 0)$ let us construct a GLc-framed manifold corresponding to the map $A+k$ with respect to the definitition of the degree.

Note, that since split $E = A(F_I) \times F_2$, the map A is transversal to the sub-space F_2. Moreover $A^{-I}(F_2) = F_2 \times R^I$ and using the notations of Section 2, $M^{n+I} = (F_2 \times R^I \cap B) = \bar{B}^2$ and the reduced map $h = (A+k)|_{M^{n+I} = \bar{B}^2} : \bar{B}^2 \longrightarrow F_2$ is homotopic to the map $\Sigma^r \chi_m$, $(r = \dim F_2 - 3)$.

With reference to Lemma I, we may assume that the reduced map coincides with $\Sigma^r \chi_m$.

Every point $c \in F_2$, $c \neq 0$, is a regular value for $h = \Sigma^r \chi_m$ (Ref. [II]), and $h^{-I}(c)$ is a one-dimensional sphere S^I.

Let us identify sub-space F_2 with a finite-dimensional Euclidean space of the corresponding dimension. If point c has coordinates $c = (c_I, c_2, c_3, 0, \dots, 0)$ with respect to some basis in F_2, then S^I belongs to the spase formed by coordinates $(x_I, x_2, x_3, x_4, 0, \dots, 0)$. which we shall denote by $R^4 \subset F_2 \times R^I$. Let R^k be a complemented subspace to R^4 in $F_2 \times R^I$.

Let us construct the framing of manifold S^I first in B^2, then "enrich" it with GLc-framing of B^2 in an induced Φ-structure $B_\Phi = \{B, A\}_\Phi$.

The framing of S^I in B^2 will be as follows:

$$\tau_k : TS^I \oplus (S^I \times F^2) = TS^I \oplus (S^I \times R^3) \oplus (S^I \times R^k) \xrightarrow{I_{TS^I} \oplus \xi^{-I} \oplus I_{S^I \times R^k}}$$

$$\longrightarrow TS^I \oplus \nu_{R^4} S^I \oplus (S^I \times R^k) \approx TS^I \oplus \nu_{R^{k+4}} S^I \approx TR^{k+4}|_{S^I} \approx S^I \times R^{k+4};$$

where $\xi : \nu_{R^4} S^I \xrightarrow{D^* \chi} S^I \times R^3$ is the trivialization of a normal bundle to S^I in R^4 created by map χ.

In order to write down the GLc-framing B^2 in the structure $B_\Phi = \{B, A\}_\Phi$, note that operator $\bar{A} : F_I \xrightarrow{A|_{F_I}} A(F_I) \times F_2 \xrightarrow{q} A(F_I)$, where q is the projection on the first factor, is a linear isomorphism, therefore the structure B_Φ can be defined by a single chart

$$\varphi : B \subset F_I \times F_2 \times R^I \longrightarrow A(F_I) \times F_2 \times R^I \qquad (7)$$

$$\varphi(x_I, x_2, t) \longrightarrow (A x_I, x_2, t)$$

It is easy to see that $M^{n+I}=(F_2 \times R^I) \cap B$ is a sub-manifold of the structure B_Φ. The normal bundle $\nu_B B^2$ to B^2 in B_Φ coincides with $B^2 \times F_I$. The trivializing map for $\nu_B B^2$ looks simple enough, because it has been induced by a linear map (7), i.e.

$$\tau'': B^2 \times F_I \longrightarrow B^2 \times A(F_I)$$

$$(x, v) \longrightarrow (x, \bar{A} v)$$

Then the map ξ_∞ (see Section 2) has the form:

$$\xi_\infty: \nu_B B^2 \xrightarrow{D^*A} B^2 \times A(F_I) \times F_2 \xrightarrow{I_{B^2} \times q} B^2 \times A(F_I)$$

$$\xi_\infty \circ \tau''^{-I}: (x,v) \longrightarrow (x, \bar{A}^{-I} v) \longrightarrow (x,v)$$

So, the degree $\tilde{d}(A + k, B, 0)$ is defined by the class of triplet (S^I, i, τ_{S^I}), where $i : S^I \longrightarrow B$ is an identical embedding and GLc-framing

$$\tau_{S^I} : TS^I \oplus (S^I \times F_2) \oplus (S^I \times A(F_I)) \xrightarrow{I_{TS^I} \oplus \xi^{-I} \oplus \xi_\infty^{-I}}$$

$$TS^I \oplus \nu_{B^2} S^I \oplus \nu_B B^2 \big|_{S^I} \approx TB_\Phi \big|_{S^I} = S^I \times A(F_I) \times F_2 \times R^I$$

is an infinite-dimensional suspension over the framing $I_{TS^I} \oplus \xi$.

We have to demonstrate yet that the degree $d(A + k, B_\Phi, o)$ is zero if m is even, and it is egual to unity if m is odd.

a) m is even.

Then the suspension of finite multiplicity over the Hopf map corresponds to zero in Z_2 and, consequently, to trivial class in Π^I_{l+I}, where $l = \dim F_2$. The last conclusion means that there exists a two-dimensional compact manifold W^2 in $R^{l+I} \times [0,I]$ with boundary $\partial W^2 = S^I$ where R^{l+I} is (l+I)-demensional Euclidean space, which we are going to identify with $F_2 \times R^I$. Besides, boundary $\partial W^2 = S^I$ lies in $F_2 \times R^I \times \{o\}$, and there is a trivialization of a normal bundle to W^2 in space $F_2 \times R^I \times R^I$:

$$\xi_I : \nu_{F_2 \times R^I \times R^I} W^2 \longrightarrow W^2 \times F_2,$$

coinciding on boundary $\partial W^2 = S^I$ with the suspension over ξ.

Without loss of generality, we may assume that $W^2 \subset B^2 \times [0,I] \subset B \times [0,I]$.

Then we suppose $i : W^2 \to B^2 \times [0,I] \subset B \times [0,I]$ to be an identical embedding, and define GLc-framing W^2 in $F_2 \times R^I \times [0,I]$ by the following formula:

$$\tau_{W^2} : TW^2 \oplus (W^2 \times F_2) \oplus (W^2 \times A(F_I)) \xrightarrow{I_{TW^2} \oplus \xi_I^{-I} \oplus I_{W^2} \times A(F_I)}$$

$$TW^2 \oplus \nu_{F_2 \times R^I \times R^I} W^2 \oplus (W^2 \times A(F_I)) \approx T(B_\Phi \times [0,I])\Big|_{W^2}$$

Triplet (W^2, i, τ_{W^2}) defines the GLc-framed bordism of the triplet (S^I, i, τ_{S^I}) and of the empty manifold in $F_I(B_\Phi)$. So, $[S^I, i, \tau_{S^I}] = 0$ in $F_I(B_\Phi)$ and in $\widetilde{F}_I(B_\Phi)$.

b) m is odd.

Let us assume the opposite, i.e. the class of triplet (S^I, i, τ_{S^I}) in $F_I(B_\Phi)$ is zero. This means that there exists a triple (W^2, j, τ) where W^2 is a compact two-dimensional manifold with boundary $\partial W^2 = S^I$, $j : W^2 \to B \times [0,I]$ is a continuous map coinciding at boundary with an identical embedding and GLc-framing $\tau : TW^2 \oplus (W^2 \times E) \to j^*(T(B_\Phi \times [0,I]))$ coinsides with τ_{S^I} on $TS^I \oplus (S^I \times E)$.

Our intention is to replace W^2 by $\widetilde{W}^2$ which lies in a finite-dimensional subspace of $F \times [0,I]$, and to replace GLc-framing τ by an infinite-dimensional suspension over a finite-dimensional framing (See Item I of this Section).

First, note that map $j : W^2 \to B \times [0,I]$ is continuously homotopic to embedding $j_I : W^2 \to B \times [0,I]$ such that the image $j_I(W^2)$ is in $F^n \times [0,I]$ where F^n is a finite-demensional sub-space in F. Next, we will use the following lemma:

<u>Lemma 2</u>. (On deformation of framing). Let (M, i_0, τ_0) be some triplet from $S_q(X_\Phi)$. Let the map $i_0 : M \to X$ be continuously homotopic to the map $i_I : M \to X$. Then the singular manifold (M, i_I) has GLc-framing τ_I in the structure X_Φ and triplets (M, i_2, τ_0) and (M, i_I, τ_I) are GLc-framed bordant in structure X_Φ.

<u>Proof of Lemma 2</u>. Let $i : M \times [0,I] \to X$ be a continuous homotopy connecting i_0 and i_I. For the bundle $i^*(TX_\Phi)$ over the compact manifold $M \times [0,I]$ (according to the covering homotopy property for bundles with a structure group), there exists the following isomorphism: $u : i^*(TX_\Phi) \to i_0^*(TX_\Phi) \times [0,I]$. Without loss of generality we can assume that $u|_{M \times \{0\}}$ is an identical isomorphism. Then triplet $(M \times [0,I], J, \tau)$, where

$J : M \times [0,I] \to X \times [0,I]$, $J(x,t)=(i(x,t),t)$ and τ has the form

$$T(M\times[0,I]) \oplus (M \times [0,I] \times E) = \{(TM \oplus (M\times E)) \times [0,I]\} \oplus \{(M \times [0,I])\times R^I\}$$

$$\xrightarrow{(\tau_0 \times I_{[0,I]}) \oplus I_{(M \times [0,I]) \times R^I}} (i_0^*(TX_\Phi) \times [0,I]) \oplus \{(M \times [0,I]) \times R^I\}$$

$$\xrightarrow{u^{-I} \oplus I_{(M \times [0,I]) \times R^I}} J^* (T (X_\Phi \times [0,I]));$$

defines a GLc-framed bordism between triplets (M, i_0, τ_0) and ($M, i_I, \tau_I = \tau|_{T(M \times \{I\}) \oplus (M\times E)}$) .Lemma 2 is proved.

Remark. We can make a new GLc-framing τ_I coincide at boundary ∂M with the old framing τ , provided homotopy i is constant on ∂M .

So, we can state that $j : W^2 \to B \times [0,I]$ is an embedding whose image lies in $F^n \times [0,I]$, $\dim F^n = n < \infty$.
Now let us deal with framing τ of W^2 .

Lemma 3. Let X_Φ be an Φ -structure on domain X of Banach space F . Let the model space $E \times R^q$ of this structure have the property of finite-dimensional approximation, and let the tangent bundle TX_Φ be trivial. Let there exist a split $E = E^n \times E^{\infty - n}$ and two GLc-framed bordant triplets (M_0, i_0, τ_0) and (M_I, i_I, τ_I), whose framings τ_0 and τ_I are infinite-dimensional suspensions over finite-dimensional framings τ_0^n and τ_I^n . Then the triplet (Y, i, τ) linking triplets (M_0, i_0, τ_0) and (M_I, i_I, τ_I) can be chosen so that for some split $E = E^{n_I} \times E^{\infty - n_I}, E^n \subset E^{n_I}, E^{\infty - n_I} \subset E^{\infty - n}$ the framing τ will be an infinite-dimensional suspension over framing τ^{n_I} in $E^{n_I} \times R^q \times [0,I]$.

Proof of Lemma 3. The bundles in which τ acts are layer-bundles. Using the finite-dimensional approximation property for their fibres, it is possible to replace isomorphism τ by layer-isomorphism τ' so that over ∂Y τ' coincides with τ . Let $\{(U_i, \varphi_i)\}_{i \in I}$ be an atlas on Y , and $\{(U_i, \alpha_i)\}_{i \in I}$ a respective set of trivializing maps for TY. Then in local presentation, τ' has the form:

$$\tau' \circ (\alpha_i \times I_{Y\times E}) : U_i \times R^{q+I} \times E \longrightarrow U_i \times R^{q+I} \times E$$

$$(x,v) \longrightarrow (x, v + a(x)\, v)$$

where $a(x)$ is a locally finite-dimensional operator. Then, there exists such covering $\{V_i^j\}$ of chart U_i and a set of finite-dimensional sub-spaces $F_{nj}^i \subset E \times R^{q+I}$ such, that $Ima(x) \subset F_{nj}^i$ for $x \in V_i^j$. From covering $\{V_i^j\}$ of compact manifold Y one can choose a finite covering and consequently, a finite-dimensional sub-space $E^k \subset E \times R^{q+I}$ such that $Ima(x) \subset E^k$ for all $x \in Y$.

Without loss of generality, we assume that $E^k = E^{n_1} \times R^{q+I}$ for some E^{n_1}. It is easy to check that the restriction

$$\tau^{n_I} = \tau|_{TY \oplus (Y \times E^{n_1})} : TY \oplus (Y \times E^{n_1}) \longrightarrow i(Y) \times E^{n_I} \times R^{q+I}$$

is an isomorphism of finite-dimensional vector bundles and $\tau'' = \tau^{n_1} \oplus I_{(Y \times E^{\infty - n_1})}$ (where $E^{\infty - n_1}$ is a closed complemented subspace to E^{n_1} in E) is GLc-framing that we search for.

Lemma 3 is proved.

So, using Lemma 2 and 3 for (W^2, j, τ), we get two-dimensional compact manifold $j_I W^2 = \widetilde{W}^2$, $\partial \widetilde{W}^2 = S^I$, lying in $F^{n_1} \times R^I \times [0,I]$ where F^{n_I} is a finite-dimensional sub-space in F, containing F_2 and framing $\tau^{n_I} : T\widetilde{W}^2 \oplus (\widetilde{W}^2 \times F^{n_1}) \to \widetilde{W}^2 \times F^{n_1} \times R^2$ coinciding over S^I with framing $I_{TS^I} \oplus (\Sigma\xi)^{-I} : TS^I \oplus (S^I \times F^n) \longrightarrow S^I \times F^n \times R^I$ (the suspesion $\Sigma\xi$ here is taken with a respective multiplicity).

All we have to do now is to bring framing τ^{n_1} to a correct form.

<u>Lemma 4.</u> Let M be a q-dimensional sub-manifold in Euclidean (n+q)-dimensional space $E^{n+q} = E^n \times E^q$ and $\tau : TM \oplus (M \times E^n) \longrightarrow M \times E^{n+q}$ its framing. We shall consider E^{n+q} as a sub-space in $E^{2(n+q)}$. Then the framing $\tau \oplus I_{M \times E^{n+q}}$ of manifold M in $E^{2(n+q)}$ can be replaced (within the class of framed bordisms) by a correct one. Moreover, should the framing be correct at the boundary ∂M, the new framing at the boundary ∂M will be a suspension over the old one.

<u>Proof of Lemma 4.</u> First, we shall prove that a framing τ can be replaced by such framing τ^I, for which image $\tau^I(M \times E^n)$ is orthoganal to image τ^I (TM).

Since τ is an isomorphism, the following split takes place: $M \times E^{n+q} = \tau(TM) \oplus \tau(M \times E^n)$. We denote by $\tau(TM)^{\perp}$ the bundle being orthogonal to $\tau(TM)$, and by $\tau(TM)_x^{\perp}$ - its fibre in point x. Let $p_x : E^{n+q} \longrightarrow \tau(TM)^{\perp}$ be an orthogonal projection.

The map

$$p : M \times E^{n+q} \longrightarrow \tau(TM)^{\perp}$$

$$p(x,v) = (x,\ p_x v)$$

defines the morphism of bundles, and the restriction $p_I = p|_{\tau(M\times E^n)}$: $\tau(M\times E^n) \to \tau(TM)^{\perp}$ defines the isomorphism of bundles.

The framing τ^I defined by formula

$$\tau^I : TM \oplus (M\times E^n) \xrightarrow{\tau} \tau(TM) \oplus \tau(M\times E^n) \xrightarrow{I_{\tau(TM)} \oplus p_I} \tau(TM) \oplus \tau(TM)^{\perp} = M\times E^{n+q}$$

will map $M\times E^n$ into the bundle $\tau^I(TM)^{\perp}$.

It is easy to check that a manifold M with framing τ will remain in the same class of framed bordisms.

Let us consider a Grassmann manifold G_q^{n+q} and a canonical bundle $\pi : \gamma_q^{n+q} \to G_q^{n+q}$. Since TM and $\tau(TM)$ are sub-bundles in $M \times E^{n+q}$, then the maps $f_0 : M \to G_q^{n+q}$, $f_0(x) = TM_x$ and $f_I : M \to G_q^{n+q}$, $f_I(x) = \tau(TM)_x$ are defined, TM and $\tau(TM)$ being induced bundles: $TM = f_0^*(\gamma_q^{n+q})$, $(TM) = f_1^*(\gamma_q^{n+q})$. The maps f_0 and f_I can be non-homotopic, but if $j : G_q^{n+q} \to G_q^{2(n+q)}$ is a natural embedding, then maps $jf_0, jf_I : M \to G_q^{2(n+q)}$ are homotopic [I2]. Let $h : M\times[0,I] \to G_q^{2(n+q)}$ be a homotopy, linking jf_0 and jf_I, then the bundle $h^*(\gamma_q^{2(n+q)})$ coincides with TM over $M\times 0$ and with $\tau(TM)$ over $M\times\{I\}$. Note that $h^*(\gamma_q^{2(n+q)})$ can be considered as a sub-bundel in $M\times E^{2(n+q)}$. Let us denote $\eta = h^*(\gamma_q^{2(n+q)})$ and use $\eta^{\perp}$ for denoting a respective orthogonal complementary sub-bundle in $M\times E^{2(n+q)}$. Then $\eta^{\perp}|_{M\times\{0\}} = TM^{\perp}$, $\eta^{\perp}|_{M\times\{I\}} = \tau(TM)^{\perp} = \tau(M\times E^n) \oplus (M\times E^{n+q})$.

Since η and $\eta^{\perp}$ are bundles over $M\times[0,I]$, the following isomorphisms take place:

$$g_0 : TM\times[0,I] \longrightarrow \eta,\quad g_I : [\tau(M\times E^n) \oplus (M\times E^{n+q})]\times[0,I] \longrightarrow \eta^{\perp}$$

such that $g_0|_{M\times\{0\}} = I_{TM}$, $g_0|_{M\times\{I\}} = \tau$ and $g_I|_{M\times\{I\}} = I_{\tau(TM)^{\perp}}$

Now we shall define a framing of $M\times[0,I]$ in $E^{2(n+q)}\times[0,I]$ by the formula

$$u : T(M\times[0,I]) \oplus (M\times[0,I]\times E^{2n+q}) = (TM\times[0,I]) \oplus (M\times[0,I]\times E^{2n+q}) \oplus$$

$$\oplus (M\times[0,I]\times R^I) \xrightarrow{g_0 \oplus (\tau|_{M\times E^n} \oplus I_{M\times E^{n+q}})\times I_{[0,I]} \oplus I_{M\times[0,I]\times R^I}}$$

$$\longrightarrow \eta\oplus[\tau(M\times E^{n})\oplus(M\times E^{n+q})]\times[0,I]\oplus(M\times[0,I]\times R^{I})\xrightarrow{I_{\eta}\oplus g_I\oplus I_{M\times[0,I]}\times R^{I}}$$

$$\longrightarrow \eta\oplus\eta^{\perp}\oplus(M\times[0,I]\times R^{I})=M\times[0,I]\times E^{2(n+q)}+I .$$

Since $u|_{M\times\{I\}}=\tau\oplus I_{M\times E^{n+q}}$ and $u|_{M\times\{0\}}=I_{TM}\oplus g_I\oplus(\tau|_{M\times E^{n}}\oplus I_{M\times E^{n+q}})$ i.e. the latter has a correct form, then the framed manifolds (M, $\Sigma\tau$) and (M, τ''), where $\tau''=u|_{M\times\{I\}}$, lie in one class of framed bordisms, the framing τ'' being correct.

If we consider the case of manifold M with a boundary we should additionally check that framing on the boundary (if it is correct) does not change. The Lemma is proved.

Applying, finally, Lemma 4 to sub-manifold $\widetilde{W}^2$ in $F^{n_1}\times R^{I}\times[0,I]$ $\subset F^{n_I}\times R^2$ we can state that its framing τ_2 in space $F^{l_I}\times R^2$ (where $l_I = 2n_I+2$) is correct, and at the boundary $\partial\widetilde{W}^2 = S^I$ it still coincides with the suspension over framing ξ, induced by Hopf map. This contradicts the statement that the Hopf map corresponds to a non-trivial class in Π^{I}_{k+I} for all $k \geqslant 3$, which is a proof of Item b) of Theorem I.

6. Application to Bifurcation Problem for Non-Linear Elliptic Boundary Value Problem

P.Rabinowitz [8] , generalizing the Krasnoselsky's theorem on bifurcations, has established one fact of global behaviour of solutions to Leray-Schauder type operator equations. This fact has been generalized by many authors both from the point of view of covering a wider class of equations, and obtaining more complete information on a respective branch of solutions. In this section by using the degree defined above, we will obtain an analogue of P. Rabinowitz's result for one non-linear elliptic boundary problem with a complex parameter.

Let $\Omega\subseteq R^n$ be a boundary domain with C^∞ -smooth boundary $\partial\Omega$, $u(x)$ a complex-valued function on $\bar{\Omega}$ and λ a complex parameter. Let us consider the boundary problem

$$F(x,u(x),\dots,D^{2m}u(x)) - \lambda u(x)=G(x,u(x),\dots,D^{2m-I}u(x),\lambda),\ x\in\Omega, \quad (8)$$

$$B_j(x,D)\,u(x)=0,\ x\in\partial\Omega,\ j=I,\dots,m. \quad (9)$$

where $F: \overline{\Omega} \times \mathbb{C}^M \longrightarrow \mathbb{C}^I$ is a function of class C^2, $G: \overline{\Omega} \times \mathbb{C}^N \times \mathbb{C}^I \to \mathbb{C}$ a function of class C^I and $B_j(x, D)$ a linear differential operator of order $m_j \leq 2m-I$, $j = I, \ldots, m$.

The solvability of problem (8)-(9) is studied in the Sobolev space of complex-valued functions $W_p^{2m+I}(\Omega)$, $p > n$.

Condition I. It is supposed that function $F(x, \xi)$, $x \in \overline{\Omega}$, $\xi = (\xi_\alpha) \in \mathbb{C}^M$ is uniformly strong elliptic, i.e. there exists a complex constant γ such that for all $x \in \overline{\Omega}$, $\xi \in \mathbb{C}^M$

$$\operatorname{Re}\left[\gamma \cdot \sum_{|\alpha|=2m} \frac{\partial F}{\partial \xi_\alpha}(x, \xi)\, \eta^\alpha\right] \geq C|\eta|^{2m}$$

for some $C > 0$ and any $\eta = (\eta_I, \ldots, \eta_n) \in R^n \setminus \{0\}$. Besides, with respect to boundary operators $\{B_j\}_{j=I}^m$ it is supposed that the Schapiro-Lopatinsky condition is fulfilled for every linearization $A(u)v = \sum_{|\alpha|=2m} \frac{\partial F}{\partial \xi_\alpha}(x, u(x), \ldots, D^{2m}u(x)) D^\alpha v$.

Let $W_{p,\{Bj\}}^{2m+I}(\Omega) = \{u \in W_p^{2m+I}(\Omega);\ B_j(x,D)\,u(x) = 0,\ x \in \partial\Omega$, $j = I, \ldots, m\}$. So, for every $u \in W_{p,\{B_j\}}^{2m+I}(\Omega)$ operator $A(u): W_{p,\{B_j\}}^{2m+I}(\Omega) \to W_p^I(\Omega)$ is a linear Fredholm one.

Condition II. It is assumed that the index of operator $A(\theta)$ (where θ is zero of space $W_p^{2m+I}(\Omega)$) is zero, and the spectrum of this operator is descrete.

Condition III. It is assumed that $F(x, 0, \ldots, 0) = G(x, 0, \ldots, 0, \lambda) = 0$

a) $\|G(x, p, \lambda)\| = o(\|p\|)$ for $\|p\| \to 0$

b) $\left\|\frac{\partial G}{\partial x_i}(x, p, \lambda)\right\| = o(\|p\|)$ for $\|p\| \to 0$

c) $\left\|\frac{\partial G}{\partial p_\alpha}(x, p, \lambda)\right\| = O(I)$ for $\|p\| \to 0$

We shall define the maps as follows:

$$f: W_{p,\{Bj\}}^{2m+I}(\Omega) \longrightarrow W_p^I(\Omega),\quad f(u) = F(u, \ldots, D^{2m}u),$$

$$g: W_{p,\{Bj\}}^{2m+I}(\Omega) \longrightarrow W_p^I(\Omega),\quad g(u) = G(u, \ldots, D^{2m-I}u, \lambda).$$

Then the problem of existence of solutions to (8)-(9) in space $W_p^{2m+I}(\Omega)$ is equivalent to a similar problem for operator equation

$$f(u) - \lambda u = g(u, \lambda),\quad u \in W_{p,\{Bj\}}^{2m+i}(\Omega),\ \lambda \in \mathbb{C}^I. \tag{IO}$$

From C^2-smoothness of F and condition I-II, it follows, that f is a C^I-smooth Fredholm map of index zero. Further,

$g(u,\lambda)$ is a completely continuous map from $W^{2m+I}_{p,\{B_j\}}(\Omega)\times\mathbb{C}$ in $W^I_p(\Omega)$. It has been shown in [9] that map f in case $p=2$ is a proper one on bounded closed sub-sets in $W^{2m+1}_{p,\{B_j\}}(\Omega)$. For an arbitrary p, the proof of this fact can be transferred with no changes. So, we will suppose that a restriction of f to any closed bounded sub-set in $W^{2m+I}_{p,\{B_j\}}(\Omega)$ is a proper map.

Besides, it follows from Condition III, that $f(\theta)=\theta, g(\theta,\lambda)=\theta$ and $\|g(u,\lambda)\|_{I,p}=o(\|u\|_{2m+I,p})$ for $\|u\|_{2m+I,p}\longrightarrow 0$ where $\|u\|_{1,p}$ is the norm in $W^1_p(\Omega)$.

Let us denote $A=A(\theta)$. Then $A=f'(\theta)$ and conseguently there takes place a representation $f(u)=Au+\omega(u)$, where

$$\|\omega(u)\|_{I,p}=o(\|u\|_{2m+I,p}),\quad \text{when } \|u\|_{2m+I,p}\longrightarrow 0$$

Theorem 2. Let λ_o be an eigenvalue of operator A of multiplicity m. We will define for $r>0$ the map $\Phi_r: W^{2m+I}_{p,\{B_j\}}(\Omega)\times\mathbb{C}\longrightarrow W^I_p(\Omega)\times R^I$ by formula

$$\Phi_r(u,\lambda)=(f(u)-\lambda u-g(u,\lambda),\ \|u\|^2_{2m+I,p}-r^2)$$

and introduce $B_{r,\rho}=\{(u,\lambda): \|u\|^2_{2m+I,p}+|\lambda-\lambda_o|^2=r^2+\rho^2\}$.

Then there exist such two positive constants r_o and ρ_o, that for any r,ρ, $o<r\leqslant r_o$, $o<\rho\leqslant\rho_o$, the degree $\tilde{d}(\Phi_r,B_{r,\rho},\theta)$ is well defined and $\tilde{d}(\Phi_r,B_{r,\rho},\theta)=m \pmod 2$.

Remark. Since $W^I_p(\Omega)\times R^I$ is a real space and we are using the degree constructed earlier for real spaces, further we will consider $W^{2m+I}_{p,\{Bj\}}(\Omega), W^I_p(\Omega)$ as a space of pairs of real functions and $\mathbb{C}$ as a two-dimensional real space.

Proof of Theorem 2. Let us denote by $S_{r,\rho}=\{(u,\lambda):\|u\|^2_{2m+I,p}+|\lambda-\lambda_o|^2=r^2+\rho^2\}$ the boundary of $B_{r,\rho}$. Since λ_o is an isolated eigenvalue of operator A, $\rho_0>o$ can be chosen so that for all ρ, $o<\rho<\rho_o$, and all φ the operator $(A-(\lambda_o+\rho e^{i\varphi})I)$ is invertible.

Next, $r_o>0$ can be chosen so small that for all $t\in[o,I]$ and $o<r\leqslant r_o$, the Fredholm proper homotopy $\Phi^t_r(u,\lambda)=(Au-\lambda u+t(\omega(u)-g(u,\lambda)),\|u\|^2_{2m+I,p}-r^2)$ is non-zero on $S_{r,\rho}$. Really, if $\Phi^t_r(u,\lambda)=\theta$, then $\|u\|^2_{2m+I,p}=r^2$. Hence, if $(u,\lambda)\in S_{r,\rho}$, then $|\lambda-\lambda_o|=\rho$, i.e. $\lambda=\lambda_o+\rho e^{i\varphi}$. Since operator $A-(\lambda_o+\rho e^{i\varphi})I$ is invertible, the equation $Au-\lambda u+t(\omega(u)-g(u,\lambda))=\theta$ for small r_o (by virtue of $\|\omega(u)\|_{I,p}=o(u_{2m+I,p})$ and

$\|g(u,\lambda)\|_{I,p} = o(\|u\|_{2m+I,p})$ has only one solution. This solution is $u = \theta$.

So, having chosen $\rho_o > o$ and $r_o > o$, for all $o < r < r_o$, $o < \rho < \rho_o$ we have

$$\tilde{d}(\Phi^o_r, B_{r,\rho}, \theta) = \tilde{d}(\Phi^I_r, B_{r,\rho}, \theta),$$

where $\Phi^o_r(u,\lambda) = (Au - \lambda u, \|u\|^2_{2m+I,p} - r^2)$.

Let us demonstrate that for the chosen r and ρ

$$\tilde{d}(\Phi^o_r, B_{r,\rho}, \theta) \equiv m \pmod 2).$$

For this purpose, we will use the splits $W^{2m+I}_{p,\{B_j\}}(\Omega) = F_I \times F_2$, where $F_2 = \bigcup_{j=I}^{\infty} \mathrm{Ker}(A - \lambda_o I)^j$, F^I is some closed complement of F_2 in $W^{2m+I}_{p,\{B_j\}}(\Omega)$, and $W^I_p(\Omega) = (A - \lambda_o I)(F_I) \times F_2$. (Since $W^{2m+I}_{p,\{B_j\}}(\Omega) \subseteq W^I_p(\Omega)$, F_2 can be considered as a subspace in $W^I_p(\Omega)$). Let $u = u_I + u_2$ correspond to the split $W^{2m+I}_{p,\{B_j\}}(\Omega) = F_I \times F_2$.

Let us write down $\lambda = \lambda_o + \mu e^{i\varphi}$, where $\mu > 0$ is some number, and let us consider homotopy $G(t,u,\lambda) = (Au - \lambda_o u - t\mu e^{i\varphi}u - \mu e^{i\varphi}u_2, \|u\|^2_{2m+I,p} - r^2)$, $o \leq t \leq I$, connecting map Φ^o_r with map $H_r(u,\lambda) = (Au - \lambda_o u - \mu e^{i\varphi}u_2, \|u\|^2_{2m+I,p} - r^2)$, (note that $G(t,u,\lambda) \neq \theta$ on S_{r,ρ_2}). Let us describe H as follows: $H_r(u,\lambda) = (Au - \lambda_o u, o) + (-\mu e^{i\varphi}u_2, \|u\|^2_{2m+I,P} - r^2) = Lu + k(u,\lambda)$. Then $L: W^{2m+I}_{p,\{B_j\}}(\Omega) \times \mathbb{C}^I \longrightarrow W^I_p(\Omega) \times R^I$, $(u,\lambda) \longrightarrow (Au - \lambda_o u, o)$ is a linear Fredholm operator of index I, and $L|_{F_I}: F_I \longrightarrow L(F_I)$ is a linear isomorphism. Next, $\dim F_2 < \infty$ and operator $k(u,\lambda) = (-\mu e^{i\varphi}u_2, \|u\|^2_{2m+I,p} - r^2)$ is a finite-dimensional one and its image is contained in $F_2 \times R^I$.

Let us consider restriction

$$\phi = (L+k)\big|_{B^2_{r,\rho}} : B^2_{r,\rho} = (F_2 \times \mathbb{C}^I) \cap B_{r,\rho} \longrightarrow F_2 \times R^I$$

$$(u_2,\lambda) \longrightarrow (Au_2 - u_2, \|u\|^2_{2m+I,p} - r^2).$$

Homotopy

$$\phi_t(u,\lambda) = (Au_2 - u_2, t(\|u_2\|^2_{2m+I,p} - r^2) + (I-t)(\rho^2 - |\lambda|^2))$$

is non-zero on $B^2_{r,\rho}$ and connects map ϕ with finite-dimensional map ϕ_o: $F_2 \times \mathbb{C} \longrightarrow F_2 \times R^I$, $\phi_o(u_2,\lambda) = (Au_2 - \lambda u_2, \rho^2 - |\lambda|^2)$

By analogy with ([IO], p.86), it can be shown that map ϕ_o is homotopic, in a class of maps non-degenerated on $B^2_{r,\rho}$, to a suspension over (-m)-multiple Hopf map. The only thing to note is that $W^p_I(\Omega) \times R^I$ satisfies the propety of the finite-dimensional approximation, and the statement of Theorem 2 now follows from Theorem I.

Corollary I.

Let us suppose that multiplicity of the eigenvalue is odd. Then (θ, λ_o) is a point of bifurcation of the solutions to equation (IO), and consequently to the problem (8)-(9).

The proof follows from Theorem 2 and properties of the degree.

Let λ_o be an eigenvalue of operator $A = f'(\theta)$ of odd multiplicity. Let us denote by W the closure of the set of non-trivial solutions (u, λ) to equation (IO) (i.e. solutions (u, λ) $u \neq \theta$) and let W' be a component of set W, containing (θ, λ_o).

Theorem 3. Let conditions I to III of this Section be fulfilled. Then either

a) Component W' is not bounded in $W^{2m+I}_{p,\{B_j\}}(\Omega) \times \mathbb{C}$, or

b) W' contains a finite number of points (θ, λ_j), where λ_j are eigenvalues of operator A; the number of such points corresponding to eigenvalues of odd multiplicity (including the point (θ, λ_o)) is even.

Proof. Let us assume that component W is bounded in $W^{2m+I}_{p,\{B_j\}}(\Omega) \times \mathbb{C}$. Then, because the map f is proper on bounded sub-sets, the component W' is compact. Because of discreteness of the spectrum of operator A, set W includes at most a finite number of points (θ, λ_j), $j=0,\dots,k$, where λ_j are eigenvalues of operator A. Let V be an open set in $W^{2m+I}_{p,\{B_j\}}(\Omega) \times \mathbb{C}$ containing W, at boundary ∂V of which there are no non-trivial solutions (u, λ), $u \neq \theta$ to equation (IO) and which does not include any other point of type (θ, λ), where λ are eigenvalues of operator A, apart from points (θ, λ_j), $j = 0,\dots,k$.

Let $r > 0$. Let us consider map $\Phi_r\colon \bar{V} \longrightarrow W^I_p(\Omega) \times R^I$ specified by formula $\Phi_r(u,\lambda) = (f(u) - \lambda u - g(u,\lambda), \|u\|^2_{2m+I,p} - r^2)$

Degree $d(\Phi_r, V, \theta)$ for it is well-defined. If the value of r is large (so that V is contained in a ball of radius r), then equation $\Phi_r(u,\lambda) = \theta$ has no solution in V, therefore its degree $d(\Phi_r, V, \theta)$ is zero. On the other hand, for small r, if (u, λ) satisfies equation $\Phi_r(u,\lambda) = \theta$, then $\|u\|_{2m+I,p} = r$ and, naturally, λ is close to one of λ_j, $j = 0,\dots,k$.

By property 4 of degree additivity,

$$d(\Phi_r, V, \theta) = \sum_{i=0}^{k} \chi_i \cdot d(\Phi_r, B^i_{r,\rho}, \theta) \qquad \text{(II)}$$

where $B^i_{r,\rho}$ - is the neighbourhood of point (θ, λ_i) defined in the proof of Theorem 2. Note that $\widetilde{F}_I(B^i_{r,\rho}) = F_I(B^i_{r,\rho}) \approx Z_2$ and $\widetilde{d}(\Phi_r, B^i_{r,\rho}, \theta) = d(\Phi_r, B^i_{r,\rho}, \theta)$, $i=0,\dots,k$.

Using Theorem 2 from equation (II) we get equality

$$\sum_{j=I}^{k} m_j \equiv 0 \pmod 2$$

where m_j is the multiplicity of eigenvalue λ_j, $j=I,\dots,k$. Whence follows Statement b) of Theorem 3.

References

I. Elworthy K.D., Tromba A.J. Differential structures and Fredholm maps on Banach manifolds // Proc. Sympos. Pure Math. (Global Analysis). - I970. v.I5 -p.45-74.

2. Sapronov Yu.I. On the dergree theory for nonlinear Fredholm maps // Trudy NII matematiki VGU. - Voronezh, I973, No II, p.93-IOI (in Russian).

3. Zvyagin V.G. Investigation of topological characteristics of nonlinear operators. PhD. Thesis, Voronezh,I974 (in Russian).

4. Zvyagin V.G. On the existence of a continuous branch of eigenvalues for nonlinear elliptic boundary value problem // Diff. equations, I977, vol.I3, No 8, p.I524-I527 (in Russian).

5. Ratiner N.M. On the degree theory for Fredholm mappings of manifolds // Equations on Manifolds. Voronezh, I982, p.I26-I29 (in Russian).

6. Borisovich Yu.G., Zvyagin V.G., Sapronov Yu.I. Nonlinear Fredholm maps and Leray-Schauder theory // Uspekhi Mat. Nauk (Russian Math. Surveys), I977, vol.32, No 4, p, 3-54.

7. Zvyagin V.G., Ratiner N.M. The degree of completely continuous perturbations of Fredholm maps and its application to bifurcation of solutions // Dokl. AN Ukr. SSR, I989, No 6, p.8-II (in Russian).

8. Rabinowitz P.H. A global theorem for non-linear eigenvalue problems and applications // Contrib. Nonlinear Fcl Anal. Academic Press. - I97I, p. II-36.

9. Zvyagin V.G. On the structure of the set of solutions of a nonlinear eiliptic problem with fixed boundary conditions // Global Analysis Studies and Applications, IV. Springer-Veriag,

I990 (Lect. Notes in Mathematics, vol. I400).

I0. Nirenberg L. Topics in non-linear functional analysis. New York, I974.

II. Pontryagin L.S. Smooth manifolds and their applications in homotopy theory. Moscow, I976 (in Russin).

I2. Husemoller D. Fibre bundies. McGraw Hill, I966.

FUCHSIAN SYSTEMS WITH REDUCIBLE MONODROMY AND THE RIEMANN-HILBERT PROBLEM

A.A.Bolibruch
Steklov Mathematical Institute
of Academy of Sciences of the USSR
ul. Vavilova, 42
117966, Moscow, USSR

Consider a homomorphism

$$\chi : \pi_1 (CP^1 \setminus \{a_1,\dots,a_n\}, z_o) \longrightarrow GL(p; C) \qquad (1)$$

from the fundamental group of the complement for any points $a_1,\dots,a_n$ in Riemann sphere CP^1 into the group GL(p; C) of (p,p) invertible complex matrices.

The Riemann-Hilbert problem (Hilbert's 21st problem) consists in proving the existence of a Fuchsian system

$$df = \omega f \qquad (2)$$

with the given monodromy (1) and the given singularities $a_1,\dots,a_n$. Recall that the system (2) is called Fuchsian if the matrix differential form ω has a pole of order not greater than one at every singular point a_i and ω is holomorphic on $CP^1 \setminus \{a_1,\dots,a_n\}$. If the point ∞ does not belong to the set $a_1,\dots,a_n$, then

$$\omega = \sum_{i=1}^{n} B^i \frac{dz}{z-a_i}, \quad \sum_{i=1}^{n} B^i = 0 . \qquad (3)$$

In papers [1-3] it was shown that the Riemann-Hilbert problem has a negative solution in the general case. All representations (1) of dimension $p = 3$, that cannot be realized as representations of any Fuchsian systems (2), (3), are described in paper [2]. All these representations are unstable in the following sense: if one perturbs the singular points a_i, then the answer to the Riemann-Hilbert problem with the same monodromy can become positive (see [2]).

In §1 of this paper we construct some new series of representations (1), that give a negative solution of the Riemann-Hilbert problem. These representations are already stable under perturbations of the points $a_1, \dots, a_n$. Then we reduce the Riemann-Hilbert problem to the problem of a holomorphic triviality of some vector bundles constructed by the representation (1). We introduce here the "Fuchsian weight" of a vector bundle on Riemann sphere and give its connection with a number of apparent singularities of Fuchsian linear differential equation constructed by the representation (1).

In §2 we discuss a problem of realization of some reducible monodromy representation of the Fuchsian system (2), (3) by any Fuchsian system with reducible set of matrix coefficients $B^1, \dots, B^n$.

The author is grateful to Prof. D.V.Anosov for giving him a simple universal description of prolongations into the singular points $a_1, \dots, a_n$ of the vector bundle, constructed by representation (1) (instead of the coordinate method, that the author used before). The simple description at the beginning of Section 1 is also obtained as a result of discussions with Prof. D.V.Anosov. The author is also grateful to Prof. A.V.Chernavskiĭ for his attention to this work.

1. The weight of a vector bundle and a number of apparent singularities of a Fuchsian equation.

1. Consider the universal covering $(\tilde{S}, \widehat{CP}, \pi)$ for the space $\widehat{CP} = CP^1 \setminus \{a_1, \dots a_n\}$. Fix a point y_0 in $\tilde{S}$ so that $\pi(y_0) = z_0$. The group $\pi_1(\widehat{CP}, z_0)$ may be identified with the group Δ of deck transformations of $\tilde{S}$. The triple $R = (\tilde{S}, \widehat{CP}, \pi)$ is a principal Δ-bundle and the group Δ acts on the left upon $\tilde{S}$. Define the right action of Δ as follows: $y\sigma = \sigma^{-1}y$, $y \in \tilde{S}$, $\sigma \in \Delta$. Let $\tilde{F} = (\tilde{E}, \widehat{CP}, \tilde{\pi})$ be a vector bundle with the structure group $GL(p;C)$, associated with R by the representation (1). The bundle $\tilde{F}$ may be described in terms of constant coordinate functions (see [4],[5]). It was also proved there that $\tilde{F}$ is holomorphically trivial. So the corresponding principal $GL(p;C)$-bundle $\tilde{P}$ is trivial ($\tilde{F}$ is associated to $\tilde{P}$). The bundle $\tilde{P}$ has the following description: $\tilde{P}_E \cong \tilde{S} \times GL(p;C)/\sim$, where $(y; G) \sim (\sigma y; \chi(\sigma)G)$, $\sigma \in \Delta$. (recall that the deck transformations group Δ is identified with $\pi_1(\widehat{CP}, z_0)$). Consider the map $w : \tilde{S} \longrightarrow \tilde{P}_E$, $w(y) = \langle (y;I) \rangle$. Let $U : \widehat{CP} \longrightarrow \tilde{P}_E$ be a holomorphic section of the bundle $\tilde{P}$. Then

$$w(y) = U(z)T(y), \quad z = \pi(y), \tag{1.1}$$

where $T(y)$ is a matrix function on $\tilde{S}$.

It follows immediately from the definition of $T(y)$ that $T(\sigma y) = T(y)\chi^{-1}(\sigma)$, therefore the form $\omega = dT \cdot T^{-1}$ is single-valued on CP^1 and holomorphic on $CP^1 \setminus \{a_1,\dots,a_n\}$ and system (2) with the constructed form ω has given monodromy (1). But one can say nothing about the type of singularities of the form ω at the points $a_1, \dots, a_n$.

Take generators g_i of the group $\pi_1(\widehat{CP}, z_o)$, corresponding to "small loops" around the points $a_1,\dots,a_n$, then $g_1 \dots g_n = 1$. Denote by G_i the matrix $\chi(g_i)$ and by E_i - the matrix $\frac{1}{2\pi i} \ln G_i$ with such eigenvalues ρ_i^j that conditions

$$0 \leqslant \operatorname{Re} \rho_i^j < 1 \tag{1.2}$$

hold.

Transform the matrix E_i by a matrix S to an upper-triangular form $E_i^o = S^{-1}E_iS$, where E_i^o consists of k blocks $(E_i^o)^1,\dots,(E_i^o)^k$ of dimensions $l_1,\dots, l_k$, corresponding to the generalized eigenspaces of X.

For each block $(E_i^o)^s$ one can consider a vector μ^s with such coordinates $\mu_i^s \in Z$, $i=1,\dots, l_s$, that the following conditions

$$\mu_1^s \geqslant \dots \geqslant \mu_{l_s}^s \tag{1.3}$$

hold. Denote by λ^i the vector $(\mu^1,\dots,\mu^k)$ and by A^{λ_i} the matrix $A^{\lambda_i} = \operatorname{diag}(\mu_1^1,\dots,\mu_{l_k}^k) = \operatorname{diag}(\lambda_1^i, \lambda_p^i)$.

Extend the bundle $\tilde{F}$ up to the bundle F^λ on CP^1 by the following way. Let O_i be a small disk in $\widehat{CP}$ with centre a_i. The choice of the generator g_i fixes a component of connectedness $\tilde{O}_i^*$ in $\tilde{O}_i = \tilde{\pi}^{-1}(O_i)$. For every point $y \in \tilde{O}_i$ there exists such point $y_* \in \tilde{O}_i^*$ that $y = \sigma y_*$, $\sigma \in \Delta$. Define the map $h^{\lambda_i} : \tilde{O}_i \longrightarrow GL(p;C)$ as follows

$$y \longmapsto S \cdot (y_*^{A^{\lambda_i}} \cdot y^{E_i^o}) \cdot S^{-1} \chi^{-1}(\sigma). \tag{1.4}$$

Here the function y_*^B denotes $\exp(B \ln y_*)$, $\ln y_* = \int_\gamma \frac{dz}{z-a_i}$, where γ is a path in O_i and the covering of γ in $\tilde{O}_i^*$ is the path from some fixed point y_i^o to the point y_*.

It is easy to see that the definition of the map (1.4) is correct and the following condition holds:

$$h^{\lambda i}(\tau y) = h^{\lambda i}(y)\chi^{-1}(\tau), \quad \tau \in \Delta. \tag{1.5}$$

Therefore a definition of the map

$$\gamma_i : \tilde{E}\,\Big|_{O_i} \longrightarrow O_i \times C^p, \tag{1.6}$$

$$\gamma_i(x) = (\pi(y);\ h^{\lambda i}(y)s), \quad x = (y,s)$$

is correct too (recall that $\tilde{F}_E = \tilde{E} = \tilde{S} \times C^p/\sim$, where $(y,s) \sim (\sigma y;\ \chi(\sigma)s)$, $\sigma \in \Delta$).

Prolong the bundle $\tilde{F}$ into the point a_i by formula (1.6). Make these prolongations for all $i = 1,\dots,n$. As a result we shall construct a bundle F^λ on CP^1, where $\lambda = (\lambda^1,\dots,\lambda^n)$. We shall further use the term "admissible" for such λ. The holomorphic type of the bundle F^λ depends on the choice of the matrices S from (1.4). Further F^λ denotes either all prolongations of $\tilde{F}$ with the given λ and all S or a prolongation of $\tilde{F}$ with the given λ and some S. Further it will be clear what F^λ means in every such situation.

Proposition 1.1. The Riemann-Hilbert problem for the representation (1) has a positive solution if and only if there exists such admissible set λ that the bundle F^λ is holomorphically trivial.

Proof. Sufficiency. Let F^λ be holomorphically trivial for some λ. Then the corresponding principal GL(P,c)-bundle P^λ is trivial too (F^λ is associated with P^λ). Let U(z) be a section of P^λ. Consider the matrix function T(y) from (1.1).

In the neighbourhood O_i of the point a_i by (1.6), (1.4) we have

$$\gamma_i(w(y)) = (\pi(y);\ h^{\lambda i}(y)) = (z;\ S y_*^{A^{\lambda_i}} y_*^{E_i^o} S^{-1}\chi^{-1}(\sigma)), \quad U(z) = (z;\tilde{U}(z)),$$

where $z = \pi(y)$, $y = \sigma y_*$, $\sigma \in \Delta$, $y_* \in \tilde{O}_i^*$.

Therefore the matrix function T(y) may be presented in $\tilde{O}_i^*$ as follows:

$$T(y) = \tilde{U}^{-1}(z)\, S\, y_*^{A^{\lambda_i}}\, y_*^{E_i^o}\, S^{-1}\chi^{-1}(\sigma) \tag{1.7}$$

and a differential form ω of the corresponding system (2) has the form

$$\omega = dT\,T^{-1} = d(\tilde{U}^{-1})\tilde{U} + \frac{\tilde{U}^{-1}S}{z-a_1}\left(A^{\lambda_i} + (z-a_1)^{A^{\lambda_i}} E_i^o\, (z-a_1)^{-A^{\lambda_i}}\right)S^{-1}U. \tag{1.8}$$

Since the set λ^i is admissible, it is easy to prove that the function $(z-a_i)^{A^{\lambda_i}} E_i^o (z-a_i)^{-A^{\lambda_i}}$ is holomorphic at the point a_i. Therefore the form ω has a pole of order at most one at the point a_i and so the system (2) is Fuchsian at a_i.

Necessity. It was proved in [6] that some fundamental matrix $T(y)$ of the Fuchsian system (2) may be presented in a neighbourhood O_i of the point a_i in the form (1.7) with some admissible set λ^i. Let $\lambda = (\lambda^1, \dots, \lambda^n)$, then the bundle F^λ is holomorphically trivial.

Indeed, by (1.7), (1.6) and (1.1) we obtain the holomorphy of the map $U(y) = w(y)\,T^{-1}(y)$ from CP^1 into P_E^λ, in other words, $U(y)$ is a holomorphic section of this bundle.

Remark 1.1. Since the first Chern class $c_1(F^\lambda)$ of the bundle F^λ is equal to

$$c_1(F^\lambda) = \sum_{i=1}^{n} \mathrm{Sp}\,(A^{\lambda_i} + E_i),$$

one can investigate only the bundles with the condition $c_1(F^\lambda) = 0$.

By the Birkhoff-Grotendick theorem (see [7]) the bundle F^λ is biholomorphically equivalent to a direct sum:

$$F^\lambda \cong \mathcal{O}(-c_1^\lambda) \oplus \dots \oplus \mathcal{O}(-c_p^\lambda), \tag{1.9}$$

where $c_1^\lambda \geqslant \dots \geqslant c_p^\lambda$, $c_i^\lambda \in \mathbb{Z}$, $\mathcal{O}(-1)$ is the Hopf bundle.

Definition. The number $\gamma(\lambda) = \sum_{i=1}^{p} (c_1^\lambda - c_i^\lambda)$ is called the Fuchsian weight of the bundle F^λ.

Call the number $\gamma(0)$ the Fuchsian weight of representation (1) and denote it by $\gamma(\chi)$.

Call the numbers $\gamma_{max}(\chi) = \max\limits_{\lambda} \gamma(\lambda)$, $\gamma_{min}(\chi) = \min\limits_{\lambda} \gamma(\lambda)$ the maximal and minimal Fuchsian weight, respectively.

Let (1) be an irreducible representation. Consider Fuchsian linear differential equation

$$y^{(p)} + q_1(z)\,y^{(p-1)} + \dots + q_p(z)y = 0 \tag{1.10}$$

with monodromy (1). Such equation always exists (see [8], [9]), but its coefficients have in the general case the supplementary apparent singularities $b_1, \dots, b_{m'}$, giving no contribution to the monodromy. By changing

$$f_i(z) = \left(\frac{z - b_i}{z - a_1}\right)^{c_i} y(z), \quad i = 1, \dots, m'$$

of the depending variable one can obtain holomorphy of all solutions of this equation at the points $b_1, \dots b_{m'}$. Denote by $m(\chi)$ a number of the supplementary apparent singularities with regard to multiplicity of the equation (1.10) with the given monodromy (1). There is the following connection between the number $m(\chi)$ and a weight $\gamma(\lambda)$ of a bundle F^{λ}.

<u>Proposition 1.2.</u> The number $m(\chi)$ satisfies the following inequality

$$m(\chi) \leqslant \frac{(n-2)p(p-1)}{2} - \gamma(\lambda) + 1 - l(\lambda),$$

where $l(\lambda)$ is a number of the first equal to c_1^{λ} numbers $c_1^{\lambda},\dots,c_p^{\lambda}$ in decomposition (1.9) for the bundle F^{λ}.

<u>Proof</u> is similar to the proof of theorem 1 in [9], presented there for the case $\lambda = 0$.

It easily follows from proposition 1.2 that

$$m(\chi) \leqslant \frac{(n-2)\,p\,(p-1)}{2} + 1 - p \qquad (1.11)$$

(see also [10]).

<u>Proposition 1.3.</u> In the case of equality in the formula (1.11) the Riemann-Hilbert problem for representation (1) has a positive solution.

<u>Proof.</u> Let F^{λ} be such prolongation of the bundle $\tilde{F}$ that $c_1(F^{\lambda}) = \sum_{i=1}^{n} \mathrm{Sp}(A^{\lambda_i} + E_i) = 0$, then in decomposition (1.9): $\sum_{i=1}^{p} c_i^{\lambda} = -c_1(F^{\lambda}) = 0$ and therefore a weight $\gamma(\lambda)$ is equal to a number pc_1^{λ}, i.e. $\gamma(\lambda)$ is a non-negative number divisible by p. But it follows from proposition 1.2 that $\gamma(\lambda) \leqslant p - l(\lambda) < p$, therefore $c_1^{\lambda} = 0$, $c_2^{\lambda} = \dots = c_p^{\lambda} = 0$ and the bundle F^{λ} is holomorphically trivial. Proposition 1.3 follows now immediately from proposition 1.1.

It was proved in [9] (corollary 2 2) that if $m(\chi) = 0$, then the Riemann-Hilbert problem for representation (1) also has a positive solution. As a result we say that the Riemann-Hilbert problem for the irreducible representation (1) necessarily has a positive solution when a minimal number $m(\chi)$ of the supplementary apparent singularities of all Fuchsian equations (1.10) with monodromy (1) is either equal to zero, or has a maximal possible value defined by the right side of inequality (1.11).

At the conclusion of this part of §1 we note that maximal and minimal weights of representation (1) satisfy the following inequalities

$$0 \leqslant \gamma_{\min}(\chi) \leqslant \gamma(\chi) \leqslant \gamma_{\max}(\chi) \leqslant \qquad (1.12)$$

$$\leqslant \frac{(n-2)\,p\,(p-1)}{2} - m(\chi) + 1 - l_{\max}(\chi)$$

If p is equal to 2, then the last inequality has the form

$$m(\chi) + \gamma_{max}(\chi) = n - 2. \tag{1.13}$$

Using this formula one can obtain a number of supplementary apparent singularities of a Fuchsian equation by invariants of a bundle F^{λ}.

2. Consider representation (1) and a bundle F^{λ}, constructed by this representation.

Lemma 1.1. If a weight $\gamma(\lambda)$ of the bundle F^{λ} is equal to zero, then the Riemann-Hilbert problem for representation (1) has a positive solution.

Proof. It follows from (1.9) that the bundle F^{λ} is biholomorphically equivalent to a bundle, which can be presented by two neighbourhoods $O_{\infty} = CP^1 \setminus \{a_1\}$, $O_1 = CP^1 \setminus \{\infty\}$ and a coordinate function $\gamma_{\infty 1} = (z - a_1)^{C^{\lambda}}$, $C^{\lambda} = \text{diag}\,(c_1^{\lambda},\dots, c_p^{\lambda})$. Each equivalent bundle may be presented by the same neighbourhoods and a coordinate function $\tilde{\gamma}_{\infty 1} = \tilde{U}(z)(z-a_1)^{C^{\lambda}}\tilde{V}^{-1}(z)$, where $\tilde{U}(z)$ and $\tilde{V}(z)$ are holomorphically invertible in O_{∞} and O_1, respectively. It follows from the last observation that a bundle P^{λ} has a holomorphic out of point a_1 section $U(z)$, which may be presented in some neighbourhood of the point a_1 in the form $U(z) = V(z)\,(z-a_1)^{-C^{\lambda}}$, where $V(z)$ is a holomorphic section of P^{λ} over this neighbourhood. The matrix function $T(y)$ from (1.1) has in this case the following form (instead of (1.7))

$$T(y) = y_*^{C^{\lambda}}\tilde{V}^{-1}(z)\, S y_*^{A^{\lambda_1}} \cdot y_*^{E_1^0} S^{-1} \chi^{-1}(\sigma). \tag{1.7'}$$

It follows from condition $\gamma(\lambda) = 0$ that $C^{\lambda} = c_1^{\lambda} I$, therefore the formula (1.7') may be transformed to the form

$$T(y) = \tilde{V}^{-1}\, S\, y_*^{A^{\lambda_1} + c_1^{\lambda} I}\, y_*^{E_1^0} S^{-1} \chi^{-1}(\sigma). \tag{1.7''}$$

Consider a bundle $F^{\lambda'}$, where $(\lambda')^i = \lambda^i$, $i \neq 1$ and $(\lambda')^1 = \lambda^1 + (c_1^{\lambda},\dots,c_1^{\lambda})$. By (1.7") one concludes that the bundle $F^{\lambda'}$ is holomorphically trivial. Lemma 1.1 follows now from proposition 1.1.

Remark 1.2. Now using the construction of the bundle $F^{\lambda'}$ one can directly verify that a form ω from (1.8), constructed by $T(y)$ from (1.7"), has a pole of order at most one at the point a_1.

Consider now a bundle F^0, constructed by representation (1).

Theorem 1.1. Let every matrix $G_i = \chi(g_i)$ of monodromy representation (1) be transformed to Jordan block and let representation (1) be reducible. Then the Riemann-Hilbert problem for this representation has a positive solution if and only if $\gamma(\chi) = 0$.

Proof. Sufficiency of the condition $\gamma(\chi) = 0$ was proved in lemma 1.1. Necessity follows from remark 6.1 [2]. It may also be obtained from proposition 2.1 and corollary 2.1. of this paper (see remark 2.1).

Remark 1.3. For irreducible representation (1) the remaining conditions of theorem 1.1 are not sufficient for a positive solution of the Riemann-Hilbert problem (see example 5.1 [2]).

Proposition 1.4. Let matrices $G_1,\ldots,G_n$ have the common invariant subspace of dimension $0 < l < p$; let them satisfy the condition $G_1 \ldots G_n = I$ and let each matrix G_i be transformed to a Jordan block with eigenvalue ν_i. If a number

$$\nu = \sum_{j=1}^{n} \frac{1}{2\pi i} \ln \nu_j$$

is noninteger, then the Riemann-Hilbert problem for representation (1) with monodromy matrices $G_i = \chi(g_i)$ and any set of points $a_1,\ldots,a_n$ has a negative solution.

Proof. The number ν is integer or noninteger together with the number $\rho = \sum_{j=1}^{n} \rho_j$, where ρ_j is from (1.1). But a Fuchsian weight $\gamma(\chi)$ of representation (1) in our case is equal to the following sum

$$\gamma(\chi) = pc_1^o - \sum_{i=1}^{p} c_i^o = pc_1^{\,o} + p\rho \tag{1.13}$$

as $\sum_{i=1}^{p} c_i^o = -c_1(F^o) = -\sum_{i=1}^{p} \operatorname{Sp} E_i^o = -p\rho$.

If the Riemann-Hilbert problem for the given representation has a positive solution, then it follows from theorem 1.1 that $\gamma(\chi) = 0$ and therefore $\rho = -c_1^o$ is an integer number.

Example 1.1 Consider the matrices

$$G_1 = \begin{pmatrix} 1100 \\ 0110 \\ 0011 \\ 0001 \end{pmatrix}, \quad G_2 = \begin{pmatrix} 311-1 \\ -4-112 \\ 0031 \\ 00-4-1 \end{pmatrix}, \quad G_3 = \begin{pmatrix} -102-1 \\ 4-101 \\ 00-10 \\ 004-1 \end{pmatrix}$$

and any set of points a_1, a_2, a_3 . Note that $G_1 \cdot G_2 \cdot G_3 = I$, the matrix G_2 may be transformed to G_1 and the matrix G_3 may be transformed to a Jordan block with eigenvalue $\nu_3 = -1$. The number $\nu = \frac{1}{2\pi i}(\ln 1 + \ln 1 + \ln(-1))$ is noninteger for any choice of logarithms, therefore all conditions of proposition 1.4 hold. So the Riemann-Hilbert problem for representation (1) with the given matrices G_1, G_2, G_3 has a negative solution.

Remark 1.4 Representation (1), constructed above, is stable under perturbation of the points a_1, a_2, a_3 (see the introduction to this paper). The dimension $p=4$, for which a similar construction is possible, is minimal.

3. Consider a Fuchsian system (2), (3) in a neighbourhood of a singular point a_i. As it was noted in the proof of theorem 1.1, it follows from [6] that there always exists a fundamental matrix T(y) of a solution space X of this system which can be presented in $\tilde{O}_i^*$ in the following form (see (1.7)).

$$T(y) = U_i(z)(z - a_i)^{A^{\lambda_i}} \, y_*^{E_i^o} \quad , \tag{1.14}$$

where $U_i(z)$ is a matrix, holomorphically invertible at the point a_i, matrices A^{λ_i} and E_i^o are defined at the beginning of §1, $\lambda^i = (\varphi_i^1, \dots, \varphi_i^p)$ is an admissible set.

Definition 1.2. Numbers φ_i^j are called i-valuations and numbers $\beta_i^j = \varphi_i^j + \rho_i^j$ are called i-exponents of the solution space X at the point a_i.

Number φ_i^j characterizes the order of growth of j-column of the matrix T(y) at the point a_i (see details in [2]) and it is also denoted by $\varphi_i(t_j)$.

It easily follows from (1.14) and from the theorem of residues sum for the form $Sp\omega = d\ln\det T$ that

$$\sum = \sum_{i=1}^{n} \sum_{j=1}^{p} \beta_i^j = 0 \ . \tag{1.15}$$

If X_l is l-dimensional invariant (under monodromy action) subspace of solution space X of system (2), (3), then

$$\sum_l = \sum_{i=1}^{n} \sum_{j=1}^{l} \beta_i^j \leq 0 \tag{1.16}$$

for the sum $\sum_l$ of exponents of the subspace X_l (see lemma 3.6,[2]).

Consider now the following example of a negative solution of the Riemann-Hilbert problem. It follows from this example that the upper triangularity of the representation (1) does not ensure a positive answer to the question of a positive solution of the Riemann-Hilbert problem.

Example 1.2. Consider any set of points a_1, a_2, a_3, a_4 and representation (1), presented by matrices $G_i = \chi(g_i)$, where

$$G_1 = \begin{pmatrix} 1 & 0 & \underline{1} & 0 & 0 & 1 & 0 \\ 0 & -1 & -1 & 0 & 0 & \underline{1} & 0 \\ 0 & 0 & 1 & \underline{1} & 2 & 2 & 1 \\ 0 & 0 & 0 & 1 & \underline{1} & 1 & 0 \\ 0 & 0 & 0 & 0 & 1 & 1 & \underline{1} \\ 0 & 0 & 0 & 0 & 0 & -1 & 0 \\ 0 & 0 & 0 & 0 & 0 & 0 & 1 \end{pmatrix}, \qquad G_2 = \begin{pmatrix} 1 & 0 & 1 & \underline{1} & -1 & 1 & 0 \\ 0 & -1 & \underline{1} & 1 & -1 & -1 & 0 \\ 0 & 0 & -1 & -1 & \underline{1} & 0 & 0 \\ 0 & 0 & 0 & 1 & 1 & 0 & \underline{1} \\ 0 & 0 & 0 & 0 & -1 & \underline{1} & -1 \\ 0 & 0 & 0 & 0 & 0 & -1 & 0 \\ 0 & 0 & 0 & 0 & 0 & 0 & 1 \end{pmatrix},$$

$$G_3 = \begin{pmatrix} 1 & 0 & \underline{1} & 0 & -1 & 1 & 1 \\ 0 & -1 & -1 & \underline{1} & 0 & 1 & 0 \\ 0 & 0 & 1 & 1 & -1 & 2 & \underline{1} \\ 0 & 0 & 0 & -1 & \underline{1} & -2 & -1 \\ 0 & 0 & 0 & 0 & -1 & \underline{1} & 1 \\ 0 & 0 & 0 & 0 & 0 & -1 & 0 \\ 0 & 0 & 0 & 0 & 0 & 0 & 1 \end{pmatrix}, \qquad G_4 = \begin{pmatrix} 1 & 0 & 1 & -1 & \underline{1} & 0 & 0 \\ 0 & -1 & \underline{1} & -2 & 1 & 0 & 0 \\ 0 & 0 & -1 & \underline{1} & 0 & 0 & 0 \\ 0 & 0 & 0 & -1 & 1 & \underline{1} & 0 \\ 0 & 0 & 0 & 0 & 1 & -1 & \underline{1} \\ 0 & 0 & 0 & 0 & 0 & -1 & 0 \\ 0 & 0 & 0 & 0 & 0 & 0 & 1 \end{pmatrix}.$$

The Riemann-Hilbert problem for this representation has a negative solution.

Proof. Suppose the contrary. Let $T(y)$ be a fundamental matrix of a solution space X of a system (2), (3) with the given monodromy. Then the subspaces X_l , $1 \leq l < p$, generated by the first l columns $t_1, \ldots, t_l$ of the matrix $T(y)$ are invariant under monodromy action. Each matrix G_i may be transformed to Jordan normal form, consisting of two Jordan blocks with eigenvalues 1 and -1, respectively. Using only the upper-triangular transformations, one can transform each matrix G_i to the form, obtained by the shuffle of these two Jordan blocks. One can obtain this form by replacing by zero all numbers being above the diagonal except for the numbers marked by the sign ___ . The character of associations of the corresponding vectors of X for matrices G_i may be described in the following way:

$$e_1^1 \quad f_1^1 \quad e_2^1 \quad e_3^1 \quad e_4^1 \quad f_2^1 \quad e_5^1 \qquad - G_1$$

$$e_1^2 \quad f_1^2 \quad f_2^2 \quad e_2^2 \quad f_3^2 \quad f_4^2 \quad e_3^2 \qquad - G_2$$

$$e_1^3 \quad f_1^3 \quad e_2^3 \quad f_2^3 \quad f_3^3 \quad f_4^3 \quad e_3^3 \qquad - G_3$$

$$e_1^4 \quad f_1^4 \quad f_2^4 \quad f_3^4 \quad e_2^4 \quad f_4^4 \quad e_3^4 \qquad - G_4$$

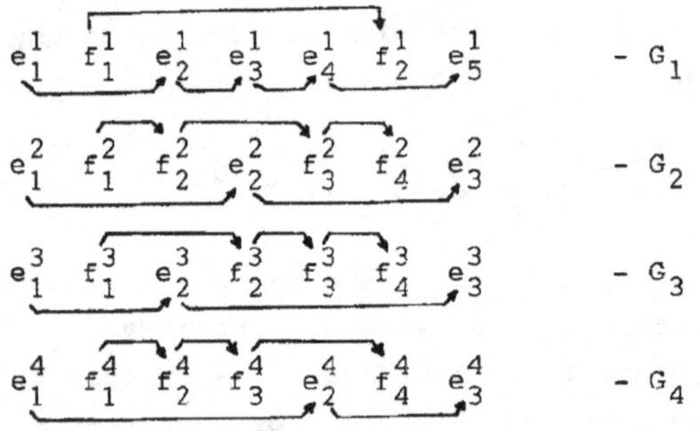

Here sequences of associated vectors, corresponding to Jordan blocks with eigenvalue 1 (vectors e_i^j) and -1 (vectors f_i^j) are denoted by the sign ⟶ . Note that any vector e_1^i, $i=1,\ldots,4$ generates the subspace X_1, vectors e_1^i, f_1^i generate the subspace X_2 and so on. Denote by a_j^i a valuation $\varphi_i(e_j^i)$ of the vector e_j^i and by b_j^i - a valuation $\varphi_i(f_j^i)$. As it follows from (1.14), definition 1.2 and the admissibility of a set $\lambda^i = (\varphi_i^1, \ldots, \varphi_i^p)$ the following inequalities

$$a_i^i \geqslant \ldots \geqslant a_5^i , \tag{1.17}$$

$$b_1^i \geqslant \ldots \geqslant b_4^i , \quad i = 1, \ldots, 4$$

hold. It follows from (1.16) that $\Sigma_1 \leqslant 0$. But using (1.15), (1.17) and (1.16), one can obtain: $0 = \Sigma_7 = \Sigma_6 + a_5^1 + a_{3_1}^2 + a_3^3 + a_3^4 \leqslant \leqslant \Sigma_6 + \Sigma_1 \leqslant 0$, therefore $\Sigma_6 = 0$ and $\Sigma_1 = a_5^1 + a_3^2 + a_3^3 + a_3^4 = 0$. It follows from this equality and (1.17) that $a_1^i = \ldots = a_5^i = a^i$, $i = 1, \ldots, 4$. One can analogously prove that $b_1^i = \ldots = b_4^i = b^i$, $i = 1, \ldots, 4$ and $\Sigma_2 = b^1 + b^2 + b^3 + b^4 + 2 = 0$. Condition (1.15) has in this case the form

$$0 = \Sigma_7 = 5a^1 + 3a^2 + 3a^3 + 3a^4 + 2b^1 + 1 + 4b^2 + + 2 + 4b^3 = 2 + 4b^4 + 2 = 2a^1 - 2b^1 - 1 \tag{1.18}$$

as $\Sigma_1 = \Sigma_2 = 0$. But it follows from (1.18) that $2a^1 - 2b^1 = 1$ for integers a^1 and b^1, which is impossible. Now the statement of example 1.2 is proved.

2. Fuchsian systems with reducible monodromy.

Consider again a Fuchsian system (2), (3) with solution space X and monodromy (1). Let a monodromy of this system be reducible and let X_l, $1 \leqslant l < p$ be a subspace of the space X, invariant under monodromy action. Denote by Σ_l a sum of exponents of this subspace.

<u>Proposition 2.1.</u> If $\Sigma_l = 0$, then there exists such Fuchsian system (2), (3) with the same monodromy that a set $\{B^i\}$ of coefficients of this system has a common l-dimensional invariant subspace.

<u>Proof.</u> Denote by T(y) a fundamental matrix of the space Y, the first l columns of which form a basis in the subspace X_l. They also form a matrix T_l. Consider a basic minor of the matrix $T_l(y)$ at the point z_o. Denote its elements by $T'(y)$ and by X' a solution space of a system $df = \omega' f$, where $\omega' = dT'(T')^{-1}$. (One can conclude without loss of generality that the lines of this basic minor belong to the first l lines of the matrix $T_l(y)$). It follows from (1.15) that $0 \geqslant \Sigma' \geqslant \Sigma_l = 0$ (see lemma 3.6 in [2]), therefore $0 = \Sigma' = \Sigma_l$ and

$\det T'(y) \neq 0$, $y \in \tilde{S}$. Let y_j be a vector consisting of the first l components of j-line of the matrix $T(y)$, $j > l$. Then

$$y_j = \sum_{i=1}^{l} c_i^j y_i \quad , \tag{2.1}$$

where y_i is an i-line of the matrix T'. Solving system (2.1) and using condition $\det T'(y) \neq 0$ we obtain: $c_i^j(z) = \dfrac{\det {}^jT'(y)}{\det T'(y)}$, $i = 1,\dots,l$, $j = l+1,\dots,p$ are singlevalued and holomorphic functions out of the points $a_1,\dots,a_n$.

It follows from condition $\Sigma' = \Sigma_l = 0$ that an inequality

$$\varphi_i(y_{jm}) \geqslant \varphi_i(t'_m)$$

for m-column t'_m of the matrix $T'(y)$ and for m-component y_{jm} of j-line of the matrix $T(y)$ holds. Therefore a sum ${}^j\Sigma'_i$ of i-exponents of a space generated by columns of the matrix ${}^jT'(y)$ and a sum Σ'_i satisfy the following inequality: ${}^j\Sigma'_i \geqslant \Sigma_i$. From this inequality and (1.15) we also obtain the holomorphy of functions $c_k^j(t) = ((z-a_i)^{{}^j\Sigma'_i - \Sigma'_i})\,0\,(1)$ at the points $a_1,\dots,a_n$. It follows from Liuville theorem that the functions $c_k^j(z)$, $1 \leqslant k \leqslant l$, $l+1 \leqslant j \leqslant p$ are constant. Consider the matrix

$$S = \left(\begin{array}{c|c} I & 0 \\ \hline C & I^{p-l} \end{array}\right)$$

where $C = \|\tilde{c}_{jk}\|$, $\tilde{c}_{jk} = -c_k^{j+p}$. It follows from (2.1) that elements $t_{km}(y)$ of the matrix $\tilde{T}(y) = ST(y)$ with indices $l+1 \leqslant k \leqslant p$, $1 \leqslant m \leqslant l$ are equal to zero for all y. Therefore the system (2), (3) with matrix form $\tilde{\omega} = d\tilde{T}\,\tilde{T}^{-1}$ has the form we need.

<u>Corollary 2.1.</u> Let monodromy matrices G_i of a Fuchsian system (2), (3) have a common invariant l-dimensional subspace X_l of solution space X and let each matrix G_i be transformed to a Jordan block. There exists such constant matrix S that by changing $g = Sf$ of the depending variable the system (2), (3) may be transformed to a system with coefficients B^i having the common invariant subspace of dimension l.

<u>Proof.</u> It is sufficient to prove equality $\Sigma_l = 0$ (see proposition 2.1). Let us transform a matrix G_i to a Jordan block. Since a set λ^i in decomposition (1.14) is admissible, we conclude that

$$\varphi_i^1 \geqslant \dots \geqslant \varphi_i^p \tag{2.2}$$

Note that the first l valuations in (2.2) are valuations of the subspace X_l at the point a_i, therefore $\sum_l \geqslant \frac{l}{p} \sum$. But it follows now from conditions (1.15), (1.16) that $\sum_l = \sum = 0$ and

$$\varphi_i^1 = \ldots = \varphi_i^p , \quad i = 1,\ldots,n . \qquad (2.3)$$

Remark 2.1. Now we can prove the necessity of theorem 1.1 without any references to other papers. It follows from (2.3) that under conditions of corollary 2.1 there exists a Fuchsian system with the given monodromy for which matrices A^{λ_i} , $i=1,\ldots,n$, from decomposition (1.14) are scalar. From this statement using proposition 1.1, we conclude that a bundle F^{λ} with the given set $\lambda = (\lambda^1,\ldots,\lambda^n)$ is holomorphically trivial. Inverting the proof of lemma 1.1 we obtain in this case the following result: the weight $\gamma(\chi)$ of hhe bundle F^{o} is equal to zero.

Reducibility of a monodromy of a Fuchsian system does not always provide the existence of such a change of the depending variable that transforms this system to a Fuchsian system with a reducible set of coefficients. More precisely, there exist representations (1) satisfying the following conditions:

1) representation (1) may be realized as a monodromy representation of a Fuchsian system (2), (3);

2) matrices G_i of monodromy representation (1) have a common invariant l -dimensional subspace;

3) there is no Fuchsian system (2), (3) with the given monodromy (1) and with coefficients having a common invariant subspace of dimension l.

Proposition 2.2. There exist such points $a_1,\ldots,a_4$ on Riemann sphere CP^1 that monodromy representation (1) with matrices of monodromy

$$G_1 = \begin{pmatrix} 1 & 1 & 1 & -1 & 0 & 0 & -1 \\ 0 & -1 & -1 & 0 & 0 & 0 & 1 \\ 0 & 0 & 1 & 1 & 2 & 2 & 2 \\ 0 & 0 & 0 & 1 & 1 & 0 & 1 \\ 0 & 0 & 0 & 0 & 1 & 1 & 1 \\ 0 & 0 & 0 & 0 & 0 & 1 & 0 \\ 0 & 0 & 0 & 0 & 0 & 0 & -1 \end{pmatrix} , \qquad G_2 = \begin{pmatrix} 1 & 1 & 0 & 1 & 1 & 1 & 0 \\ 0 & -1 & 1 & 1 & -1 & +1 & -1 \\ 0 & 0 & -1 & -1 & 1 & -1 & 0 \\ 0 & 0 & 0 & 1 & 1 & 1 & 0 \\ 0 & 0 & 0 & 0 & -1 & -1 & 1 \\ 0 & 0 & 0 & 0 & 0 & 1 & 0 \\ 0 & 0 & 0 & 0 & 0 & 0 & -1 \end{pmatrix}$$

$$G_3 = \begin{pmatrix} 1 & 0 & 1 & 0 & -1 & 0 & 0 \\ 0 & -1 & -1 & 1 & -1 & 1 & 2 \\ 0 & 0 & 1 & 1 & -1 & 1 & 2 \\ 0 & 0 & 0 & -1 & 1 & -1 & -2 \\ 0 & 0 & 0 & 0 & -1 & 1 & 1 \\ 0 & 0 & 0 & 0 & 0 & 1 & 1 \\ 0 & 0 & 0 & 0 & 0 & 0 & -1 \end{pmatrix} , \qquad G_4 = \begin{pmatrix} 1 & 0 & 1 & -1 & 1 & 1 & 0 \\ 0 & -1 & 1 & -2 & 0 & 0 & 0 \\ 0 & 0 & -1 & 1 & 0 & 0 & 0 \\ 0 & 0 & 0 & -1 & 1 & 0 & 1 \\ 0 & 0 & 0 & 0 & 1 & 1 & 0 \\ 0 & 0 & 0 & 0 & 0 & 1 & 1 \\ 0 & 0 & 0 & 0 & 0 & 0 & -1 \end{pmatrix}$$

satisfies the conditions 1) - 3).

Proof. Prove at first that there is no Fuchsian system with the given monodromy and with coefficients having a common one-dimensional invariant subspace. Suppose the opposite. Let such system exist, then it has the solution $e_1 = (f_1, 0, 0)$ with condition $\sum_{i=1}^{4} \varphi_i(e_1) = 0 = \Sigma_1$ where Σ_1 is a sum of exponents of the subspace $X_1 \subset X$. The further proof of this part identically coincides with the proof of example 1.2 after obtaining an equality $\Sigma_1 = 0$. We only need to change the scheme of association in that example for the following one

$$
\begin{array}{ccccccc}
e_1^1 & f_1^1 & e_2^1 & e_3^1 & e_4^1 & e_5^1 & f_2^1 \\
\end{array} \quad - G_1
$$

$$
\begin{array}{ccccccc}
e_1^2 & f_1^2 & f_2^2 & e_2^2 & f_3^2 & e_3^2 & f_4^2 \\
\end{array} \quad - G_2
$$

$$
\begin{array}{ccccccc}
e_1^3 & f_1^3 & e_2^3 & f_2^3 & f_3^3 & e_3^3 & f_4^3 \\
\end{array} \quad - G_3
$$

$$
\begin{array}{ccccccc}
e_1^4 & f_1^4 & f_2^4 & f_3^4 & e_2^4 & e_3^4 & f_4^4 \\
\end{array} \quad - G_4
$$

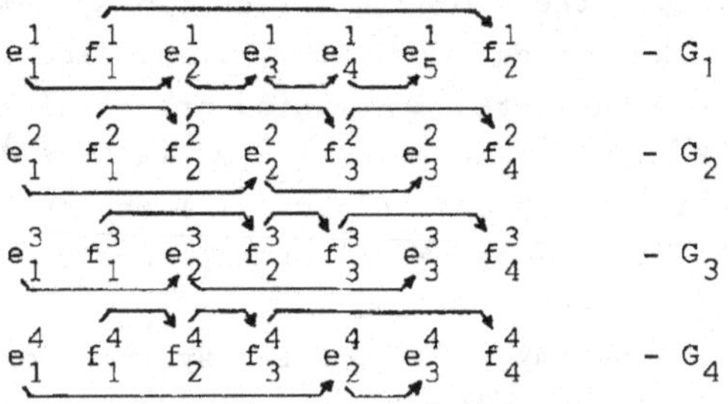

Prove now that with some choice of points $a_1, \ldots, a_4$ the given representation may be realized as a monodromy representation of a Fuchsian system (2), (3).

Let $a_1 = 0$ and let a_2, a_3, a_4 be any points.

By corollary 3.1 [2] there exists a system (2) with the given monodromy χ, Fuchsian at the points a_2, a_3, a_4 and regular at the point $a_1 = 0$ (i.e. all solutions of this system have a polynomial growth at the point a_1), a fundamental matrix $T(y) = \|t_{ij}\|$ of which has the upper triangular form. By changing $g(y) = \Gamma(z)f(y)$ of the depending variable one can transform this system to a system with a fundamental matrix $T(y)$ satisfying the following conditions:

$$\varphi_i(t_{jj}) = 0, \qquad j = 1, \ldots, 7, \quad i = 2, 3, 4,$$

$$\varphi_i(t_{11}) = \varphi_1(t_{66}) = 0, \quad \varphi_1(t_{22}) = \varphi_1(t_{77}) = \qquad (2.4)$$

$$= -2, \quad \varphi_1(t_{33}) = \varphi_1(t_{44}) = \varphi_1(t_{55}) = -1$$

(these equalities follow from the fact that a sum of exponents of a function t_{jj} is equal to zero for all j).

Consider a two-dimensional subrepresentation χ_2 of the representation χ and the coresponding subsystem $df = \omega_2 f$ of our system with a fundamental matrix T_2. By (1.14) the matrix T_2 has the following form in a neighbourhood of the point a_1:

$$T_2(y) = \begin{pmatrix} a(z) & e(z) \\ 0 & b(z) \end{pmatrix} (z-a_1)^{\begin{pmatrix} 0 & 0 \\ 0 & -2 \end{pmatrix}} y_*^{E_1^0} \qquad (2.5)$$

where $\pi(y_*) = z-a_1$, $a(a_1)\cdot b(a_1) \neq 0$. Since a_1 is a regular point for our system then $e(z) = P(\frac{1}{z-a_1}) + h(z)$, where P is a polynomial, $h(z)$ is holomorphic at $a_1 = 0$ and $h(a_1) = 0$. Change a depending variable in our system as follows: $g = \Gamma(z)f$,

$$\Gamma(z) = \left(\begin{array}{cc|c} 1 & -\frac{1}{b(z)}P(\frac{1}{z-a_1}) & 0 \\ 0 & 1 & \\ \hline & 0 & I^5 \end{array}\right) \qquad (2.6)$$

As a result we also obtain a Fuchsian at the points a_2, a_3, a_4 system, but a function $e(z)$ in (2.5) for this new system has the form

$$e(z) = c\,(z-a_1)^m(1 + 0\,(1)), \qquad (2.7)$$

where $m > 0$. Note that $c \neq 0$ because in the opposite case the system $df = \omega_2 f$ is decomposed, i.e. χ_2 is a commutative representation, but this fact contradicts the conditions of the proposition.

It follows from proposition 5.6 [2] that using any sufficiently small perturbation of the point $a_1 = 0$ one can obtain a condition

$$\gamma(\chi_2) = 0\,. \qquad (2.8)$$

But in this case by proposition 5.2 [2] the following equality

$$m = 1 \qquad (2.9)$$

holds.

Consider a system (2) with the given monodromy, some singular points $a_1,\dots,a_4$, Fuchsian at the points a_2, a_3, a_4, regular at a_1 with the upper triangular fundamental matrix $T(y)$, satisfying the conditions (2.4), (2.5), (2.7)-(2.9). As it is proved above such system always exists. Consider a change $g = \Gamma(z)f$ of the depending variable for this system, where

$$\Gamma(z) = \left(\begin{array}{cc|c} 1 & 0 & \\ -\dfrac{b(a_1)}{c(z-a_1)} & 1 & 0 \\ \hline 0 & & I^5 \end{array}\right)$$

As a result of this transformation a subsystem $df = \omega_2 f$ transforms to a system with the following decomposition (2.5):

$$T'(y) = V_1(z)\,(z - a_1)^{-I}\, y_*^{E_1^0} \tag{2.10}$$

where the matrix
$$V_1(z) = \begin{pmatrix} a(z)(z-a_1) & c(1 + O(1)) \\ -(\frac{ab}{c})(z) & d(z) \end{pmatrix}$$

is holomorphically invertible at the point a_1.

Consider a quotient representation χ/χ_5 and the corresponding quotient system. Treat them in the same manner as χ_2 above. As a result we obtain a system (2), Fuchsian at the points a_2, a_3, a_4 with a fundamental matrix $T(y)$ satisfying the following conditions: a matrix A^{λ_i} in decomposition (1.14) is equal to $-I$ and

$$U_1(z) = \left(\begin{array}{c|ccc|c} V_1(z) & & * & & * \\ \hline & u_{33} & * & * & \\ 0 & 0 & u_{44} & * & * \\ & 0 & 0 & u_{55} & \\ \hline 0 & & 0 & & V_2(z) \end{array}\right) \tag{2.11}$$

It follows from (2.4), (2.10) that matrices $V_1(z)$, $V_2(z)$ and elements u_{33}, u_{44}, u_{55} are holomorphically invertible at the point a_1. Concerning the elements marked by the sign * one can only say that they are meromorphic at the point a_1. Using a change of the depending variable, analogous to (2.6), which was described in detail in [2; p.35 and lemmas 3.1, 3.2], one can obtain holomorphy at the point a_1 of the elements marked by the sign * in (2.11). It follows now from (1.8) that the constructed system (2) is Fuchsian at the point a_1.

References

1. Bolibruch A.A. The Riemann - Hilbert problem on a complex projective line. Mat. Zametki (Math. Notes), v.46, No.3, p.118-120 (1989) (in Russian).

2. Bolibruch A.A. The Riemann - Hilbert problem. Uspekhi Mat. Nauk (Soviet mathematical surveys), v.45, No.2, p.3-47 (1990) (in Russian).

3. Anosov D.V. Hilbert's 21st problem (according to Bolibruch). IMA preprint series, #660 Inst. for Math. and its appl. Univ. of Minnesota (1990).

4. Röhrl H. Das Riemann - Hilbertsche Problem der Theorie der linearen Differentialgleichungen. Math. Ann., b.133, s.1-25 (1957).

5. Förster O. Riemannsche Flächen. Berlin e.a., Springer-Verlag, 1977.

6. Gantmacher F.R. The theory of matrices. N.Y.: Chelsea, 1977.

7. Okonek Ch., Schneider M., Spindler H. Vector bundles on complex projective spaces. Birhäuser, 1980.

8. Deligne P. Equations différentielles à points singuliers reguliers. L.N. in math., No.163, 1970.

9. Bolibruch A.A. On the construction of a Fuchsian differential equation with the given monodromy representation. Mat. Zametki (Math. Notes), v.48, No.5, p.22-34 (1990) (in Russian).

10. Ohtsuki M. On the number of apparent singularities of a linear differential equation. Tokyo J. Math., v.5, No.1, p.23-26 (1982).

FINITELY SMOOTH NORMAL FORMS OF VECTOR FIELDS IN THE VICINITY OF A REST POINT

I.U.Bronsteĭn and A.Ya.Kopanskiĭ
Institute of Mathematics and Computer Centre
Academy of Sciences of SSR Moldova
277028, Chişinău, Moldova, USSR

1. Consider the vector field

$$\dot{x} = dx/dt = F(x) \quad (x \in R^d, \ t \in R) \tag{1}$$

with the origin being a rest point (i.e., $F(0) = 0$). Assume that $F \in C^K$, $K \geq 1$. Denote $A = DF(0)$, $f(x) = F(x) - Ax$. Equation (1) can be rewritten as

$$\dot{x} = Ax + f(x), \ f(0) = 0, \ Df(0) = 0. \tag{2}$$

In what follows it is assumed that the eigenvalues of the linear operator A do not belong to the imaginary axis. In this case, the rest point $x = 0$ is said to be hyperbolic. According to the well-known Grobman-Hartman theorem (see, for example, [1]) the system (2) is topologically linearizable, that is, there exists a local homeomorphism $y = H(x)$ transferring solutions of (2) to solutions of the linear system $\dot{y} = Ay$. It should be noted that the conjugating homeomorphism H may fail to be smooth. As it was shown by Hartman [2], the system $\dot{x} = 2x$, $\dot{y} = y + xz$, $\dot{z} = -z$ does not admit a C^1-linearization.

Let k be a positive integer. Let G^k denote the group of local C^k-diffeomorphisms $\Phi : (R^d, 0) \to (R^d, 0)$ with $D\Phi(0) = id$.

In this paper we consider the problem of reducing the vector field (2) to a simpler form by the aid of coordinate changes from the group G^k. To be more precise, we discuss the following two questions:

1) Under what conditions does there exist a transformation $y = \Phi(x)$, $\Phi \in G^k$, which brings the equation (2) to the form $\dot{y} = Ay + p(y)$, where p is a polynomial (of degree more than 1)?

2) What kinds of monomials entering the polynomial p can be eli-

minated by transformations from G^k?

As to the first question, let us note that there exist smooth vector fields that cannot be reduced near a hyperbolic point to a polynomial form by any transformation of the class C^1.

Example 1. Let us show that the following C^1-system

$$\dot{x} = 2x, \ \dot{y} = y + xz(x^2 + z^2)^{-1/3}, \ \dot{z} = -z \tag{3}$$

is not locally C^1-conjugated with a polynomial vector field. Suppose the contrary. According to some results obtained by V.S.Samovol (see below) the polynomial normal form corresponding to the linear part of (3) with respect to the class of all C^1 transformations is as follows:

$$\dot{u} = 2u, \ \dot{v} = v + auw, \ \dot{w} = -w. \tag{4}$$

By the assumption, there exists a local C^1-diffeomorphism Ψ

$$x = f(u,v,w), \ y = g(u,v,w), \ z = h(u,v,w)$$

conjugating (3) and (4). It is clear that $\Psi(0,0,0) = (0,0,0)$.

Because Ψ carries the stable (unstable) manifold of the point $(0,0,0)$ with respect to the system (4) to the corresponding manifold for the system (3), one has

$$f(0,0,w) = 0, \quad g(0,0,w) = 0, \tag{5}$$

$$h(u,v,0) = 0. \tag{6}$$

Since Ψ is a local homeomorphism, it follows from (5) that

$$h(0,0,w) \neq 0 \quad \text{if} \quad w \neq 0. \tag{7}$$

Write the solutions of (3) and (4):

$$x(t) = xe^{2t}, \ y(t) = [y + xz\int_0^t (x^2e^{4t} + z^2e^{-2t})^{-1/3}dt]e^t, \ z(t) = ze^{-t};$$

$$u(t) = ue^{2t}, \ v(t) = (v + auwt)e^t, \ w(t) = we^{-t}.$$

Let u and w be fixed sufficiently small non-zero numbers. The map Ψ transfers the solution $\varphi(t) = (ue^{2t}, auwte^t, we^{-t})$ of (4) with initial condition $\varphi(0) = (u,0,w)$ to a solution of (3). Hence

$$f(ue^{2t},\ auwte^{t},\ we^{-t}) = f(u,0,w)e^{2t}; \tag{8}$$

$$g(ue^{2t},\ auwte^{t},\ we^{-t}) =$$

$$= \left\{g(u,0,w)+f(u,0,w)h(u,0,w)\int_0^t [g^2(u,0,w)e^{4t}+h^2(u,0,w)^{-2t}]^{-\frac{1}{3}}dt\right\}e^t;$$

$$h(ue^{2t},\ auwte^{t},\ we^{-t}) = h(u,0,w)e^{-t}. \tag{9}$$

Let $\{t_n\} \to +\infty$. Denote

$$u_n = ue^{-2t_n}. \tag{10}$$

Then

$$(u_n e^{2t},\ au_n wte^{t},\ we^{-t}) = (ue^{2(t-t_n)},\ auwte^{t-2t_n},\ we^{-t}).$$

Replace in the identity (9) t by t_n and u by u_n , then

$$g(u,\ auwt_n e^{-t_n},\ we^{-t_n}) =$$

$$= \left\{g(u_n,\ 0,w) + f(u_n,0,w)h(u_n,\ 0,w)\int_0^{t_n}[g^2(u_n,\ 0,w)e^{4t} + \right.$$

$$\left. + h^2(u_n,\ 0,w)e^{-2t}]^{-\frac{1}{3}}\,dt\right\} e^t . \tag{11}$$

Since $g(0,0,w) = 0$ (see (5)), one has

$$g(u_n,0,w) = g'_u(0,0,w)u_n + o(u_n). \tag{12}$$

Further,

$$R = f(u_n,\ 0,w)h(u_n,\ 0,w)e^{t_n} = f(u_n,\ 0,w)e^{2t_n}h(u_n,0,w)e^{-t_n}$$

and by using the equations (8) and (10) one obtains

$$R = f(u,au_n wt_n e^{t_n},\ we^{-t_n})h(u_n,0,w)e^{-t_n} =$$

$$= f(u,auwt_n e^{-t_n},\ we^{-t_n})\ h\ (u_n,0,w)e^{-t_n} .$$

By virtue of (12) the equality (11) gets the form

$$g(u,auwt_n e^{-t_n}, we^{-t_n}) = [g'_u(0,0,w)u_n + o(u_n)]e^{t_n} +$$

$$+ f(u,auwt_n e^{-t_n}, we^{-t_n})\, h(u_n,0,w)e^{-t_n} \times$$

$$\times \int_0^{t_n} [g^2(u_n, 0,w)e^{4t} + h^2(u_n, 0,w)e^{-2t}]^{-1/3}\, dt. \qquad (13)$$

The following estimates are valid:

$$I_n = \int_0^{t_n} [g^2(u_n,o,w)e^{4t} + h^2(u_n,0,w)e^{-2t}]^{-1/3}\, dt \leqslant$$

$$\leqslant \int_0^{t_n} [h(u_n,0,w)]^{-2/3}\, e^{2t/3} dt \leqslant \text{const} \cdot e^{2t_n/3},$$

since $\lim h(u_n,0,w) = h(0,0,w) \neq 0$ by (7) (remember that $w \neq 0$). Hence it follows from (13) as $n \to \infty$ that

$$g(u,0,o) = 0, \qquad (14)$$

therefore

$$g(u,auwt_n e^{-t_n}, we^{-t_n}) = g'_v(u,0,o)auwt_n e^{-t_n} +$$

$$+ g'_w(u,0,0)we^{-t_n} + o(|auwt_n e^{-t}| + |we^{-t}|). \qquad (15)$$

Divide both parts of the equality (13) by $t_n e^{-t_n}$, then by virtue of (15) one obtains

$$g'_v(u,0,0)auw + g'_w(u,0,0)wt_n^{-1} + o(t_n^{-1}) =$$

$$= [g'_n(0,0,w)u_n + o(u_n)]e^{2t_n} t_n^{-1} + f(u,auwt_n e^{-t_n}, we^{-t_n}) \times$$

$$\times h(u_n, 0,w)t_n^{-1} I_n. \qquad (16)$$

The left part of the last equality and the first term of the right part are bounded as $n \to \infty$ (see(10)).
Let $t \in [t_n/4, 3t_n/4]$, then by (12) and (10) one gets

$$g^2(u_n,0,w)e^{4t} + h^2(u_n,0,w)e^{-2t} \leqslant c_1 e^{4t-4t_n} + c_2 e^{-2t} \leqslant$$

$$\leqslant c_1 e^{-t_n} + c_2 e^{-t_n/2} \leqslant c_3 e^{-t_n/2}.$$

Consequently,

$$t_n^{-1} I_n \geqslant t_n^{-1} e^{t_n/6} c_4 \frac{t_n}{2} \to \infty \quad (n \to \infty).$$

It therefore follows from (16) that $f(u,0,0)h(0,0,w) = 0$. Since $w \neq 0$ one has $h(0,0,w) \neq 0$ by (7), hence $f(u,0,0) = 0$. Using (6) and (14), one concludes that $\Psi(u,0,0) = (0,0,0)$. Remembering that $u \neq 0$, $\Psi(0,0,0) = (0,0,0)$ and Ψ is locally one-to-one, we get a contradiction. This contradiction shows that the vector field (3) cannot be reduced to a polynomial form by a local C^1-diffeomorphism. This example answers affirmatively the question raised by A.D.Myshkis.

The first results concerning the reduction problem of smooth vector fields near a hyperbolic rest point to the polynomial form were obtained by Sternberg [3] and Chen [4]. Takens [5] generalized these results to the case of non-hyperbolic fixed points (in this situation, the normal form is, of course, polynomial only with respect to the hyperbolic variables, that is, in the direction transversal to the central manifold). Robinson [6] considered the problem of finitely smooth normalisation in the neighbourhood of an arbitrary smooth invariant manifold. The results of Sternberg and Chen were elaborated and extended in many directions by Belitskii [7] and Sell [8].

The classical results of Poincaré and Dulac (see, for example, [1]) combined with the above mentioned theorems of Sternberg and Chen show that when considering the second question one may assume without loss of generality that the polynomial vector field contains only resonant monomials. One should also have in mind that the polynomial resonant normal form is, in general, not unique. For example, consider the vector field $\dot{x} = x + y$, $\dot{y} = y$, $\dot{z} = 2z - y^2$. The nonlinear term entering the third equation is a resonant one. But the polynomial change of variables $u = x$, $v = y$, $w = z + 2xy - y^2$ brings the above system to the linear form. Note that the linear approximation operator of this system is not diagonalizable. It is easy to show that if the linear part is diagonal and the vector field contains only one resonant monomial of degree k, then there does not exist a polynomial change of variables that linearizes the k-germ of the vector field.

The problem of further simplification of the polynomial resonant normal form by using finitely smooth changes of variables was considered at first by Samovol [9, 10]. V.S.Samovol formulated a certain sufficient condition allowing one to delete a monomial from the normal form by a transformation of the class C^k (the so-called condition $S(k)$). In some cases the condition $S(k)$ cannot be improved. Recently Samovol

[11-15] has shown that in general the condition S(k) can be weakened. The authors [16] found an essentially more general condition, A(k), that includes all the results of Samovol [9 - 15].

The aim of this paper is to sum up the results of the investigations carried out in the above mentioned two directions and to formulate some unsolved problems.

2. Let us present the main statements answering the first question.

Let $\nu_1, \dots, \nu_d$ denote the eigenvalues of the operator A and $\theta_1, \dots, \theta_n$ be all distinct numbers contained in the set $\{\mathrm{Re}\,\nu_i : i = 1, \dots, d\}$. By the hyperbolicity assumption, $\theta_i \neq 0$ $(i = 1, \dots, n)$. Introduce new notations for the numbers $\theta_1, \dots, \theta_n$ as follows

$$-\lambda_\ell < \dots < -\lambda_1 < 0 < \mu_1 < \dots < \mu_m \quad (m + \ell = n).$$

The space R^d can be represented as a direct sum of A-invariant linear subspaces $E_1, \dots, E_n$ such that the eigenvalues of the operator $A|E_i$ satisfy the condition $\mathrm{Re}\,\nu_i = \theta_i$ $(i = 1, \dots, n)$. Set

$$Q_o(k) = [\frac{\lambda_\ell}{\lambda_1} + k\,(\frac{\mu_m}{\lambda_1} + 1)] + [\frac{\mu_m}{\mu_1} + k\,(\frac{\lambda_\ell}{\mu_1} + 1)] + 2;$$

$$M = [k\,\lambda_\ell/\lambda_1] + 1, \qquad N = [k\,\mu_m/\mu_1] + 1,$$

$$Q_s(k) = M + N + \max\{M, N\}; \qquad Q_B(k) = kn + 1$$

(the square brackets denote the integer part of a number). If $m > 0$ and $\ell > 0$, then the rest point $x = 0$ is said to be a saddle point and if $m = 0$ or $\ell = 0$, then $x = 0$ is called a node. Consider at first the case of a saddle point.

<u>Theorem 1</u>. Let K and k be positive integers. Assume that the vector field (1) is of class C^K, $x = 0$ is a hyperbolic saddle point and $A = DF(0)$. If $K \geq \min\{Q_o(k), Q_s(k), Q_B(k)\}$ then the vector field (1) near the point 0 by a transformation $y = \varphi(x)$, $\varphi \in G^k$, can be reduced to the polynomial resonant normal form

$$\dot{y} = Ay + \sum_{|\tau|=2}^{N} p_\tau y^\tau. \tag{17}$$

Here $\tau \in \mathbb{Z}^n_+$, p_τ denotes a multihomogeneous polynomial from $P_\tau(E_1,\dots,E_n;\ E_1 \oplus \dots \oplus E_n)$, $p_\tau = (p^1_\tau, \dots, p^n_\tau)$ and $p^i_\tau \neq 0$ implies $\Theta_i = (\tau, \Theta) \equiv \tau^1 \Theta_1 + \dots + \tau^n \Theta_n$ (by the resonance condition). Below we shall give an upper estimate for N.

In fact, theorem 1 includes three statements. In the case K $Q_o(k)$ it improves the results by Sternberg [3], Chen [4] and Takens [5]. If K $Q_s(k)$, we get a theorem slightly stronger than the results announced by Sell [8]. In the case: K $Q_B(k)$, theorem 1 implies theorem 5.14. from the book by Belitskii [7]. Examples show that neither of these three statements is a consequence of two others, i.e. $\min\{Q_o(k), Q_s(k), Q_B(k)\}$ can be attained by each of these tree quantities.

Example 2.

$$\dot{x}_i = -(16,4 + 3,6)x_i \qquad (i = 1,\dots,11);$$

$$\dot{y}_j = (-11,8 + 21,8j)y_j \qquad (j = 1,\dots,10);$$

$$\dot{y}_{11} = 228y_{11} + \Big(\sum_{i=1}^{11} x_i^2 + \sum_{i=1}^{11} y_i^2\Big)^{148}\Big(\sum_{i=1}^{11} |x_i| + \sum_{i=1}^{11} |y_i|\Big)^{\frac{1}{2}}.$$

Here $n = 22$, $\ell = 11$, $m = 11$, $\lambda_1 = 20$, $\lambda_{11} = 56$, $\mu_1 = 10$, $\mu_{11} = 228$, $K = 296$. Set $k = 14$, then $Q_o(14) = 292 < K$; $Q_s(14) = 680$; $Q_B(14) = 309$.

Example 3.

$$\dot{x}_i = -(0,75 + 0,25i)x_i \qquad (i = 1,\dots,5);$$

$$\dot{y}_j = (75 + 25j)y_j \qquad (j = 1,\dots,4);$$

$$\dot{y}_5 = 200y_5 + \Big(\sum_{i=1}^{5} x_i^2 + \sum_{i=1}^{5} y_i^2\Big)^{8}\Big(\sum_{i=1}^{5} |x_i| + \sum_{i=1}^{5} |y_i|\Big)^{\frac{1}{2}}.$$

Here $n = 10$, $\ell = 5$, $m = 5$, $\lambda_1 = 1$, $\lambda_5 = 2$, $\mu_1 = 100$, $\mu_5 = 200$, $K = 16$. Let $k = 2$. Then $Q_o(2) = 410$, $Q_5(2) = 15$ K, $Q_B(2) = 21$.

Example 4.

$$\dot{x}_1 = -x_1, \quad \dot{x}_2 = -4x_2, \quad \dot{y}_1 = 100y_1,$$

$$\dot{y}_2 = 200y_2 + (x_1^2 + x_2^2 + y_1^2 + y_2^2)^6(|x_1| + |x_2| + |y_1| + |y_2|)^{\frac{1}{2}}.$$

For this system $n = 4$, $\ell = 2$, $m = 2$, $\lambda_1 = 1$, $\lambda_2 = 4$, $\mu_1 = 100$, $\mu_2 = 200$, $K = 12$. For $k = 2$ one has $Q_o(2) = 412$; $Q_s(2) = 23$; $Q_B(2) = 9 \quad 12$.

Remark. Subject to arbitrarily small C^1-disturbances of the vector field (1), the number n of distinct values of real parts of eigenvalues of the operator $A = DF(0)$ can become equal to the dimension d of the phase space. Therefore it seems at first glance that the estimate of the smoothness order, $K \geq Q_B(k) = kn + 1$, that enters in the third statement, is unstable under perturbations. But it follows from the proof that this estimate is in fact valid for all sufficiently C^1-close vector fields.

In each of these three cases, the proof of theorem 1 is carried out in its own way. It would be desirable to get a strengthening of theorem 1 that includes the three mentioned statements and to work out a unified proof.

In the nodal case one has the following

Theorem 2. Let K and k be positive integers. Assume that the vector field (1) is of class C^K, $x = 0$ is a nodal rest point (for definiteness, of the attracting type, i.e., $m = 0$). If $K \geq \min\{[\lambda_\ell / \lambda_1]+1, k\ell+1\}$, then the vector field (1) near the origin by a change of variables from G^k can be reduced to the polynomial resonant normal form

$$\dot{y} = Ay + \sum_{|\tau|=2}^{N_1} p_\tau y^\tau .$$

Theorem 2 strengthens the results by Sternberg [17] and Belitskii [7]. Theorem 2 includes in fact two statements. It is easy to give examples showing that none of these two statements is a consequence of another.

3. Let us consider now the second question. We shall use the notation introduced above. Let $\dim E_i = m_i$ $(i = 1,\dots,n)$. Denote the coordinates in E_i by $x_{i,1},\dots, x_{i,m_i}$. Remember that $R^d = E_1+\dots+E_n$. For each multiindex $\omega \in \mathbb{Z}^d_+$ we establish its correspondence with the multiindex $\tau = \tau(\omega) \in \mathbb{Z}^n_+$ defined by the rule $\tau^i = \omega_{i,1}+\dots+\omega_{i,m_i}$ (it is assumed that $\omega = (\omega^1,\dots,\omega^d) = (\omega_{1,1},\dots,\omega_{1,m_1},\dots, \omega_{n,1},\dots\dots\dots, \omega_{n,m_n}))$. Given a polynomial

$$p(x) = \Big(\sum_{|\omega|=2}^{N} p^1_\omega x^\omega, \dots, \sum_{|\omega|=2}^{N} p^d_\omega x^\omega\Big) \qquad (18)$$

in scalar variables $x_1, \ldots, x_d$, we denote by $\tilde{p}$ the corresponding polynomial in vector variables $z_1, \ldots, z_n$,

$$\tilde{p}(z) = \Big(\sum_{|\tau|=2}^{N} \tilde{p}^1_\tau z^\tau , \ldots, \sum_{|\tau|=2}^{N} \tilde{p}^n_\tau z^\tau \Big), \tag{19}$$

where $\tau \in \mathbb{Z}^n_+$, $\tilde{p}^i_\tau \in P_\tau(E_1, \ldots, E_n ; E_i)$ $(i = 1, \ldots, n)$. The polynomial (19) is said to be weakly resonant if $\tilde{p}^i_\tau \neq 0$ implies $\Theta_i = \tau^1 \Theta_1 + \ldots + \tau^n \Theta_n$. For $\tau \in \mathbb{Z}^n_+$ we shall use the notation $\tau = (\alpha, \beta) = (\alpha^\ell, \ldots, \alpha^1, \beta^1, \ldots, \beta^m)$ which corresponds to the rearrangement $(\Theta_1, \ldots, \Theta_n) = (-\lambda_\ell, \ldots, -\lambda_1, \mu_1, \ldots, \mu_m)$.

Following Samovol [9], we shall say that the multiindex $\omega \in \mathbb{Z}^d_+$ and the corresponding multiindex $\tau(\omega) \in \mathbb{Z}^n_+$ satisfy the condition S(k) with respect to the operator A if at least one of the following inequalities

$$\alpha^1 \lambda_1 + \ldots + \alpha^r \lambda_r > k \lambda_r \qquad (1 \leq r \leq \ell)$$

$$\beta^1 \mu_1 + \ldots + \beta^s \mu_s > k \mu_s \qquad (1 \leq s \leq m)$$

holds.

Samovol [10] has shown that if the multiindex $\omega \in \mathbb{Z}^d_+$ satisfies the condition S(k), then the weakly resonant monomial x^ω can be deleted from the polynomial normal form by a change of variables belonging to the group G^k. In particular, the normal C^1-form contains only weakly resonant quadratic terms of the form $x_i x_j$ such that $\operatorname{Re} \nu_i < 0$, $\operatorname{Re} \nu_j > 0$ (all other monomials satisfy the condition S(1)).

To illustrate Samovol's theorem let us consider the following model example

$$\dot{z} = \Theta z + x_\ell^{\alpha^\ell} \ldots x_1^{\alpha^1} y_1^{\beta^1} \ldots y_m^{\beta^m} ; \tag{20}$$

$$\dot{x}_i = -\lambda_i x_i \quad (i = 1, \ldots, \quad); \quad \dot{y}_j = \mu_j y_j \quad (j = 1, \ldots, m),$$

where $\lambda_\ell > \ldots > \lambda_1 > 0$; $\mu_m > \ldots > \mu_1 > 0$; $\Theta = -\alpha^\ell \lambda_\ell - \ldots - \alpha^1 \lambda_1 + \beta^1 \mu_1 + \ldots + \beta^m \mu_m$, i.e. $x^\alpha y^\beta = x_\ell^{\alpha^\ell} \ldots x_1^{\alpha^1} y_1^{\beta^1} \ldots y_m^{\beta^m}$ is a resonant monomial. Suppose that the multiindex $(\alpha^\ell, \ldots, \alpha^1, \beta^1, \ldots, \beta^m)$ satisfies the condition S(k). For definiteness, assume that the inequality $\alpha^1 \lambda_1 + \ldots + \alpha^r \lambda_r > k \lambda_r$ holds. Select a number M such that $M > k \lambda_r$. The change of variables

$$z = w + \frac{1}{M} u^{\alpha} v^{\beta} \ln\Big(\sum_{i=1}^{\tau} |u_i|^{M/\lambda_i}\Big);$$

$$u_i = x_i \quad (i = 1, \dots, \ ); \quad v_j = y_j \quad (j = 1, \dots, m)$$

is of class C^k and it conjugates (20) with the linear system.

Lemma 1. If $\tau \in \mathbb{Z}^n_+$ and $|\tau| > k\big(\ln \frac{\lambda_\ell \mu_m}{\lambda_1 \mu_1} + 2\big)$, then τ satisfies the condition $S(k)$.

Proof. Suppose the contrary:

$$\alpha^1 \lambda_1 + \dots + \alpha^r \lambda_r \leqslant k \lambda_r \quad (1 \leqslant r \leqslant \ell);$$

$$\beta^1 \mu_1 + \dots + \beta^s \mu_s \leqslant k \mu_s \quad (1 \leqslant s \leqslant m).$$

Consider the problem on maximization of the linear functional $|\alpha| = \alpha^1 + \dots + \alpha^\ell$ on the convex polyhedral domain in R^ℓ, defined by the system of inequalities $\alpha^1 \leqslant k$, $\alpha^1 \lambda_1 + \alpha^2 \lambda_2 \leqslant k \lambda_2, \dots, \alpha^1 \lambda_1 + \dots + \alpha_\ell \lambda_\ell \leqslant k \lambda_\ell$. This domain has a unique vertex $\alpha^1 = k$, $\alpha^2 = k - k\frac{\lambda_1}{\lambda_2}, \dots, \alpha^\ell = k - k\frac{\lambda_{\ell-1}}{\lambda_\ell}$. Therefore

$$|\alpha| \leqslant \alpha(\lambda_1, \dots, \lambda_\ell) = k\ell - k\Big(\frac{\lambda_1}{\lambda_2} + \frac{\lambda_2}{\lambda_3} + \dots + \frac{\lambda_{\ell-1}}{\lambda_\ell}\Big).$$

Let us investigate how the quantity $\alpha(\lambda_1, \dots, \lambda_\ell)$ depends upon the numbers $\lambda_1, \dots, \lambda_\ell$ whenever the quotient $\lambda_\ell / \lambda_1 = \rho$ is assumed to be fixed. Denote $\rho_i = \lambda_i / \lambda_{i+1}$ $(i = 1, \dots, \ell - 1)$, then

$$\alpha(\lambda_1, \dots, \lambda_\ell) \leqslant k\ell - \min\{k(\rho_1 + \dots + \rho_{\ell-1}) : 0 < \rho_i < 1,$$
$$\rho_1 \cdots \rho_{\ell-1} = 1/\rho\}.$$

It is not difficult to show that the minimum is attained at the point $\rho_i = \rho^{-\frac{1}{\ell-1}}$ $(i = 1, \dots, \ell - 1)$, i.e.

$$\alpha(\lambda_1, \dots, \lambda_\ell) \leqslant \alpha(\ell) = k\ell - (\ell-1)k\rho^{-\frac{1}{\ell-1}}.$$

The function $\alpha(\ell)$ is monotonically increasing and $\lim_{\ell\to\infty} \alpha(\ell) = (1 + \ln\rho)k$. Thus

$$|\alpha| \leqslant k\Big(\ln\frac{\lambda_\ell}{\lambda_1} + 1\Big).$$

Similarly,

$$|\beta| \leq k\ (\ln \frac{\mu_m}{\mu_1} + 1).$$

Our assertion follows from the last two inequalities.

We shall say that the multiindex $\tau \in \mathbb{Z}^n_+$, $|\tau| \geq 2$, satisfies the condition A(K) if there exist a number $p \geq 1$, vectors $\sigma_i \in R^n_+(k) = \{a \in R^n_+ : \text{either } a^j = 0 \text{ or } a^j \geq k\ (j = 1,\dots,n)\}$ $(i = 1,\dots,p)$ and a number δ equal either to 1 or to -1 such that for each pair of multiindices $(\gamma, \rho) \in \Gamma_k(\sigma) = \{(\gamma,\rho) : \gamma \in \mathbb{Z}^n_+,\ \rho \in \mathbb{Z}^p_+,\ 0 \leq |\rho| \leq |\gamma| \leq k,\ \tau^i + \rho^1\sigma^i_1 + \dots + \rho^p\sigma^i_p \geq \gamma^i\ (i = 1,\dots,n)\}$ one can find real numbers $A_1,\dots,A_p$ that satisfy the system of inequalities

$$A_i \geq 0 \qquad (i = 1,\dots,p);$$

$$A_1\sigma^i_1 + \dots + A_p\sigma^i_p \leq \tau^i + \rho^1\sigma^i_1 + \dots + \rho^p\sigma^i_p - \gamma^i \quad (i = 1,\dots,n);$$

$$\delta\,[\sum_{i=1}^{p}(A_i - \rho^i)\sum_{j=1}^{n}\sigma_i^j\,\Theta_j\,] > 0. \qquad (21)$$

In this paper we have no possibility to explain the origin and essence of the cumbersome technical condition A(k). We only note that the condition A(k) is preserved under small C^1-perturbations of the vector field. Indeed, the numbers $\Theta_1,\dots,\Theta_n$ enter only the last (strong) inequality of (21).

Let us explain the sense of the condition A(k) by considering the model example (20). Let p be some positive integer; $\varepsilon_i = (\varepsilon^\ell_i,\dots,\varepsilon^1_i)$ and $\varkappa_i = (\varkappa^1_i,\dots,\varkappa^m_i)$ be vectors with non-negative entries such that all the numbers $\Theta^*_i = -(\varepsilon_i,\lambda) + (\varkappa_i,\mu)$ have the same sign $(i = 1,\dots,p)$. Let M be some sufficiently large positive number. It can be directly verified that the change of variables

$$z = w + \frac{1}{M}\,u^\alpha v^\beta \ln(\sum_{i=1}^{p} |u^{\varepsilon_i} v^{\varkappa_i}|^{M/|\Theta|_i}); \qquad (22)$$

$$x_i = u_i \quad (i = 1,\dots,\ell); \qquad y_j = v_j \quad (j = 1,\dots,m)$$

conjugates (20) with its linear part. The identification of the smoothness class of the transformation (22) entails considerable difficulties.

Condition A(k) arose in fact as a sufficient condition put on the vectors $\sigma_i = (\varepsilon^\ell_i,\dots,\varepsilon^1_i, \varkappa^1_i,\dots,\varkappa^m_i)$ $(i = 1,\dots,p)$ in order to ensure C^k-smoothness of the map (22).

Let us indicate some connections between S(k) and A(k).

Lemma 2. $S(k) \Rightarrow A(k)$.

Lemma 3. In the nodal case, $S(k) \Leftrightarrow A(k)$.

Example 5. Consider the vector field

$$\dot{x}_2 = -690x_2 + x_1^5 y_1^{10} y_2^4, \quad \dot{x}_1 = -300x_1, \quad \dot{y}_1 = y_1, \quad \dot{y}_2 = 200\, y_2 .$$

It is easy to verify that the monomial $x_1^5 y_1^{10} y_2^4$ that corresponds to the multiindex $\tau = (0,5,10,4)$ satisfies the condition S(9), but does not satisfy conditions S(k) for $k > 9$. At the same time τ satisfies the condition A(13) with $\sigma_1 = (0,0,200,0)$, $\sigma_2 = (0,166,0,250)$. Thus, in the saddle case, the condition A(k) is weaker than S(k).

By using the duality principle in the theory of linear inequalities one can prove the following assertion.

Lemma 4. The multiindex τ satisfies the condition A(k) iff there exist vectors $\sigma_1, \ldots, \sigma_p \in R_+^n(k)$ and a number $\theta^* \neq 0$ such that $\sum_{j=1}^n \sigma_i^j \theta_j = \theta^* (i = 1,\ldots,p)$ and for each $(\gamma, \rho) \in \Gamma_k(\sigma)$ the conditions

$$u_j \geq 0 \quad (j = 1,\ldots,n), \quad \sum_{j=1}^n u_j \sigma_i^j \geq 1 \quad (i = 1,\ldots,p) \tag{23}$$

imply

$$\sum_{j=1}^n (\tau^j + \rho^1 \sigma_1^j + \ldots + \rho^p \sigma_p^j - \gamma^j) u_j > \rho^1 + \ldots + \rho^p. \tag{24}$$

Since (24) is a linear inequality for each $(\gamma, \rho) \in \Gamma_k(\sigma)$ and the conditions (23) determine a convex polyhedral domain D, it suffices to verify the inequalities (24) at the vertices of D only. But it should be noted that the search of vertices is a non-trivial task and we were able to solve it only in some particular cases (i.e., for special choices of the vectors $\sigma_1, \ldots, \sigma_p$). Along this way one obtains

Lemma 5. Each of the two conditions formulated below implies A(k).

1) There exist an integer τ, $1 \leq \tau \leq \ell - 1$, and positive numbers $\varepsilon_1 \leq \ldots \leq \varepsilon_m$ such that the following inequalities hold

$$\sum_{i=1}^{\tau} \alpha^i \lambda_i + \sum_{j=1}^m \frac{\beta^j \mu_j}{\varepsilon_j} > k \max\left\{ \lambda_\tau ; \frac{\mu_j}{\varepsilon_j} \right. \quad (1 \leq j \leq m) \quad ;$$

$$\sum_{i=1}^{\tau} \alpha^i \lambda_i + \sum_{i=\tau+1}^{\ell} \frac{\alpha^i \lambda_i}{1+\varepsilon_1} > k \max\left\{ \lambda_\tau ; \frac{\lambda_\ell}{1+\varepsilon_1} \right\} ;$$

$$\sum_{i=1}^{\tau} \alpha^i \lambda_i + \sum_{i=\tau+1}^{\ell} \frac{\alpha^i \lambda_i}{1+\varepsilon_q} + \sum_{j=1}^{q-1} \frac{\beta^j \mu_j (\varepsilon_q - \varepsilon_j)}{\varepsilon_j (1+\varepsilon_q)} >$$

$$> k \max\left\{ \lambda_\tau ; \frac{\lambda_\ell}{1+\varepsilon_q} ; \frac{\mu_j (\varepsilon_q - \varepsilon_j)}{\varepsilon_j (1+\varepsilon_q)} \ (1 \le j \le q-1) \right\} \quad (q = 2,\dots,m).$$

2) There exist positive numbers $\varepsilon_1 \le \dots \le \varepsilon_\ell$ and an integer s, $1 \le s \le m-1$, such that

$$\sum_{j=1}^{s} \beta^j \mu_j + \sum_{i=1}^{\ell} \frac{\alpha^i \lambda_i}{\varepsilon_i} > k \max\left\{ \mu_s ; \frac{\lambda_i}{\varepsilon_i} \ (1 \le i \le \ell) \right\} ;$$

$$\sum_{j=1}^{s} \beta^j \mu_j + \sum_{j=s+1}^{m} \frac{\beta^j \mu_j}{1+\varepsilon_1} > k \max\left\{ \mu_s ; \frac{\mu_m}{1+\varepsilon_1} \right\} ;$$

$$\sum_{j=1}^{s} \beta^j \mu_j + \sum_{j=s+1}^{m} \frac{\beta^j \mu_j}{1+\varepsilon_q} + \sum_{i=1}^{q-1} \frac{\alpha^i \lambda_i (\varepsilon_q - \varepsilon_i)}{\varepsilon_i (1+\varepsilon_q)} >$$

$$> k \max\left\{ \mu_s ; \frac{\mu_m}{1+\varepsilon_q} ; \frac{\lambda_i (\varepsilon_q - \varepsilon_i)}{\varepsilon_i (1+\varepsilon_q)} \ (1 \le i \le q-1) \right\} \quad (q=2,\dots,\ell).$$

The following theorem can be regarded as the main result of the theory of finitely smooth normal forms.

<u>Theorem 3.</u> Suppose that the conditions of theorem 1 are fulfilled. Then by a transformation belonging to G^k the vector field (1) can be reduced to the normal form (17) containing only such weakly resonant multihomogeneous polynomials $p^i_\tau y^\tau$ that τ does not satisfy $A(k)$. Similarly, one may assume that in theorem 2 $p^i_\tau \ne 0$ implies that τ does not satisfy $S(k)$. Moreover, by lemma 1 one may put $N = [\, k(\ln \frac{\lambda_\ell \mu_m}{\lambda_1 \mu_1} + 2)]$ in theorem 1 and $N_1 = [k(\ln \frac{\lambda_\ell}{\lambda_1} + 1)]$ in theorem 2.

From theorem 3 and lemma 5 one can deduce the recent results obtained by Samovol [11 - 15].

It remains an open question whether the estimate of smoothness of the transformation killing the polynomial $p^i_\tau y^\tau$ as given by A(k) can be improved.

Define the support of the C^k-normal form (17) as the set of all multiindices $\tau \in \mathbb{Z}^n_+$ such that the term $p^i_\tau y^\tau$ cannot be eliminated by a transformation of the class C^k.

Let us show that the support of the C^k-normal form is not, in general, a convex subset of $\mathbb{Z}^n_+$. The following example answers the question put by A.D.Bryuno.

<u>Example 6.</u> Consider the vector field

$$\dot{x}_4 = -34x_4 + x_1^{18}x_2^8 \quad , \quad \dot{x}_3 = -18x_3 + x_1^{18}x_2^4y^4 \quad ,$$

$$\dot{x}_2 = -2x_2 + x_1^1y^8 \quad , \quad \dot{x}_1 = -x_1 \quad , \quad \dot{y} = 2y .$$

It is not difficult to show that the multiindices $\tau_1 = (0,0,18,8,0)$ and $\tau_2 = (0,0,18,0,8)$ corresponding to the monomials $x_1^{18}x_2^8$ and $x_1^{18}y^8$ satisfy the condition S(17) but do not satisfy A(k) for $k > 17$. The monomial $x_1^{18}x_2^4y^4$ corresponds to the multiindex $\tau_3 = \frac{1}{2}(\tau_1 + \tau_2)$. Applying condition 1) of lemma 5 in the case $\tau = 1$, $\ell = 2$, $m = 1$, $\varepsilon_1 = 2$, $\alpha^1 = 18$, $\alpha^2 = 4$, $\beta = 4$, $\lambda_1 = 1$, $\lambda_2 = 2$, $\mu = 2$ one gets the following system of inequalities:

$$18 + \frac{8}{1+2} > k \max \left\{ 1, \frac{2}{1+2} \right\} ,$$

$$18 + \frac{8}{2} > k \max \left\{ 1, 1 \right\} .$$

Hence it follows that the multiindex τ_3 satisfies the condition A(20). Thus the support of the C^{20}-normal form contains the multiindices τ_1 and τ_2 but does not contain their half-sum.

It would be interesting to give a geometric description of C^k-normal form supports. In particular, what can be said about the set of all multiindices $\tau \in \mathbb{Z}^n_+$ that do not satisfy the condition A(k) (for a fixed linear operator A)?

All the results presented in this review have their counterparts in the theory of local diffeomorphisms near a hyperbolic fixed point. Some results similar to theorems 1-3 were obtained for vector fields in the vicinity of a non-hyperbolic rest point. In such a case one gets, of course, normal forms which are polynomials in the hyperbolic variables only, but the coefficients are functions defined on the central manifold. Moreover, for a fixed order of smoothness, k, the support of the normal form (with respect to the hyperbolic variables) contains only multiindices that do not satisfy the condition A(k). The detailed proofs of the above mentioned and many other results can be found in the book recently prepared by the authors [18]. To find a conjugating transformation one needs to solve a certain system of functional equations in some special functional spaces. This, in turn, is based on some essential notions and results of global analysis (in particular, jets of vector bundle sections, the fixed point principle for bundle morphisms and the theorem on a smooth invariant section).

References

1. Arnol'd V.I., Il'yashenko Yu.S. Ordinary differential equations // Encyclopaedia of Math. Sciences. Vol.1. 1988.
2. Hartman Ph. On local homeomorphisms of Euclidean Spaces // Bol.Soc. Mat. Mexicana. 1960. V.5. P.220-241.
3. Sternberg S. On the structure of local homeomorphisms of Euclidean n-space.II. // Amer. J. Math. 1958. V.80, No.3. P.623-631.
4. Chen K.T. Equivalence and decomposition of vector fields about an elementary critical point // Amer. J. Math. 1963. V.85, No.4. P.693-722.
5. Takens F. Partially hyperbolic fixed points // Topology. 1971. V.10, No.2. P.133-147.
6. Robinson R.C. Differentiable conjugacy near compact invariant manifolds // Bol. Soc. Brasil. Math. 1971. V.2, No.1. P.33-44.
7. Belitskiǐ G.R. Normal forms, invariants and local mappings. Kiev, 1979 (in Russian).
8. Sell G.R. Smooth linearization near a fixed point // Amer. J. Math. 1985. V.107, No.5. P.1035-1091.
9. Samovol V.S. Linearization of systems of differential equations in the vicinity of invariant toroidal manifolds // Trudy Mosk. Mat. obshch. 1979. V.38. P.187-219 (in Russian).
10. Samovol V.S. Equivalence of systems of differential equations in the neighbourhood of a rest point // Trudy Mosk. Mat. obshch.1982. V.44. P.213-234(in Russian).
11. Samovol V.S. Linearization of an autonomous system in the neighbourhood of a hyperbolic rest point // Diff. uravn. 1987. V.23, No.6. P.1098-1099 (in Russian).
12. Samovol V.S. On smooth linearization of systems of differential equations in the neighbourhood of a saddle rest point // Uspekhi mat. nauk. 1988. T.43. No.4. P.223-224 (in Russian).
13. Samovol V.S. On some conditions sufficient for smooth linearlization of an autonomous system in the vicinity of a rest point // Izv.AN KazSSR. 1988, No.3. P.41-44 (in Russian).
14. Samovol V.S. Linearization of a system of ordinary differential

equations in the neighbourhood of a rest point of the saddle type // DAN UkrSSR. Ser A. 1989. No.1. P.30-33 (in Russian).

15. Samovol V.S. On a necessary and sufficient condition for smooth linearization of an autonomous system on the plane in the vicinity of a rest point // Mat. Zametki. 1989. V.46, No.1. P.67-77 (in Russian).

16. Bronsteǐn I.U., Kopanskiǐ A.Ya. Finitely smooth polynomial normal forms of C^∞-diffeomorphisms in the neighbourhood of a fixed point // Funk. analiz i ego prilozh. 1990. V.24, No.2. P.79-80. (in Russian).

17. Sternberg S. Local contractions and a theorem of Poincaré// Amer.J. Math. 1957. V.79, No.5. P.809-824.

18. Bronsteǐn I.U., Kopanskiǐ A.Ya. Invariant manifolds and normal forms. Kishinev. 1992 (in Russian).

GENERALIZED DEGREE OF MULTI-VALUED MAPPINGS

B.D.Gel'man
Department of Mathematics
Voronezh State University
394693, Voronezh, USSR

Topological invariants of multi-valued mappings have been studied by many authors (see, for example, surveys [1-3] and references in them).

In Voronezh the problem of the investigation of topological invariants of multi-valued mappings was posed by Yu.G.Borisovich in 1968. This theme has been widely developed by him and his former students (the author of the present paper is one of them) in their research.

The present paper introduces the notion of the generalized degree for a broad class of multi-valued mappings. Naturally this class comprises many classes of multi-valued mappings, for which the degree has been constructed earlier, and it allows one to consider them all from the same point of view. Note that the value of the generalized degree for acyclic and generalized-acyclic multi-valued mappings coincides with the value of the degree constructed earlier (see [1,5]).

This paper also studies the questions connected with the invariance of the generalized degree under linear homotopies and proves some theorems on the calculation of the generalized degree. On the basis of one variant of the theorem on the product of degrees it is shown that the generalized degree of the multi-valued mapping $F(z) = \sqrt[n]{z}$ is equal to $\frac{1}{n}$.

The paper closes with the definition of the generalized rotation of the multi-valued vector field and the proofs of some theorems of fixed points of multi-valued mappings.

0. Main definitions.

Let X and Y be metric spaces; the multi-valued mapping (m-mapping) F of space X into space Y is correspondence which associates every point $x \in X$ with a non-empty subset $F(x) \subset Y$.

Denote by $K(Y)$ a set of all non-empty compact subsets in Y. Further we shall consider m-mappings with compact images (though it is not always essential). Such m-mappings will be denoted by $F : X \longrightarrow K(Y)$.

The set $\Gamma_X(F) \subset X \times Y$; $\Gamma_X(F) = \{(x,y) \mid x \in X,\ y \in F(x)\}$ is called a graph of m-mapping F over the set X.

Consider continuous mappings $p : \Gamma_X(F) \longrightarrow X$, $q : \Gamma_X(F) \to Y$ which are restrictions of the natural projections $pr_X : X \times Y \longrightarrow X$; $pr_Y : X \times Y \longrightarrow Y$ on $\Gamma_X(F)$. It is evident that for any $x \in X$ we have the equality $F(x) = q \circ p^{-1}(x)$.

Thus, any m-mapping $F : X \longrightarrow K(Y)$ defines five objects $(X, Y, \Gamma_X(F), p, q)$. The reverse also holds, namely, if a quintuple (X,Y,Z,f,g) is given, where X,Y,Z are metric spaces, $f : Z \longrightarrow X$, $g : Z \longrightarrow Y$ are some continuous mappings, f is surjective and a pre-image of any point is a compact in Z, then the equality $F(x) = g \circ f^{-1}(x)$ defines the m-mapping $F : X \longrightarrow K(Y)$.

It is easy to see that the same m-mapping can be defined by different quintuples.

<u>0.1. Definition</u>. The quintuples (X,Y,Z_0,f_0,g_0) and (X,Y,Z_1,f_1,g_1) are called equivalent if they define the same m-mapping.

In the set of all equivalent quintuples, giving the m-mapping F, a special role is played by one of them.

<u>0.2. Definition</u>. Representation of the m-mapping $F : X \longrightarrow K(Y)$ in the form of a quintuple $(X, Y, \Gamma_X(F), p, q)$ is called a canonical representation of the m-mapping F.

Let Y be a metric vector space, $X \subset Y$, $F : X \longrightarrow K(Y)$.

The point $x_0 \in X$ is called a fixed point of the m-mapping F if $x_0 \in F(x_0)$.

If $F(x_0) \ni 0$, then the point x_0 is called a singular point of the m-mapping F.

The definition of the upper semicontinuous m-mapping and its properties are considered to be known (see, e.g. [4]).

In this paper the Alexander-Cech cohomologies are used (see, e.g. [10]).

1. Splitting mappings.

Let G_1, G_2 be additive abelian groups and $\tau : G_1 \longrightarrow G_2$ an arbitrary homomorphism.

1.1. Definition. We say that τ is a splitting homomorphism if for τ there exists a left inverse homomorphism δ, i.e. $\delta : G_2 \longrightarrow G_1$ and the composition $\delta \circ \tau : G_1 \longrightarrow G_1$ is an identical automorphism.

1.2. Lemma. Homomorphism τ is said to be splitting iff the following conditions are satisfied:

a) τ is a monomorphism;

b) group G_2 is isomorphic to a direct sum $\mathrm{Im}\,\tau \oplus G$, where G is some abelian group.

The proof is obvious.

1.3. Corollary. If G_1 and G_2 are modules over field P, $\tau : G_1 \longrightarrow G_2$ is a homomorphism of the modules and Ker $\tau = \{0\}$, then τ is a splitting homomorphism.

1.4. Lemma. Let $\tau_1 : G_1 \longrightarrow G_2$, $\tau_2 : G_2 \longrightarrow G_3$ be splitting homomorphisms, then the composition $\tau = \tau_2 \circ \tau_1 : G_1 \longrightarrow G_3$ is also a splitting homomorphism.

The proof is obvious.

The statement of lemma 1.4 is transferred to the composition of any finite number of splitting homomorphisms.

Let X and Y be metric spaces; $A \subset X$, $B \subset Y$ be some subsets. Let $f : (X,A) \longrightarrow (Y,B)$ be a continuous mapping of a pair of spaces.

1.5. Definition. We say that the mapping f is splitting in dimension k with respect to the group of coefficients G over the space (Y,B) (for brevity, (k,G)-splitting) if the homomorphism $f^* : H^k(Y,B,G) \longrightarrow H^k(X,A,G)$ is a splitting homomorphism.

1.6. Example. Let $f : (X,A) \longrightarrow (Y,B)$ be a continuous mapping of pairs of spaces having a continuous right inverse. Then f is a splitting mapping for any dimension and any group of coefficients.

The proof is obvious.

1.7. Example. ((k,G)-Vietoris mappings). Let f be a continuous mapping of a metric space X into a metric space Y. Consider a set $M_k(f) \subset Y$ of those points in which a k-acyclicity of pre-images is violated, i.e.

$$M_o(f) = \{ y \mid y \in Y,\ H^o(f^{-1}(y),\ G) \neq G \},$$

$$M_k(f) = \{ y \mid y \in Y,\ H^k(f^{-1}(y),\ G) \neq 0 \},\ k > 0.$$

Denote by $d_k(f)$ a relative dimension of the set $M_k(f)$ in Y, i.e. $d_k(f) = rd_Y(M_k(f))$. If $M_k(f) = \emptyset$, then we say that $d_k(f) = -\infty$.

The mapping $f : X \quad Y$ will be called (k,G)-Vietoris (see [1]) if the following conditions are satisfied:

a) f is proper and surjective;

b) $d_k(f) \leqslant n-2-k$ for all $k \geqslant 0$.

Let $f : X \longrightarrow Y$ be a (k,G)-Vietoris mapping, B a subset in Y, $A = f^{-1}(B)$. Let us prove then that $f : (X,A) \longrightarrow (Y,B)$ is a $(k+1,G)$-splitting mapping.

Proof. Consider the following commutative diagram:

$$\begin{array}{ccccccccc}
H^k(Y,G) & \xrightarrow{i^*_k} & H^k(B,G) & \xrightarrow{\delta} & H^{k+1}(Y,B,G) & \xrightarrow{j^*_{k+1}} & H^{k+1}(Y,G) & \xrightarrow{i^*_{k+1}} & H^{k+1}(B,G) \\
\downarrow{\scriptstyle f^*_1} & & \downarrow{\scriptstyle f^*_2} & & \downarrow{\scriptstyle f^*} & & \downarrow{\scriptstyle f^*_3} & & \downarrow{\scriptstyle f^*_4} \\
H^k(X,G) & \xrightarrow{\hat{i}^*_k} & H^k(A,G) & \xrightarrow{\hat{\delta}} & H^{k+1}(X,A,G) & \xrightarrow{\hat{j}^*_{k+1}} & H^{k+1}(X,G) & \xrightarrow{\hat{i}^*_{k+1}} & H^{k+1}(A,G)
\end{array}$$

In the diagram the horizontal lines are exact and f^*_i, $i=1,2,3,4$ are isomorphisms by virtue of the theorem of Vietoris-Begle-Sklyarenko [8] . Then f^*, by virtue of the lemma on five homeomorphisms [9], is also an isomorphism, i.e. the splitting homomorphism.

If f is a $(1,G)$-Vietoris mapping, then we shall simply call it Vietoris mapping (if the group of coefficients G is known). Vietoris mappings have a number of useful properties:

a) let $f : X \longrightarrow Y$ be a Vietoris mapping; then for any set $B \subset Y$ the mapping $f|_{f^{-1}(B)} : f^{-1}(B) \longrightarrow B$ is also Vietoris;

b) let $f_1 : X \longrightarrow Y$, $f_2 : Y \longrightarrow Z$ be Vietoris mappings; then their composition is also a vietoris mapping;

c) let $f : X \longrightarrow Y$ be a Vietoris mapping and B some subset in Y, $A = f^{-1}(B)$; then f is a (k,G)-splitting mapping for any k over (Y,B).

The proof of properties a) and b) is contained in, for example, [5] , the proof of property c) follows from 1.7.

1.8. Example. Let Z be some metric space, $\{(X_\alpha, A_\alpha)\}_{\alpha \in J}$ a family of compact pairs in Z directed with respect to the inclusion and let $(X,A) = (\bigcap_\alpha X_\alpha, \bigcap_\alpha A_\alpha)$. Consider a family of continuous mappings $\{f_\alpha\}_{\alpha \in J}$, $f_\alpha : (X_\alpha, A_\alpha) \longrightarrow (Y,B)$ such that the diagram

$$\begin{array}{ccc}
(X_\alpha, A_\alpha) & \xrightarrow{i^\beta_\alpha} & (X_\beta, A_\beta) \\
 & {\scriptstyle f_\alpha} \searrow \quad \swarrow {\scriptstyle f_\beta} & \\
 & (Y, B) &
\end{array}$$

is commutative for any $\alpha \leqslant \beta$, where i^β_α is an inclusion mapping.

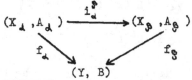

Denote $f : (X,A) \longrightarrow (Y,B)$, $f = f_\alpha$, for an arbitrary $\alpha \in \mathfrak{I}$.

If for any $\alpha \in \mathfrak{I}$ there exists a continuous mapping $g_\alpha : (Y,B) \longrightarrow (X_\alpha, A_\alpha)$, which is a right inverse mapping with respect to the mapping f_α , so that for any α and β , $\alpha \leq \beta$, the mappings

$$g_\beta : (Y,B) \longrightarrow (X_\beta, A_\beta) \quad \text{and} \quad i_\alpha^\beta \circ g_\alpha : (Y,B) \rightarrow (X_\alpha, A_\alpha) \rightarrow (X_\beta, A_\beta)$$

are homotopic as the mappings of the pairs of spaces, then the mapping $f : (X,A) \longrightarrow (Y, B)$ is (k,G)-splitting for any k and G.

Proof. By virtue of continuity of Alexander-Cech cohomology theory [10] we have:

$$\varinjlim H^k(X_\alpha, A_\alpha, G) = H^k(X,A,G)$$

We have the following commutative diagram:

$$\begin{array}{ccc} H^k(X_\alpha, A_\alpha, G) & \xrightarrow{\;i_\alpha^*\;} & H^k(X,A,G) \\ {\scriptstyle g^*}\downarrow & \nwarrow^{f_\alpha^*} & \uparrow{\scriptstyle f^*} \\ H^k(Y,B,G) & = & H^k(Y,B,G) \end{array}$$

By virtue of the above assumptions, in these diagrams it is possible to pass on to the limit with respect to α , then the limit homomorphism $g^* = \varinjlim g_\alpha^*$ is left inverse to the homomorphism f^*, i.e. $f : (X,A) \longrightarrow (Y,B)$ is a splitting.

1.9. Lemma. Let $f : (X,A) \longrightarrow (Y,B)$ be a certain continuous mapping. If there exists a continuous mapping $g : (Z,C) \longrightarrow (X,A)$ such that the composition $f \circ g : (Z,C) \longrightarrow (Y,B)$ is a (k,G)-splitting mapping, then f is also a (k,G)-splitting mapping.

Proof. Consider the commutative diagram:

$$\begin{array}{ccccc} H^k(Y,B,G) & & \xrightarrow{\;f^*\;} & & H^k(X,A,G) & \\ & {\scriptstyle g^*\circ f^*}\searrow & & \swarrow{\scriptstyle g^*} & & \\ & & H^k(Z,C,G) & \xrightarrow[\;\sigma\;]{} & H^k(Y,B,G) & , \end{array}$$

where σ is a left inverse homomorphism to $g^* \circ f^*$. Then $\sigma_1 = \sigma \circ g^* : H^k(X,A,G) \longrightarrow H^k(Y,B,G)$ is left inverse to the homomorphism f^* .

1.10. Corollary. Let $f : (X,A) \longrightarrow (Y,B)$ be a continuous mapping and let $(Z,C) \subset (X,A)$ and $\hat{f} = f|_{(Z,C)}$ be a (k,g)-splitting mapping, then f is a (k,G)-splitting mapping.

Proof. By way of the mapping $g : (Z,C) \longrightarrow (X,A)$ let us consider the inclusion, then this statement follows from lemma 1.9.

1.11. Lemma. Let $f : (X,A) \longrightarrow (Y,B)$ and $g : (Y,B) \longrightarrow (Z,C)$ be continuous (k,G)-splitting mappings, then their composition $g \circ f : (X,A) \longrightarrow (Z,C)$ is also a (k,G)-splitting mapping.

The proof follows from lemma 1.4.

It is also evident that the composition of any finite number of (k,G)-splitting mappings is also (k,G)-splitting.

Let B be a closed disk in a finite-dimensional space $\mathbb{R}^n$, $S = \partial B$. Let X be a metric space and $f : X \longrightarrow B$ a continuous mapping.

1.12. Lemma. If there exists such a field P that the following conditions are satisfied;

1) inclusion $f^{-1}(S) \xrightarrow{i} X$ induces a zero homomorphism $i^* : H^{n-1}(X,P) \longrightarrow H^{n-1}(f^{-1}(S),P)$;

2) homomorphism $f_s^* : H^{n-1}(S,P) \longrightarrow H^{n-1}(f^{-1}(S),P)$ is non-zero, where $f_s = f|_S$;

then the mapping f is (n,P)-splitting over (B,S).

Proof. By virtue of corollary 1.3 it is sufficient to prove that the homomorphism $f^* : H^n(B,S,P) \longrightarrow H^n(X,f^{-1}(S),P)$ is a monomorphism. For this consider the commutative diagram:

$$\begin{array}{ccccccc}
\ldots \to H^{n-1}(X,P) & \xrightarrow{i^*} & H^{n-1}(f^{-1}(S),P) & \xrightarrow{\delta} & H^n(X,f^{-1}(S),P) & \to \ldots \\
\uparrow f_B^* & & \uparrow f_S^* & & \uparrow f^* & \\
\ldots \to H^{n-1}(B,P) & \xrightarrow{i_1^*} & H^{n-1}(S,P) & \xrightarrow{\delta_1} & H^n(B,S,P) & \to \ldots
\end{array}$$

The lines of this diagram are exact sequences of the corresponding pairs of spaces. By virtue of the lemma's conditions δ and f_s^* are monomorphisms and δ_1 is an isomorphism. Then f^* is also a monomorphism, which proves the statement.

2. Admissible multi-valued mappings.

Let X, Y be metric spaces and let A be a closed subset in X. Let $F : X \longrightarrow K(Y)$ be an arbitrary m-mapping. Denote by $\Gamma_X(F)$, $\Gamma_A(F)$ the graphs of m-mapping F over X and A, respectively, i.e.

$$\Gamma_X(F) = \{(x,y) \mid x \in X,\ y \in F(x)\}, \quad \Gamma_A(F) = \{(x,y) \mid x \in A,\ y \in F(x)\}.$$

Then the projection $P : (\Gamma_X(F), \Gamma_A(F)) \longrightarrow (X,A)$, given by the condition $P(x,y) = x$, is well-defined.

<u>2.1. Definition</u>. We call m-mapping F (n,G)-admissible if projection p is a (n,G)-splitting mapping over (X,A). Denote by $Dop^n_G(X,A)$ the set of all (n,G)-admissible m-mappings F.

Consider some examples of (n,G)-admissible m-mappings.

<u>2.2. Example</u>. Let $F : X \longrightarrow K(Y)$ be an m-mapping having a continuous selection f (i.e. $f : X \quad Y$, $f(x) \quad F(x)$ for any $x \quad X$). Then $F \in Dop^n_G(X,A)$ for any group G and any n.

<u>Proof</u>. If f is a continuous selection of m-mapping F, then the continuous mapping $\hat{f} : (X,A) \longrightarrow (\Gamma_X(F), \Gamma_A(F))$, $f(x) = (x,f(x))$ is right inverse to mapping P. Consequently, by virtue of example 1.6 P is an (n,G)-splitting mapping.

<u>2.3. Example</u>. Let F be upper semicontinous m-mapping such that p is (n,G)-Vietoris mapping, then $F \in Dop^{n+1}_G(X,A)$.

The proof follows from example 1.7.

We shall call such m-mappings (n,G)-acyclic.

If F is (1,G)-acyclic m-mapping, then an image of every point $x \in X$ is acyclic. We shall call such m-mappings G-acyclic (or simply acyclic).

<u>2.4. Example</u>. Let F be such an m-mapping that there exist a metric space Z and continuous mappings $f : Z \longrightarrow X$ and $g : Z \longrightarrow Y$, satisfying the following conditions:

a) f is a proper surjective mapping;

b) $g(f^{-1}(x)) \subset F(x)$ for any $x \in X$;

c) f is (n,G)-Vietoris mapping.

Then $F \in Dop^{n+1}_G(X,A)$.

<u>Remark</u>. We shall call such m-mappings generalized (n,G)-acyclic.

<u>Proof</u>. Holds the following commutative diagram:

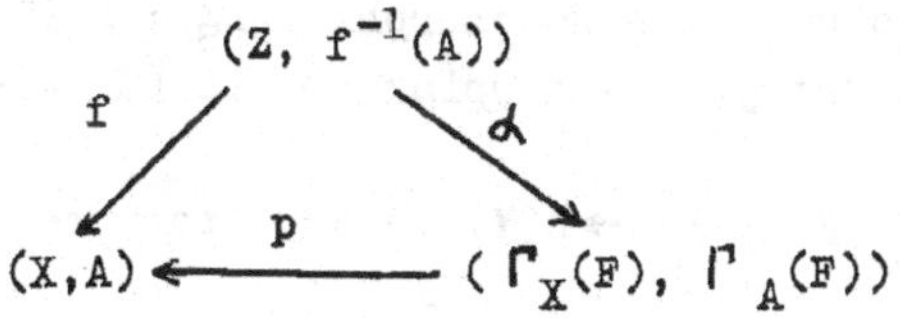

where the mapping α is defined by the condition $\alpha(z) = (f(z),g(z))$.

Since the composition $p \circ \alpha = f : Z \quad X$ is (n,G)-Vietoris mapping, then the latter is (n+1,G)-splitting mapping. Then, by virtue of lemma 1.9 the mapping p is also (n+1,G)-splitting.

<u>2.5. Example</u>. Let X,Y be compact metric spaces and $F : X \longrightarrow K(Y)$ upper semicontinuous m-mapping. We shall say that the continuous map-

ping $f : X \longrightarrow Y$ is an ε-approximation of F if $\Gamma_X(f) \subset U_\varepsilon(\Gamma_X(F))$, where $\Gamma_X(f)$ is a graph of the mapping f and $U_\varepsilon(\ {}_X(F))$ is an ε - neighbourhood of the set $\Gamma_X(F)$ in the space $X \times Y$. Let $Z_\varepsilon = U_\varepsilon(\Gamma_X(F))$, then we define the continuous mapping $p_\varepsilon : Z_\varepsilon \longrightarrow X$, where p_ε is a natural projection of Z_ε on X. Let A be a closed subset in X, $B_\varepsilon = p_\varepsilon^{-1}(A)$.

Theorem. Let there exist a monotonically decreasing sequence of positive numbers $\{\varepsilon_n\}$, $\lim \varepsilon_n = 0$, such that

a) for any ε_n there exists a continuous ε_n-approximation f_n of m-mapping F;

b) for any ε_n and ε_m, $n \leq m$, the mappings

$$\hat{f}_n,\ \hat{f}_m : (X,A) \longrightarrow (Z_{\varepsilon_m}, B_{\varepsilon_m}),\quad \hat{f}_j(x) = (x, f_j(x)),\ j = n,m,$$

are homotopic as the mappings of the pairs of spaces.

Then $F \in \mathrm{Dop}^n_G(X,A)$ for any group G and any n.

Proof. The validity of this theorem follows from the properties of upper semicontinuous m-mappings (see, for example, [4]) and from the example 1.8.

2.6. Example. Let B be a closed unit circle in the complex plane $\mathbb{C}$, $S = \partial B$. Let $F : B \to \mathbb{C}$ be an m-mapping defined by the condition $F(z) = \sqrt[n]{z}$.

Then $F \in \mathrm{Dop}^2_Q(B,S)$, where Q is the field of rational numbers.

Proof. Let us show that $p : \Gamma_B(F) \longrightarrow B$ satisfies the conditions of lemma 1.12 if we consider the field Q of rational numbers as the field of coefficients. Indeed, since the set $\Gamma_B(F)$ is contractible in itself to the point, then the inclusion $i : p^{-1}(S) \longrightarrow \Gamma_B(F)$ induces a zero homomorphism in dimension 1, i.e. $i^* : H^1(\Gamma_B(F),Q) \longrightarrow H^1(\Gamma_S(F),Q)$ is zero. Further, it is evident that the set $\Gamma_S(F)$ is homeomorphic to circle S, the homomorphism $P_S^* : H^1(S,Q) \to H^1(\Gamma_S(F),Q)$ being non-zero. Consequently, the mapping p is $(2,Q)$-splitting over (B,S).

2.7. Theorem. Let $F_1, F_2 : X \longrightarrow K(Y)$ be arbitrary m-mappings satisfying the conditions:

a) $F_1(x) \subset F_2(x)$ for any $x \in X$;

b) $F_1 \in \mathrm{Dop}^n_G(X,A)$;

then F belongs to $\mathrm{Dop}^n_G(X,A)$.

Proof. By virtue of the theorem's conditions $(\Gamma_X(F_1), \Gamma_A(F_1)) \subset (\Gamma_X(F_2), \Gamma_A(F_2))$ and holds the following commutative diagram:

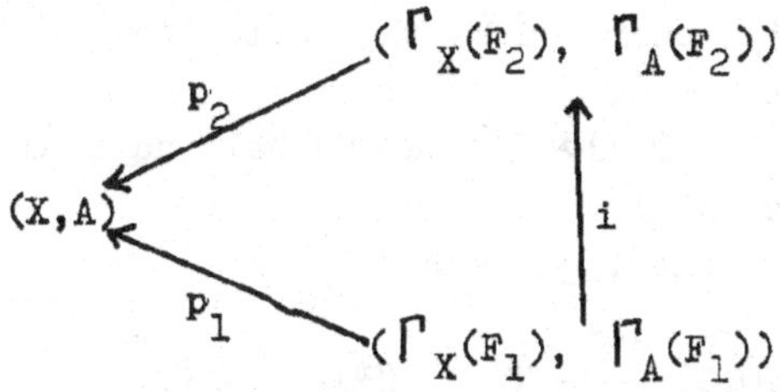

where i is the inclusion's mapping, p_1 and p_2 are projections from the respective graphs. Since p_1 is an (n,G)-splitting mapping, then by virtue of corollary 1.10 the mapping p_2 is also (n,G)-splitting.

<u>2.8. Theorem</u>. Let m-mapping $F : X \longrightarrow K(Y)$ be given by quintuple (X,Y,Z,f,g). If f is (n,G)-splitting mapping over (X,A), then $F \in \mathrm{Dop}^n_G(X,A)$.

<u>Proof</u>. Consider the commutative diagram:

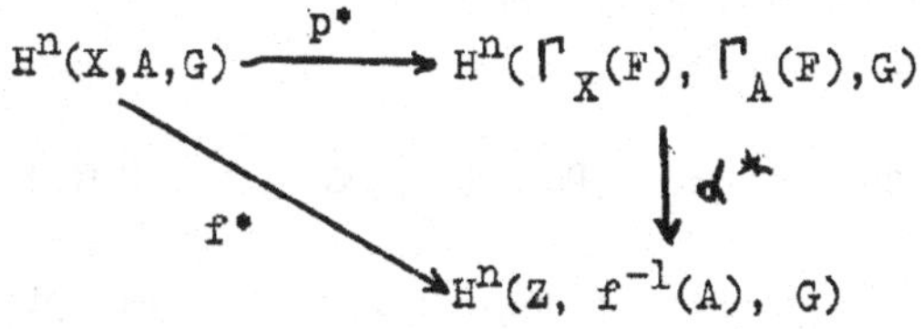

where α^* is a homomorphism generated by the continuous mapping $\alpha(z) = (f(z),\ g(z))$. Since f^* is a splitting homomorphism, then p^* is also splitting.

3. Definition of generalized degree of multi-valued mapping.

Let B be a closed unit disk in a finite-dimensional space $\mathbb{R}^n$, $S = \partial B$. Let $\varphi : B \longrightarrow K(\mathbb{R}^n)$ be a certain m-mapping, satisfying the conditions:

a) $\varphi(x) \not\ni 0$ for any $x \in S$;

b) $\varphi \in \mathrm{Dop}^n_G(B,S)$, where G is one of the following groups: Z is a group of integers; Z_p is a field of residue classes modulo p (p is a prime number); Q is a field of rational numbers.

Denote by $\delta(\varphi,G)$ a set of all homomorphisms which are left-inverse to the homomorphism $P^* : H^n(B,S,G) \longrightarrow H^n(\Gamma_B(\varphi),\ \Gamma_S(\varphi),G)$. Choose in $H^n(\mathbb{R}^n,\ \mathbb{R}^n \setminus 0,G)$ and $H^n(B,S,G)$ the compatible orientations, i.e. such generators $z_1 \in H^n(\mathbb{R}^n,\ \mathbb{R}^n\ 0,G)$, $z_2 \in H^n(B,S,G)$ that the homomorphism i^*, generated by the inclusion $i : (B,S) \to (\mathbb{R}^n, \mathbb{R}^n \setminus 0)$, transfers z_1 into z_2.

Let $q : (\Gamma_B(\varphi), \Gamma_S(\varphi)) \to (\mathbb{R}^n, \mathbb{R}^n \setminus 0)$ be a natural projection of the graph $\Gamma_B(\varphi)$ in $\mathbb{R}^n$.

Consider the following diagram:

$$\begin{array}{ccccc} H^n(\mathbb{R}^n, \mathbb{R}^n \setminus 0, G) & \xrightarrow{q^*} & H^n(\Gamma_B(\varphi), \Gamma_S(\varphi), G) & & \\ & & p^* \nearrow \quad \searrow \delta & & \\ & & H^n(B,S,G) = H^n(B,S,G) & & \end{array}$$

where $\delta \in \mathfrak{S}(\varphi, G)$.

Then $\delta \circ q^*(z_1) = k \cdot z_2$, where $k \in G$.

3.1. Definition. The number $k \in G$ is called a degree of the admissible m-mapping φ over the disk B with respect to the homomorphism δ and is denoted by $\deg_G(\varphi, B, \delta)$.

3.2. Definition. The set

$$\mathrm{Deg}\,(\varphi, B, G) = \{ \deg_G(\varphi, B, \delta) \mid \delta \in \mathfrak{S}(\varphi, G) \}$$

is called a generalized degree of the admissible m-mapping over the disk B.

Let us consider some properties of a generalized degree.

3.3. Proposition. Let $\varphi \in \mathrm{Dop}^n_G(B,S)$ and $\varphi(x) \not\ni 0$ for any $x \in B$, then $\mathrm{Deg}\,(\varphi, B, G) = \{0\}$.

Proof. Since $\varphi(x) \not\ni 0$ for any $x \in B$, then projection q can be represented in the following form:

$$q : (\Gamma_B(\varphi), \Gamma_S(\varphi)) \xrightarrow{\hat{q}} (\mathbb{R}^n \setminus 0, \mathbb{R}^n \setminus 0) \xrightarrow{j} (\mathbb{R}^n, \mathbb{R}^n \setminus 0)$$

where $\hat{q} : (\Gamma_B(\varphi), \Gamma_S(\varphi)) \to (\mathbb{R}^n \setminus 0, \mathbb{R}^n \setminus 0)$ is a projection from the graph and j is an inclusion. Then, since $H^n(\mathbb{R}^n \setminus 0, \mathbb{R}^n \setminus 0, G) = 0$, $q^* = \hat{q}^* \circ j^* = 0$. Consequently, $\delta \circ q^*(z_1) = 0$ for any homomorphism $\delta \in \mathfrak{S}(\varphi, G)$.

3.4. Corollary. If $\mathrm{Deg}\,(\varphi, B, G) \neq 0$, then there exists a point $x_0 \in \mathrm{Int}\, B$ such that $\varphi(x_0) \ni 0$.

3.5. Proposition. Let $\varphi_1, \varphi_2 : B \to K(\mathbb{R}^n)$ be m-mappings satisfying the following conditions:

a) $\varphi_1(x) \subset \varphi_2(x)$ for any $x \in B$;

b) $\varphi_2(x) \not\ni 0$ for any $x \in S$;

c) $\Phi_1 \in \mathrm{Dop}_G^n(B,S)$;
then $\Phi_2 \in \mathrm{Dop}_G^n(B,S)$, for Φ_1 and Φ_2 the generalized degrees are defined and $\mathrm{Deg}(\Phi_1, B, G) \subset \mathrm{Deg}(\Phi_2, B, S)$.

<u>Proof</u>. The admissibility of m-mapping Φ_2 follows from theorem 2.7. By virtue of the fact that Φ_2 does not contain singular points on S, Φ_1 does not contain them either. Consequently, $\mathrm{Deg}(\Phi_1, B, G)$ and $\mathrm{Deg}(\Phi_2, B, G)$ are defined. In order to clarify the connection between them let us consider the following diagram:

$$\begin{array}{ccccc} & & H^n(\Gamma_B(\Phi_1), \Gamma_S(\Phi_1), G) & & \\ & \overset{q_1^*}{\nearrow} & \uparrow i^* & \overset{p_1^*}{\nwarrow} & \\ H^n(\mathbb{R}^n, \mathbb{R}^n \setminus 0, G) & & & & H^n(B,S,G) \\ & \underset{q_2^*}{\searrow} & & \underset{p_2^*}{\swarrow} & \\ & & H^n(\Gamma_B(\Phi_2), \Gamma_S(\Phi_2), G) & & \end{array}$$

where q_1^*, q_2^*, p_1^*, p_2^* are homomorphisms generated by the corresponding projections from the graphs and i^* is a homomorphism generated by the inclusion. Let $\delta_1 \in \mathfrak{S}(\Phi_1, G)$, then by virtue of the diagram's commutativity $\delta_2 = \delta_1 \circ i^* \in \mathfrak{S}(\Phi_2, G)$.

Let $k = \deg_G(\Phi_1, B, \delta_1)$, then $\delta_1 \circ q_1^*(z_1) = k \cdot z_2$, $\delta_1 \circ i^* \circ q_2^*(z_1) = k \cdot z_2$, whence $\delta_2 \circ q_2^*(z_1) = k \cdot z_2$, i.e. $k = \deg_G(\Phi_2, B, \delta_2)$. Consequently, $\mathrm{Deg}(\Phi_1, B, G) \subset \mathrm{Deg}(\Phi_2, B, G)$.

<u>3.6. Proposition</u>. Let $\Phi : B \to K(\mathbb{R}^n) - (n-1, Z)$ be an acyclic m-mapping, $\Phi(x) \not\ni 0$ for any $x \in S$. Then $\Phi \in \mathrm{Dop}_Z^n(B,S)$ and $\mathrm{Deg}(\Phi, B, Z) = \{k\}$, where k is a degree of m-mapping Φ (see, for example, [1]).

<u>Proof</u>. Consider the following commutative diagram:

$$\begin{array}{ccccc} H^n(\mathbb{R}^n, \mathbb{R}^n \setminus 0, Z) & \xrightarrow{q^*} & H^n(\Gamma_B(\Phi), \Gamma_S(\Phi), Z) & \xleftarrow{p^*} & H^n(B,S,Z) \\ \uparrow \delta_1 & & \uparrow \delta_2 & & \uparrow \delta_3 \\ H^{n-1}(\mathbb{R}^n \setminus 0, Z) & \xrightarrow{q_1^*} & H^{n-1}(\Gamma_S(\Phi), Z) & \xleftarrow{p_1^*} & H^n(S, Z) \end{array}$$

where vertical homomorphisms are connecting homomorphisms from exact sequencies of the pairs (they are isomorphisms) and q_1^* and p_1^* are generated by the restrictions of the mappings q and p on the corresponding sets. Let $z_1 \in H^n(\mathbb{R}^n, \mathbb{R}^n\ 0, Z)$, $z_2 \in H^n(B,S,Z)$ be generators giving the compatible orientations. Then $z_1' = \delta_1^{-1}(z_1)$,

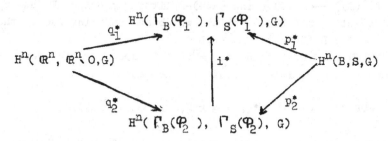

$z_2' = \delta_3^{-1}(z_2)$ give the compatible orientations in $H^{n-1}(\mathbb{R}^n \setminus 0, Z)$ and $H^{n-1}(S,Z)$, respectively. By virtue of the fact that P^* is an isomorphism, we have $(P^*)^{-1} \circ q^*(z_1) = k \cdot z_2$, i.e. $\mathrm{Deg}(\varphi,B,Z) = \{k\}$. Then by virtue of the diagram's commutativity $\delta_3 \circ (P_1^*)^{-1} \circ q_1^* (\delta_1)^{-1}(z_1) = kz_2$, or $(P_1^*)^{-1} q_1^*(z_1') = kz_2'$, i.e. $k = \deg(\varphi, B)$. This proves the statement.

Let $\varphi \in \mathrm{Dop}_G^n(B,S)$ have no singular points on S and let there exist representation φ in the form of a quintuple $\mu = (B, R^n, Z, f, g)$, $f : (Z, L = f^{-1}(S)) \longrightarrow (B,S)$ being (n,G)-splitting mapping. Denote by $\mathcal{S}_\mu(\varphi, G)$ a set of all homomorphisms which are left inverse to the homomorphism $f^* : H^n(B,S,G) \longrightarrow H^n(Z,L,G)$.

If $z_1 \in H^n(\mathbb{R}^n, \mathbb{R}^n \setminus 0, G)$ and $z_2 \in H^n(B,S,G)$ are compatible orientations, then it is possible to consider the set

$$\mathrm{Deg}_\mu(\varphi, B, G) = \{k \mid \delta \circ g^*(z_1) = kz_2,\ \delta \in \mathcal{S}_\mu(\varphi, G)\}.$$

<u>3.7. Theorem</u>. Under the above assumptions the set $\mathrm{Deg}_\mu(\varphi,B,G) \subset \mathrm{Deg}(\varphi,B,G)$.

<u>Proof</u>. Consider the following commutative diagram:

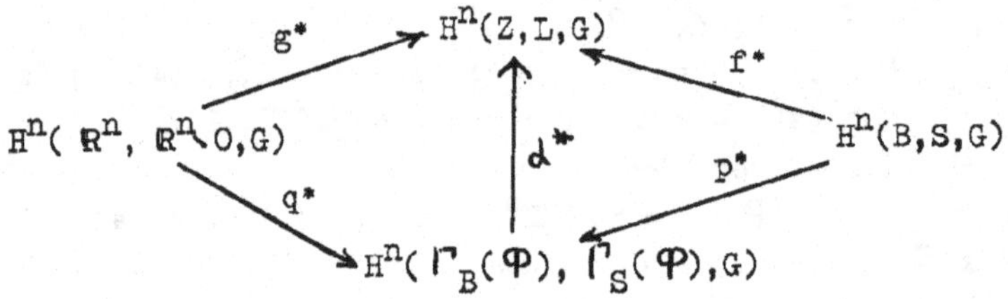

where α^* is a homomorphism generated by the mapping

$$\alpha = (f,g) : (Z,L) \longrightarrow (\Gamma_B(\varphi), \Gamma_S(\varphi)).$$

Let $k \in \mathrm{Deg}_\mu(\varphi,B,G)$, then there exists $\delta \in \mathcal{S}_\mu(\varphi,G)$ such that $\delta \circ g^*(z_1) = k\, z_2$. Then $\delta \circ \alpha^* \circ q^*(z_1) = k\, z_2$, since the homomorphism $\delta \circ \alpha^*$ is left inverse to the homomorphism p^*, i.e. $\delta_1 = \delta \circ \alpha^* \in \mathcal{S}(\varphi,G)$, then $k \in \mathrm{Deg}(\varphi,B,G)$.

<u>3.8. Corollary</u>. If m-mapping $\varphi : B \rightarrow K(\mathbb{R}^n)$ is generalized $(n-1,Z)$-acyclic and has no singular points on S, then $\varphi \in \mathrm{Dop}_Z^n(B,S)$ and $\mathrm{Deg}(\varphi,B,Z) \supset \mathrm{Deg}(\varphi)$, where $\mathrm{Deg}(\varphi)$ is a degree of a generalized $(n-1,Z)$-acyclic m-mapping.

The proof of this statement follows from the definition of $\mathrm{Deg}(\varphi)$ (see, for example, [1]), theorem 3.7 and proposition 3.5.

3.9. Theorem. Let $\Phi_1 \in \mathrm{Dop}^n_G(B,S)$ and let there exist an upper semicontinuous m-mapping $\Phi_2 : S \to K(\mathbb{R}^n\setminus 0)$ satisfying the following conditions:

1) $\Phi_1(x) \subset \Phi_2(x)$ for any $x \in S$;

2) the set $\Phi_2(x)$ is G-acyclic for any $x \in S$. Then $\mathrm{Deg}(\Phi_1,B,G) = \{k\}$, where k is some number.

Proof. Denote by p_2, q_2 projections of the graph $\Gamma_S(\Phi_2)$ in S and $\mathbb{R}^n\setminus 0$, respectively. Then there takes place the following commutative diagram:

$$\begin{array}{ccccc}
 & \overset{q_2^*}{\nearrow} & H^{n-1}(\Gamma_S(\Phi_2),G) & \overset{p_2^*}{\nwarrow} & \\
H^{n-1}(\mathbb{R}^n\setminus 0,G) & \xrightarrow{\hat{q}^*} & H^{n-1}(\Gamma_S(\Phi_1),G) & \xleftarrow{\hat{p}^*} & H^{n-1}(S,G) \\
\downarrow \delta_1 & & \downarrow \delta & & \downarrow \delta_3 \\
H^n(\mathbb{R}^n,\mathbb{R}^n\setminus 0,G) & \xrightarrow{q^*} & H^n(\Gamma_B(\Phi_1),\Gamma_S(\Phi_1),G) & \xleftarrow{p^*} & H^n(B,S,G)
\end{array}$$

Here $\delta, \delta_1, \delta_3$ are connecting homomorphisms of exact sequences of the pairs, i^* is a homomorphism induced by the inclusion, $p^*, q^*, \hat{p}^*, \hat{q}^*$ are projections from the corresponding graphs.

Let $z_1 \in H^n(\mathbb{R}^n, \mathbb{R}^n\setminus 0, G)$, $z_2 \in H^n(B,S,G)$ be generators giving the compatible orientations. Then for any $\varphi \in \mathfrak{S}(\Phi_1,G)$ we obtain:

$\varphi \circ q^*(z_1) = k \cdot z_2$, but $q^* = \delta \circ i^* \circ q_2^* \circ (\delta_1)^{-1}$. Then

$$\varphi \circ \delta \circ i^* \circ q_2^* \circ (\delta_1)^{-1}(z_1) = \delta_3 \circ (p_2^*)^{-1} \circ q_2^* \circ (\delta_1)^{-1}(z_1) = k \cdot z_2 .$$

Consequently, $(p_2^*)^{-1} \circ q_2^*(z_1') = k \cdot z_2'$, where $z_1' = (\delta_1)^{-1}(z_1)$ and $z_2' = (\delta_3)^{-1}(z_2)$. Since k does not depend on the choice of homomorphism φ, $\mathrm{Deg}(\Phi_1,B,G) = \{k\}$, where k is defined by the relation

$$(p_2^*)^{-1} \circ q_2^*(z_1') = k \cdot z_2' .$$

4. On the calculation of generalized degree.

Let, as before, B be a closed unit disk in the finite-dimensional space $\mathbb{R}^n$, $S = \partial B$. Let $\Phi_0, \Phi_1 : B \to K(\mathbb{R}^n)$ be upper semicontinuous m-mappings satisfying the conditions:

a) Φ_0 is m-mapping with G-acyclic images;

b) $\Phi_1 \in Dop^n_G(B,S)$.

4.1. Definition. We say that Φ_1 is linearly homotopic to Φ_o if m-mapping $\Psi : [0,1] \times B \longrightarrow K(\mathbb{R}^n)$, defined by the condition $\Psi(\lambda,x) = (1-\lambda)\Phi_o(x) + \lambda\Phi_1(x)$, has no singular points on $[0,1] \times S$, i.e. $\Psi(\lambda,x) \not\ni 0$ for any $\lambda \in [0,1]$, $x \in S$.

4.2. Theorem. If Φ_1 is linearly homotopic to Φ_o, then $Deg(\Phi_1,B,G) = Deg(\Phi_o,B,G) = \{k\}$.

Proof. Let us introduce the following notation: $[a,b]$ is a segment connecting points a and b, where $a,b \in \mathbb{R}^n$. Consider the set $X \subset B \times \mathbb{R}^n \times \mathbb{R}^n \times \mathbb{R}^n$ defined by the condition:

$$X = \{(x,y,z,u) \mid x \in B,\ y \in \Phi_o(x),\ z \in \Phi_1(x),\ u \in [y,z]\}.$$

Analogously the following set is defined:

$$X_1 = \{(x,y,z) \mid x \in B,\ y \in \Phi_o(x),\ z \in \Phi_1(x)\} \subset B \times \mathbb{R}^n \times \mathbb{R}^n .$$

Consider continuous mappings which are projections on the corresponding sets:

$$\alpha_1 : X \longrightarrow X_1, \quad \alpha_1(x,y,z,u) = (x,y,z);$$

$$\alpha_2 : X \longrightarrow \Gamma_B(\Phi_1), \quad \alpha_2(x,y,z) = (x,z);$$

$$\beta : X \longrightarrow \Gamma_B(\Phi_o), \quad \beta(x,y,z,u) = (x,y);$$

$$f : X \longrightarrow \mathbb{R}^n, \quad f(x,y,z,u) = u;$$

$$t : X \longrightarrow B \quad t(x,y,z,u) = x.$$

Note that α_1 is Vietoris mapping, since a pre-image of every point is a closed segment in $\mathbb{R}^n$. Analogously, α_2 is Vietoris mapping, since a pre-image of each point is homeomorphic to $\Phi_o(x)$, i.e. it is acyclic. Denote $\alpha = \alpha_2 \circ \alpha_1 : X \longrightarrow \Gamma_B(\Phi_1)$, then α is also Vietoris mapping as a composition of Vietoris mappings.

Analogously it is possible to consider the sets:

$$Y = \{(x,y,z,u) \mid x \in S,\ y \in \Phi_o(x),\ z \in \Phi_1(x),\ u \in [y,z]\},$$

$$Y_1 = \{(x,y,z) \mid x \in S,\ y \in \Phi_o(x),\ z \in \Phi_1(x)\} .$$

Obviously, $\alpha_1 : (X,Y) \longrightarrow (X_1,Y_1)$; $\alpha_2 : (X_1,Y_1) \longrightarrow$

$$(\Gamma_B(\varphi_1),\ \Gamma_S(\varphi_1));\quad \beta : (X,Y) \longrightarrow (\Gamma_B(\varphi_0),\ \Gamma_S(\varphi_0));$$

$f : (X,Y) \longrightarrow (\mathbb{R}^n, \mathbb{R}^n \setminus 0)$; $t ; (X,Y) \longrightarrow (B,S)$.

Then holds the following commutative diagram:

$$\begin{array}{ccccc}
 & & H^n(\Gamma_B(\varphi_0),\ \Gamma_S(\varphi_0),G) & & \\
 & \nearrow^{q_0^*} & \downarrow \beta^* & \nwarrow^{p_0^*} & \\
H^n(\mathbb{R}^n, \mathbb{R}^n \setminus 0,G) & \xrightarrow{f^*} & H^n(X,Y,G) & \xleftarrow{t^*} & H^n(B,S,G) \\
 & \searrow_{q_1^*} & \uparrow \alpha^* & \swarrow_{p_1^*} & \\
 & & H^n(\Gamma_B(\varphi_1),\ \Gamma_S(\varphi_1),G) & &
\end{array}$$

where p_i, q_i are projections from the corresponding graphs, $i = 0,1$. This diagram is commutative since the mappings $q_1 \circ \alpha$, $q_0 \circ \beta$, $f : (X,Y) \longrightarrow (\mathbb{R}^n, \mathbb{R}^n \setminus 0)$ are linearly homotopic.

Let $\mathrm{Deg}(\varphi_0,B,G) = \{k\}$, then $(p_0^*)^{-1} \circ q_0^*(z_1) = k \cdot z_2$, i.e. $q_0^*(z_1) = k\, p_0^*(z_2)$. By virtue of the diagram's commutativity $\alpha^* \circ q_0^*(z_1) = k \beta^* \circ p_0^*(z_2)$, or $f^*(z_1) = k \cdot t^*(z_2)$. Then $\alpha^* \circ q_1^*(z_1) = k \alpha^* \circ p_1^*(z_2)$, since α^* is an isomorphism, then $q_1^*(z_1) = k\, p_1^*(z_2)$. Consequently, for any $\sigma \in \sigma(\varphi_1,G)$ we have $\sigma \circ q_1^*(z_1) = k\, z_2$; i.e. $\mathrm{Deg}(\varphi_1,B,G) = \{k\}$.

Let us consider some corollaries from this theorem.

4.3. Theorem. Let m-mapping $\varphi : B \longrightarrow K(\mathbb{R}^n)$, $\varphi \in \mathrm{Dop}_G^n(B,S)$. If there exists such a vector $y_0 \in \mathbb{R}^n \setminus 0$ that for any $x \in S$, $\lambda \geq 0$ the condition

$$\lambda y_0 \cap \varphi(x) = \emptyset$$

is satisfied, then $\mathrm{Deg}(\varphi,B,G) = \{0\}$.

Proof. It is evident that from the theorem's conditions there follows the absence of singular points of φ on the set S, i.e. $\varphi(x) \not\ni 0$ for any $x \in S$. Consequently, the degree $\mathrm{Deg}(\varphi,B,G)$ is defined. Consider a continuous mapping $\psi(x) = -y_0$. It is obvious that φ is linearly homotopic to the mapping ψ. Then $\mathrm{Deg}(\varphi,B,G) =$

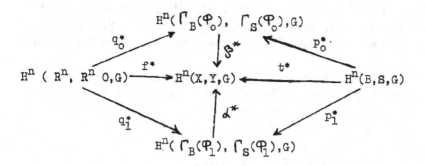

$= \mathrm{Deg}(\mathcal{S},B,G)$, but the mapping $\mathcal{S}$ does not equal zero at any point $x \in B$. Consequently, by virtue of proposition 3.3 $\mathrm{Deg}(\mathcal{S},B,G) = \{0\}$, which proves the statement.

4.4. Theorem. Let $\Phi : B \to K(\mathbb{R}^n)$ be (n,G)-admissible m-mapping without singular points on S. If m-mapping $F : B \to K(\mathbb{R}^n)$, $F(x) = x - \Phi(x)$ is such that $F(S) \subset B$, then $\mathrm{Deg}(\Phi,B,G) = \{1\}$.

Proof. Consider a mapping $\mathcal{S} : B \to \mathbb{R}^n$, $\mathcal{S}(x) = x$. Let us show that Φ is linearly homotopic to this mapping. Let $\Psi(\lambda,x) = (1-\lambda)x + \lambda\Phi(x) = x - \lambda F(x)$. For $\lambda = 1$, $\Psi(1,x) = \Phi(x)$ has no singular points on S. If $0 \leq \lambda < 1$, then $\|x\| > \lambda\|y\|$ for any $x \in S$, $y \in F(x)$, i.e. $\Psi(\lambda,x) \not\ni 0$ for $x \in S$, $\lambda \in [0,1]$. Then $\mathrm{Deg}(\Phi,B,G) = \mathrm{Deg}(\mathcal{S},B,G) = \{1\}$.

Let us consider some analogues of the theorem of odd field.

4.5. Theorem. Let m-mapping $\Phi_1 : B \to K(\mathbb{R}^n)$ be (n,Z)-admissible m-mapping. If there exists an upper semicontinuous m-mapping $\Phi_2 : S \to K(\mathbb{R}^n \setminus 0)$, satisfying the conditions:

a) $\Phi_1(x) \subset \Phi_2(x)$ for any $x \in S$;

b) $\Phi_2(x)$ is Z-acyclic set for any $x \in S$;

c) $\lambda\Phi_2(x) \cap \Phi_2(-x) = \emptyset$ for any $x \in S$ and $\lambda \geq 0$.

Then $\mathrm{Deg}(\Phi_1,B,Z) = \{k\}$, where $k \neq 0$.

Proof. By virtue of theorem 3.9 $\mathrm{Deg}(\Phi_1,B,Z) = \{k\}$, where k is defined by the relation $(p_2^*)^{-1} \circ q_2^*(z_1') = k \cdot z_2'$, where z_1', z_2' are generators giving the compatible orientations. It is known (see, for example, [5]) that $k \neq 0$. This proves the theorem.

4.6. Corollary. Let an upper semicontinuous m-mapping $\Phi_1 : B \to K(\mathbb{R}^n)$ satisfy the following conditions:

a) $\Phi_1 \in \mathrm{Dop}_Z^n(B,S)$;

b) for every point $x \in S$ there exists a linear functional $\mathcal{S}_x : \mathbb{R}^n \to \mathbb{R}^1$, strictly dividing the sets $\Phi_1(x)$ and $\Phi_1(-x)$.

Then $\mathrm{Deg}(\Phi_1,B,Z) = \{k\}$, where $k \neq 0$.

Proof. Consider an upper semicontinuous m-mapping $\Phi_2 : S \to K(\mathbb{R}^n \setminus 0)$ defined by the condition $\Phi_2(x) = \overline{co}\,\Phi_1(x)$. It is easy to check that Φ_2 satisfies the conditions of theorem 4.5, which proves the statement.

4.7. Theorem. Let m-mapping $\Phi_1 \in \mathrm{Dop}_Z^n(B,S)$ and let the following conditions be satisfied:

a) $\Phi_1(x) \not\ni 0$ for any $x \in S$;

b) there exists an upper semicontinuous m-mapping $\Phi_2 : B \to K(\mathbb{R}^n)$ such that $\Phi_2(x) \subset \Phi_1(x)$ for any $x \in B$;

c) the set $\Phi_2(x)$ is Z-acyclic for any $x \in B$;

d) $\Phi_2(x) \cap \lambda\Phi_2(-x) = \emptyset$ for any $x \in S$ and $\lambda \geq 0$, then $\mathrm{Deg}\,(\varphi_1,B,Z)\ni k$, where $k \neq 0$.

Proof. By virtue of proposition 3.5 and 3.6 we have an inclusion $\mathrm{Deg}\,(\varphi_1,B,Z)\supset \mathrm{Deg}(\varphi_2,B,Z) = \{k\}$. Since $k \neq 0$ (see the proof of theorem 4.6), $\mathrm{Deg}\,(\varphi_1, B, Z) \ni k \neq 0$.

5. Theorem on product of degrees.

Let $\varphi : B \longrightarrow K(\mathbb{R}^n)$ be (n,G)-admissible m-mapping without singular points on S and let $g : (\mathbb{R}^n,\mathbb{R}^n\setminus 0) \longrightarrow (\mathbb{R}^n, \mathbb{R}^n\setminus 0)$ be a continuous mapping. Then m-mapping $\Psi = g\circ\varphi$,given by the condition

$$\Psi(x) = \bigcup_{y\in\varphi(x)} g(y)$$

is defined.

5.1. Lemma. M-mapping Ψ is (n,G)-admissible m-mapping without singular points on S.

Proof. The fact that Ψ has no singular points on S is obvious. Let us show its (n,G)-admissibility. Since φ is defined by the quintuple $(B, \mathbb{R}^n, \Gamma_B(\varphi),p,q)$, the quintuple $(B, \mathbb{R}^n, \Gamma_B(\varphi),p,g\circ q)$ defines m-mapping Ψ . Since P is (n,G)-splitting mapping over (B,S), by virtue of theorem 2.8 m-mapping Ψ is (n,G)-admissible.

Let $z_1\in H^n(\mathbb{R}^n, \mathbb{R}^n\setminus 0,G)$ be a generating element, then for $g^* : H^n(\mathbb{R}^n, \mathbb{R}^n\setminus 0, G) \longrightarrow H^n(\mathbb{R}^n, \mathbb{R}^n\setminus 0,G)$ the degree $\deg(g)$, $g^*(z_1) = \deg(g)\cdot z_1$ is defined.

5.2. Theorem. Under the above assumptions

$$\mathrm{Deg}(\Psi,B,G) \supset \deg\,(g)\cdot \mathrm{Deg}(\varphi,B,G).$$

Proof. Consider a commutative diagram

$$\begin{array}{ccccc}
(\mathbb{R}^n, \mathbb{R}^n\setminus 0) & \xleftarrow{\ g\ } & (\mathbb{R}^n, \mathbb{R}^n\setminus 0) & \xleftarrow{\ q_\varphi\ } & (\Gamma_B(\varphi), \Gamma_S(\varphi)) \\
 & & & & \downarrow p_\varphi \\
 & & (\Gamma_B(\Psi), \Gamma_S(\Psi)) & \xrightarrow{\ p_\Psi\ } & (B,S)
\end{array}$$

(with $q_\Psi : (\Gamma_B(\Psi), \Gamma_S(\Psi)) \to (\mathbb{R}^n, \mathbb{R}^n\setminus 0)$ and $\alpha : (\Gamma_B(\varphi), \Gamma_S(\varphi)) \to (\Gamma_B(\Psi), \Gamma_S(\Psi))$)

where p_φ, q_φ, p_Ψ, q_Ψ are projections from the corresponding graphs and $\alpha\,(x,y) = (x,g(y))$.

In cohomologies we obtain the following diagram:

$$
\begin{array}{ccccc}
H^n(\mathbb{R}^n, \mathbb{R}^n \setminus 0, G) & \xrightarrow{g^*} & H^n(\mathbb{R}^n, \mathbb{R}^n \setminus 0, G) & \xrightarrow{q^*_\varphi} & H^n(\Gamma_B(\varphi), \Gamma_S(\varphi), G) \\
 & \searrow^{q^*_\psi} & & \nearrow^{\alpha^*} & \uparrow p^*_\varphi \\
 & & H^n(\Gamma_B(\psi), \Gamma_S(\psi), G) & \xleftarrow{p^*_\psi} & H^n(B,S,G)
\end{array}
$$

Let $\sigma \in \mathcal{G}(\varphi, G)$, i.e. $\sigma : H^n(\Gamma_B(\varphi), \Gamma_S(\varphi), G) \longrightarrow H^n(B,S,G)$ and it is left inverse to P^*_φ. Then $\sigma \circ \alpha^*$ is left inverse to the homomorphism P^*_ψ. Let $\gamma \in \mathrm{Deg}(\varphi, B, G)$, where γ is defined by the relation $\sigma \circ q^* \ (z_1) = \gamma \cdot z_2$. Then $\sigma \circ q^*_\varphi \circ g(z_1) = \deg(g) \cdot \sigma \circ q^*_\varphi (z_1) =$
$= \deg(g) \cdot \gamma \cdot z_2$.

By virtue of the diagram's commutativity

$$\sigma \circ \alpha^* \cdot q^*_\psi (z_1) = \sigma \circ q^*_\varphi \circ g(z_1) = \deg(g) \cdot \gamma \cdot z_2$$

Then $\deg(g) \cdot \gamma \in \mathrm{Deg}(\psi, B, G)$, Q.E.D.

As a corollary from this theorem we can calculate the degree of m-mapping in example 2.6.

Let B be a closed unit space in a complex plane $\mathbb{C}$, $\varphi : B \to \mathbb{C}$ be defined by the condition $\varphi(z) = \sqrt[n]{z}$, where n is integer, $n > 0$.

<u>5.3. Theorem</u>. $\mathrm{Deg}(\varphi, B, Q) = \{\frac{1}{n}\}$.

<u>Proof</u>. Consider $g : (\mathbb{C}, \mathbb{C} \setminus 0) \longrightarrow (\mathbb{C}, \mathbb{C} \setminus 0)$, defined by the condition $g(z) = z^n$. It is known that $\deg(g) = n$. Then m-mapping $\psi(z) = g \circ F(z) =$
$= z$. Consequently, $\mathrm{Deg}(\psi, B, Q) = 1$. On the other hand, by virtue of the proved theorem $\mathrm{Deg}(\psi, B, Q) \supset n \cdot \mathrm{Deg}(\varphi, B, Q)$. Consequently, $\mathrm{Deg}(\varphi, B, Q) = \frac{1}{n}$.

6. Rotation of multi-valued vector field and fixed points.

Let $F : B \to K(\mathbb{R}^n)$ be some m-mapping. M-mapping $\varphi = i - F : B \to K(\mathbb{R}^n)$, defined by the relation

$$\varphi(x) = x - F(x)$$

is called a multi-valued vector field (mv-field), generated by F.

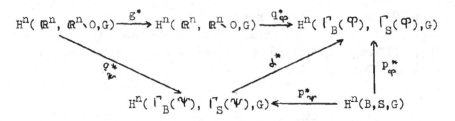

Let $(B, \mathbb{R}^n, \Gamma_B(F), r, s)$ be a canonical representation of m-mapping F, then the quituple of objects $(B, \mathbb{R}^n, \Gamma_B(F), r, r-s)$ is a representation of mv-field φ . Let $(B, \mathbb{R}^n, \Gamma_B(\varphi), p, q)$ be a canonical representation of φ, then holds the following statement.

6.1. Lemma. The quituples $(B, \mathbb{R}^n, \Gamma_B(F), r, r-s)$ and $(B, \mathbb{R}^n, \Gamma_B(\varphi), p, q)$ are homeomorphic, i.e. there exists a homeomorphism $\alpha : \Gamma_B(F) \longrightarrow \Gamma_B(\varphi)$ such that the diagram

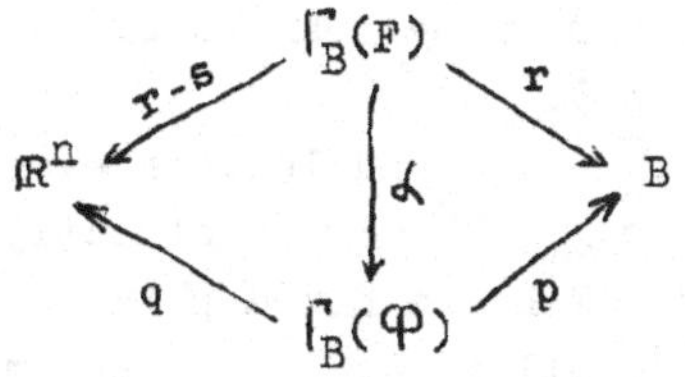

is commutative.

Proof. Mapping α is defined by the relation $\alpha(x,y) = (x, x-y)$.

Let m-mapping F have no fixed points on $S = \partial B$, i.e. $F(x) \not\ni x$ for any $x \in S$. Then mv-field φ has no singular points on S and in cohomologies we have the following commutative diagram:

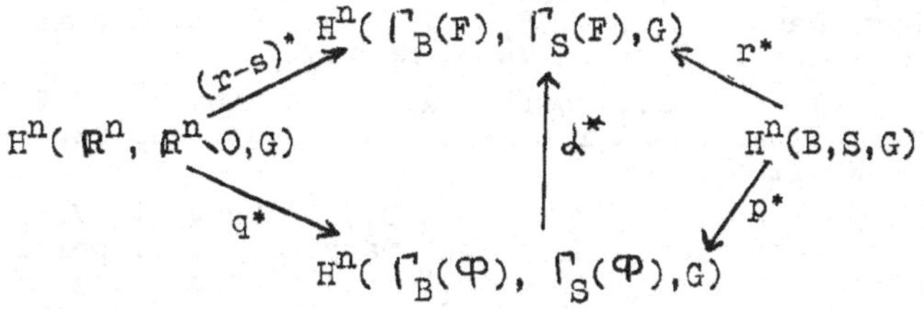

where α^* is an isomorphism.

Denote by $\mathfrak{S}(r)$ and $\mathfrak{S}(p)$ the sets of homomorphisms which are left inverse to the homomorphisms r^* and p^* respectively. It is obvious that the isomorphism α^* establishes bijection between them.

6.2. Lemma. If $F \in \mathrm{Dop}^n_G(B,S)$, then $\varphi \in \mathrm{Dop}^n_G(B,S)$ and $\mathrm{Deg}(\varphi, B, S) = \{k \mid \sigma \circ (r-s)^*(z_1) = k \cdot z_2, \sigma \in \mathfrak{S}(r)\}$.

The proof of this lemma is obvious.

6.3. Definition. If $F \in \mathrm{Dop}^n_G(B,S)$ and F has no fixed points on S, then the generalized degree of φ is called a rotation of mv-field $\varphi = i - F$.

The following statement follows from the properties of the generalized degree.

6.4. Theorem. If $\mathrm{Deg}(i-F, B, G) \neq 0$, then F has a fixed point in Int B.

Thus, if the rotation of mv-field φ is known, it is possible to prove fixed points theorems. Let us consider some statements.

6.5. Theorem. Let $F \in \mathrm{Dop}_G^n(B,S)$ and $F(S) \subset B$, then m-mapping F has a fixed point in the disk B.

Proof. If F has a fixed point on S, then the theorem is proved. Otherwise (see theorem 4.4) $\mathrm{Deg}(\varphi, B, G) = \{1\}$. Consequently, F has a fixed point in Int B.

Realizing this theorem for different classes of (n,G)-admissible m-mappings, we obtain a number of well-known fixed points theorems (see, for example, [1.5,6,7]).

6.6. Theorem. Let $F \in \mathrm{Dop}_Z^n(B,S)$ and let for every point $x \in S$ there exists a linear functional $\varphi_x : \mathbb{R}^n \to \mathbb{R}^1$, strictly dividing the sets $x - F(x)$ and $-x-F(-x)$. Then F has a fixed point.

The proof follows from corollary 4.6 and theorem 6.4.

It is easy to obtain some more fixed points theorems.

References

1. Borisovich Yu.G., Gel'man B.D., Myshkis A.D.,Obukhovskii V.V. Topological methods in the theory of fixed points of multi-valued mappings. Uspekhi Mat. Nauk, 1980, v.35, No.1, 59-126 (in Russian ; see English translation in Russian Math. Surveys).
2. Borisovich Yu.G., Gel'man B.D., Myshkis A.D., Obukhovskii V.V. Multi-valued mappings. Itogi nauki i tekhn. VINITI Mat. analiz, 1982, V.19, 151-211 (in Russian).
3. Borisovich Yu.G., Gel'man B.D., Myshkis A.D., Obukhovskii V.V. On new results in the theory of multi-valued mappings. I. Topological characteristics and solvability of operator relations. Itogi nauki i tekhn. VINITI. Mat. analiz, 1987, V.25, 121-195 (in Russian).
4. Borisovich Yu.G., Gel'man B.D., Myshkis A.D., Obukhovskii V.V. Introduction to the theory of multi-valued mappings. Voronezh University Press, 1986, 104 p. (in Russian).
5. Górniewicz L., Homological methods in fixed-point theory of multivalued maps. Rozpr. mat., 1976, No.129, 66 pp.
6. Dzedzej Z., Fixed point index theory for a class of nonacyclic multivalued maps. Rozpr. mat., 1985, No.253, 58 pp.
7. Siegberg H.W., Skordev G., Fixed point index and chain approximations. Pacif.J.Math., 1982, V.102, No.2, pp.455-486.
8. Sklyarenko E.G. On some applications of the theory of sheaves in general topology. Uspekhi Mat. Nauk, 1964, V.19, No.6, 47-70 (in Russian; see English translation in Russian Math. Surveys).
9. Dold A. Lectures on algebraic topology. Springer, 1972.
10. Spanier E.H. Algebraic topology. McGraw-Hill, 1966.

ON FREDHOLMIAN ASPECTS OF LINEAR TRANSMISSION PROBLEMS

G.N.Khimshiashvili
A.Razmadze Mathematical Institute
Georgian Academy of Sciences
Z.Rukhadze street,1
380093, Tbilissi, Georgia, USSR

The paper describes Fredholmian properties of certain group-theoretical generalizations of the classical linear transmission problem (LTP) [1] (also known as the Riemann boundary value problem or the Riemann-Hilbert problem) introduced by the author in [2], [3]. Some considerations are extended for the first time beyond the framework of the Riemann sphere and compact groups of coefficients, which leads to new problems and perspectives.

The main attention is given to the conditions under which the problems under consideration are of the Fredholm type and to the connections with the global analysis, in particular with the theory of Fredholm structures [4]. In the original and basic case when the coefficients belong to a compact Lie group our approach largely rests on the geometric theory of loop groups from the book of A.Pressley and G.Segal [5], and also on B.Boyarski ideas about an abstract scheme of the LTP [6], [7] which are very close to manipulations with Fredholm Grassmanian of the polarized Hilbert space used in [5].

We start with describing the essence of our interpretation of the LTP which leads to fairly general formulations. The so-called generalized LTP (GLTP) proper which is our main concern in this paper is treated in [2]. Note that from the very beginning our considerations are restricted to the GLTP for the Riemann sphere and a compact group of coefficients because then the above-mentioned connection with the theory of Fredholm structures can be traced most clearly.

Other variants of the GLTP with a one-dimensional domain of definition are briefly discussed in the concluding sections of the paper. where the above-mentioned particular case revealing completely the meaning we assign to the term "Fredholmian aspects" is used as a guiding model. In the same section we also introduce the necessary functional spaces and linear operators, including generalized Toeplitz operators (GTO) in the spirit of [6], and present some necessary facts from the theory of compact Lie groups [8].

§3 contains the results and constructions which enable us to use the generalized Birkhoff factorization (GBF) [5] to investigate the GLTP on the Riemann sphere. The partial indices of loops and the so-called Birkhoff strata are also introduced here in complete analogy with the corresponding classical notions investigated in quite a number of works of which we attach particular importance to [7] and [9]. The connection of the GBF with the theory of holomorphic bundles and one application of partial indices within the latter theory are also described. It should be emphasized that it is due to the availability of the GBF that in the case under consideration it becomes possible to develop a meaningful theory including the analogues of all classical results [1] which is presented in the next section.

One of the key points in §4 is the proof of the Fredholm property of the family of GTO's (theorem 2) and calculation of their indices and $\mathbb{Z}_2$ -indices [10] in terms of partial indices of the coefficient loop. It should be noted that even in the case of the classical LTP, corresponding to the unitary group of coefficients, this leads to new results if we take various representations differing from the standard representation used in the LTP.

The next section is in a certain sense the central one, since it summarizes all notions discussed in the paper and gives - for the first time - a detailed description of the previously announced [11] construction of Fredholm structures on the coefficient loop group using our GLTP theory. Here the essential use is also made of arguments in the spirit of global analysis from [4] and [12]; in particular, we resort to the fundamental diagram from [4] and give an alternative way of introducing a Fredholm structure using the triviality of the tangent bundle of a Hilbert manifold [12]. With these things in mind, the main result (theorem 3) immediately follows from the above-mentioned properties of the GTO. All this, in turn, enables one to have a new look at the geometry of loop groups and Birkhoff strata, since the latter turn out to be Fredholm submanifolds with respect to the constructed structures. In particular, standard constructions from the Fredholm structures theory make it possible to define characteristic classes of a loop group [14] and cohomological fundamental classes of Birkhoff strata [9] treated in §6.

The contents of these two sections is closely connected with D.Freed's recent paper [13] in which he constructs similar structures corresponding, as may be shown, to the particular case of the adjoint representation of the classical Lie group. Next, omitting the detailed comments, we give the first results towards calculation of the constructed cohomological classes in terms of the well known Bott description of cohomologies of a loop group [14] based on the reduction to the situations from [9] and [13].

As it was already mentioned, many of the results presented are also valid for GLTPs on Riemann surfaces of non-zero genus and this is briefly discussed in § 7. The reason for the theory being incomplete and for the discussion being brief is that the corresponding geometric problems of the holomorphic bundles theory [15] are considerably more difficult here [16].

The concluding § 8 contains a brief discussion of the results and perspectives.

1. Generalized Riemann Boundary Value Problem.

A few words should be said about the terminology because the classical problem which we are interested in has many formulations, modifications, generalizations and names [1], [17]. Without going into particulars, we would like to note that for convenience of our exposition it is sufficient to use two names - LTP and Riemann boundary value problem (RBVP), assuming that in the LTP an emphasis is placed on the definition domain being one-dimensional and the conjugation condition being linear, whereas in the RBVP an emphasis is made on the complex analyticity of the situation and on the presence of two submanifolds with the common boundary. Taking all this into account, we may briefly say that we are not interested in the generalization of the LTP as such, since quite a satisfactory generalization variant under the name of abstract linear transmission problem (ALTP) was formulated by B.Boyarski [6] in terms of Fredholm pairs of subspaces, but we are interested in what may be called complex-analytically meaningful forms of the RBVP, which fits the title of the section.

To make the above clear, let us recall the formulation of the RBVP [1]. Let $\mathbb{P} = \overline{\mathbb{C}}$ be the Riemann sphere represented in the form of the union $\mathbb{P} = B_0 \cup \mathbb{T} \cup B_\infty$, where B_0 is the unit circle, $\mathbb{T}$ the unit circumference, B_∞ a complementary domain containing the point at infinity $\{\infty\}$. Let $A(z)$ be a $n \times n$ matrix-function from Hölder's or Sobolev's class. For open subsets U, V of some complex manifolds we shall denote by $A(U,V)$ a set of all continuous mappings of the closure of U in V which are holomorphic inside U. Then the RBVP with the coefficient A can be formulated as the problem of the existence and number of pairs $(X_0, X_\infty) \in A(B_0, \mathbb{C}^n) \times A(B_\infty, \mathbb{C}^n)$ with the normalization $X_\infty(\infty) = 0$, satisfying on $\mathbb{T}$ the Riemann condition or the linear transmission condition

$$X_0(z) = A(z)X_\infty(z) \ , \quad \forall z \in \mathbb{T}. \tag{1}$$

Because of a remarkable simplicity of condition (1) it can be generalized in every possible way. It is natural to use for this purpose representation spaces of operator algebras playing the part of the algebra of matrix functions in the RBVP formulation. This seems to be quite sensible, but since our attention will be focused on Fredholmian aspects which ordinarily imply the reversibility of the coefficient, our consideration here will be restricted directly to the group-theoretical scheme.

Namely, let us assume that we are given a complex (possibly infinite-dimensional) manifold M represented as a combination of two submanifolds $M = M_0 \cup S \cup M_\infty$ with the common boundary S and the marked point $\infty \in M_\infty$. Let, besides, we be given a complex action r (not necessarily linear) of some Lie group G on the manifold V which in what follows will be assumed for the sake of simplicity to be a vector space. Then under the generalized Riemann boundary value problem we shall understand the problem of the existence and number of pairs $(X_0, X_\infty) \in A(M_0, V) \times A(M_\infty, V)$ with the norming $X_\infty(\infty) = 0$, satisfying on S the conjugation condition

$$X_0(z) = r(f(z))X_\infty(z). \tag{2}$$

From [2] and [18] it follows that such a generality is not excessive as it might seem at first glance Indeed, in the case when r is the action of the Lie group on its homogeneous space there is a deep connection with the differential geometry and physics [18]. Moreover, sometimes it is sensible to give up holomorphic mappings, replacing them by the solutions of a suitable elliptic system of the Vekua or Moisel-Theodorescu type [19], but we are not going to dwell on this here.

Indeed, in what follows it will always be assumed that M is one-dimensional, G is compact and the action r is linear.

2. Generalized Linear Transmission Problem (GLTP).

Thus let G be a compact Lie group [8] of rank d with the Lie algebra A. For convenience G will be assumed to be connected. It is well known that every group of such kind has the complexification $G_{\mathbb{C}}$ with $A_{\mathbb{C}}$ as a Lie algebra [5], this fact is very important, since it provides the existence of a complex structure on a group of marked loops which is needed for introducing GTOs to be considered further.

By LG we shall always denote a group of continuous marked loops (i.e. the loops which map the number $1 \in \mathbb{T}$ into the unit of the group G) on G. LG is assumed to be provided with pointwise multiplication of loops and the ordinary compact-open topology [5]. We shall also need some conditions for loop regularity and that is why by L_pG we shall denote Sobolev loops with the index $p \geqslant 1/2$ [5]. Indeed, we can work everywhere with Hölder loops or simply with continuosly differentiable loops. Note that all loop groups appearing in our discussion are complex Banach manifolds [5] and hence it is sensible to speak of holomorphic mappings.

Let it further be assumed that we are given some linear representation r of the group G in a complex vector space V. According to [8] it can be assumed without loss of generality that simultaneously we are also given a representation of the complexification G in the space V which will be denoted by the same letter r.

In this general description of the GLTP we omit the indication to the class of loop regularity and simply write LG. Assuming further that we are given a loop $f \in LG$, we can formulate the (homogeneous) GLTP with the coefficient f [3] denoted by P_f^r as the problem of the existence and number of pairs $(X_0, X_\infty) \in A(B_0, V) \times A(B_\infty, V)$ with the norming $X_\infty(\infty) = 0$, satisfying on T the following linear conjugation condition

$$X_0 = r(f)X_\infty . \qquad (3)$$

Note that it makes sense to speak of boundary values of components of the solution (X_0, X_∞), since, quite similarly to the classical LTP[1], they are sought for in Bergman-type spaces selected for this purpose. Next, for each loop h in V we can introduce a similar problem $P_{f,h}^r$ (nonhomogeneous, with the right-hand side h), replacing the conjugation condition (3) by

$$X_o(z) - r(f(z))X_\infty(z) = h(z), \ \forall z \in \mathbb{T}. \quad (4)$$

After this we can say that we are interested in the kernel and the co-kernel (i.e. in the solution and conditions of solvability) of the natural linear operator R_f^r given by the left side of (4) and acting from the above-mentioned space of pairs (piecewise holomorphic mappings) into a space of loops on V (usually of the Hölder or the Sobolev class). Using the exact matrix representation [8] and the classical LTP theory [17], we can show that such operators are singular integral operators with the Cauchy kernel [17], which makes it possible to construct their Fredholm theory by classical means. However, for us it is more convenient to use operators similar to the classical Toeplitz operators [1], [5].

That is why we also introduce a separable Hilbert space $\mathbb{H}=L^2(\mathbb{T},V)$ polarized by the decomposition $\mathbb{H} = H_+ + H_-$, where $H_+ = H^2(\mathbb{T},V)$ is the ordinary Hardy space [1], [5], consisting of L^2 - integrable (quadratically summable on $\mathbb{T}$) boundary values of holomorphic mappings from B_o into V, and H_- is its orthogonal complement in H. Denote by P_+ the corresponding orthogonal projection and by M_f^r the multiplication operator in the representation r given by the formula $M_f^r(g) = r(f)g$. Now the GTO with the symbol $f \in LG$ can be defined by the equality

$$T_f^r = P_+M_f^r \ : \ H_+ \longrightarrow H_+ . \quad (5)$$

Now the connection of these operators with the Fredholm Grassmanian from [5] becomes clear. Indeed, it is not difficult to verify that if the subspace W belongs to the Fredholm Grassmanian Gr($\mathbb{H}$), then the pair of subspaces (H_+, W) gives rise to an abstract Fredholm linear transmission problem in the Boyarski sense [7] and the index of this problem coincides with the virtual dimension of W. Conversely, if the operator T_f^r has Fredholm properties, then there arises a Fredholm pair (H_+, $M_f^r(H_+)$); therefore $M_f^r(H_+) \in$ Gr and again the index is equal to the virtual dimension. Thus we have three equivalent ways for describing our problem, which will be useful later on. At the same time it is clear that the Fredholm properties of the operators R_f^r and T_f^r are equivalent and hence the LTP is called the Fredholm problem if the above-mentioned operators have Fredholm properties, and the general value of the index of these operators is called the index of the GLTP. In what follows we shall use the language of GTOs, since many of their properties follow immediately from the properties of the Grassmanian

Gr($\mathbb{H}$).

Remark 1. In the particular case when $G = U(n)$ is a unitary group [8] we have $G_{\mathbb{C}} = GL(n,\mathbb{C})$ is a complete linear group. If now the standard representation s of the group $U(n)$ on $\mathbb{C}^n$ is taken as r, then eqs (3) and (1) will coincide and our GLTP will become the classical GTO. Note that the GLTP leads to new situations even for the unitary group, since representations differing from s can be taken as r.

It is understood that the same observation holds for other compact groups, too, and as a result we have quite a series of problems. But, as a matter of fact, we are interested first of all in irreducible representations of simple compact groups. For simplicity we do not consider exclusive groups from Cartan's list [8] and show this by using the word combination "classical simple groups". In other words, we shall speak only of groups of the series $U(n)$, $O(n)$ and $Sp(n)$.

3. Generalized Birkhoff Factorization.

Our purpose is to introduce some numerical loop invariants similar to partial indices of matrix-functions [17] which allow one to describe solutions of the GLTP. Following [5] let us choose the maximum torus T^d in G and a system of positive roots. In what follows it will be assumed that we consider the group of Sobolev loops $L_{1/2}G$ with index 1/2 [5].

Let us define nilpotent subgroups $N_0^{\pm}$ in $G_{\mathbb{C}}$ whose Lie algebras are generated by root vectors in $A_{\mathbb{C}}$ corresponding to positive (negative, resp.) roots. We also introduce subgroups L^+ in $L_{1/2}G_{\mathbb{C}}$, consisting of loops which are boundary values of holomorphic mappings of the domain B_0 (B_∞, resp.) into the group $G_{\mathbb{C}}$ and subgroups N^+, consisting of loops from L^+ (L^-, resp.) such that $f(0)$ lies in N_0^+ ($f(\infty)$ lies in N_0^-). Without going into details we denote by B_K an orbit of a subspace $K.H_+$ under the natural action of the subgroup N^- on the Grassmanian where $K \in \mathbb{Z}^d$ is a multiindex and $K.H$ denotes a subspace obtained by applying the diagonal action of the maximum torus T^d of the group G to V [5]. In other words, we assume that the multiindex determines a point in the Lie algebra of the torus T^d and we apply the exponential mapping of the torus to this point and then the obtained element acts onto the subspace H_+. In this notation according to theorem 8.5 we obtain from [5] the decomposition

$$L_{1/2}\, G_{\mathbb{C}} = \bigcup_K B_K \,. \qquad (6)$$

Recalling the classical case, we see that this reminds us of the theorem on factorization of matrix-functions with respect to contour [1] and hence we can also introduce a similar notion for the loop group LG.

Namely, by the term "the generalized Birkhoff factorization (GBF) of an arbitrary loop $f \in LG$" we shall mean the representation of the latter in the form

$$f = f_+ H f_- \ , \tag{7}$$

where f_+ belongs to a subgroup of the same name from subgroups $L^{\pm} G_{\mathbb{C}}$, and H is a homomorphism from $\mathbb{T}$ into T^d.

Note that the set of all such homomorphisms H is numbered just by multi-indices $K \in \mathbb{Z}^d$, and we can associate with the loop $f \in LG$ the multi-index corresponding to H from (7). The fact below proves that this is possible and correct.

<u>Generalized Birkhoff Theorem</u>. Each loop f from $L_{1/2}G$ has factorization (7), where the loop f defines the homomorphism H uniquely.

For the detailed proof of this important result see [5]. Note that we could also define the factorization with a different order of co-factors $f_{\pm}$ and the main result would remain valid. Our choice of factorization is suitable for the GLTP formulation.

Numbers k_i making up the multi-index K from (7) are called (right) partial indices of the loop f and are denoted by $k_i(f)$. We sometimes refer to them as loop ranks. It is understood that the left ranks are also defined and there arises a question of comparing them for one and the same group. For completeness of our discussion we shall answer this question by constructing explicitly loops with an arbitrary number of left ranks provided that the right ranks are given in the sense of the classical construction from [8] and [9].

<u>Proposition 1</u>. The sum of ranks does not depend on a factorization type. As for the rest properties, the left and the right ranks are independent.

The obtained analytical loop invariants are very useful in two respects: they can be used to describe the solvability of the GLTP and to reveal the connection of the GLTP with the theory of holomorphic main $G_{\mathbb{C}}$ - fiberings over the Riemann sphere [15].

Indeed, taking into account the fact that $L^+G_{\mathbb{C}}$ groups are exactly isomorphisms of morphisms of trivial $G_{\mathbb{C}}$ - fiberings over the contractible pieces B_0 and B_{∞} of the Riemann sphere, we immediately arrive at the following generalization of Grothendieck's theorem [15], [16].

Proposition 2 (see [5]). A point from LG is a class of an isomorphism of pairs (E, t), where E is the holomorphic main $G_{\mathbb{C}}$ -fibering on $\mathbb{P}$ and t is its trivialization over B_{∞}. Classes of an isomorphism of such fiberings are numbered by non-ordered multi-indices $K \in \mathbb{Z}^d$ given as sets of partial indices of the loop that expresses the fibering sewing function for the fixed partitioning of $\mathbb{P}$.

We shall return to fiberings in §6 but already now we can give one application of the connection obtained.

Theorem 1. Let E be the main $G_{\mathbb{C}}$ - fibering over $\mathbb{P}$ with the sewing function f. Then the dimension of its versal deformation base 15 is expressed by the formula

$$\sum_{k_i > k_j} (k_i - k_j - 1), \tag{8}$$

where $k_i = k_i(f)$ are partial indices of the sewing function.

To prove the theorem it is enough to use the geometric properties of decomposition (6) established in [5]. Indeed, it is shown there that each subset B_K is a smooth submanifold of the group of loops with the complex structure mentioned in the introduction. Besides, it becomes immediately evident that the type of isomorphism fibering remains unchanged only along B_K. Therefore, according to [16], the dimension of the versal deformation base is simply equal to the codimension of B_K in LG, which, again according to [5], can be calculated by formula (8).

Remark 2. It is understood that in the classical case we obtain ordinary right partial indices [17]. As is known, their sum is equal to the Fredholm index of the RBVP with any choice of factorization type, which explains the interdependence of the ranks from Proposition 1. Following [5], ranks B_K will be called Birkhoff strata or B-strata. According to [5], they give a complex-analytical stratification of the group of loops.

Corollary 1. The holomorphic $G_{\mathbb{C}}$ -fibering is holomorphically stable iff all pairwise differences of the ranks of its sewing function does not exceed 1.

Indeed, the stability is equivalent to the fact that the dimension of the versal deformation base is equal to zero and any pairwise difference of ranks exceeding 1 makes a nonzero contribution to the formula [8].

The established connection can be made deeper at the expense of a purely algebraic description of main fiberings in terms of non-abelian cohomologies (see, e.g., [20]). We have the bijection between $H^1(\mathbb{P}, G\ )$ and the classes of an isomorphism of main holomorphic $G_{\mathbb{C}}$ - fiberings over $\mathbb{P}$ [20] . On the other hand, we have the well known isomorphism $H^1(\mathbb{P}, T^d)/W(T^d) \cong H^1(\mathbb{P}, G_{\mathbb{C}})$, where T^d is the maximal torus and $W(T)$ is the Weyl group. From the theory of roots [8] it is not difficult to conclude that $H^1(P, T^d) \cong \mathrm{Hom}(\mathbb{C}^*, T^d)$ is identified with the lattice of weights for the dual system of the chosen system of roots $R(G, T^d)$. Now we must take into account the fact that for the connected semi-simple complex group G of the adjoint type (with the trivial centre) $H^1(\mathbb{P}, G)$ is described as a set of dominant coweights for $R(G, T^d)$. Therefore all holomorphic properties of the given $G_{\mathbb{C}}$ - fibering can be restored by the corresponding dominant coweight and after that it is easy to interpret the integers ascribed to simple roots as pairwise differences of sewing function ranks. By virtue of Corollary 1 we obtain the following description of holomorphically stable fiberings from [21] .

Corollary 2. For the classical simple group G all holomorphic stable $G_{\mathbb{C}}$ -fiberings over $\mathbb{P}$ correspond to the simple roots (marked by ⊠) in the following way:

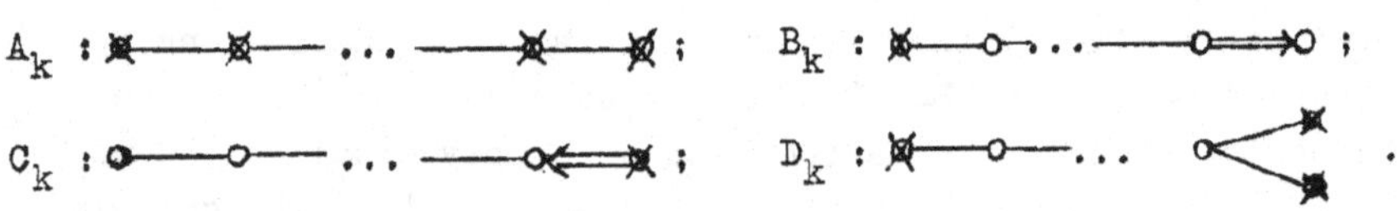

4. Fredholm Theory of the GLTP.

By the virtue of the equivalence between the operators R_f^r and T_f^r mentioned in §2, it is convenient to derive the main result in terms of the Fredholm Grassmanian [5] .

Theorem 2. Let G be a classical simple compact Lie group. Then the GLTP P_f^r has the Fredholm properties for any loop f from L_pG with $p > 1/2$ and for any linear representation r.

Proof. It is clear that it is enough to assume that r is irreducible. As it was already mentioned, the desired conclusion is equivalent to the fact that a pair of subspaces $(H_+, M_f^r H_+)$ is Fredholmian in the sense of T.Kato [6] . Note also that, according to the definition of the bounded group of the polarized Hilbert space $\mathbb{H}$ [5] , the pair (H_+, AH_+) is Fredholmian for any $A \in GL_r(H)$, and its index can be calculated as virtual dimension of AH_+. In our case the latter reduces to ind $(P_+|AH_+)$ which for $A = M_f^r$ coincides with $\mathrm{ind} T_f^r$. According to Proposition 6.3.3 from [5], the operator of multiplication by $f \in L_pG$ in the representation r lies in $GL_p(H)$ for $p > 1/2$, which completes the proof.

Indeed, the imbedding of L_pG into $GL_p(H)$ is proved in [5] only in the case corresponding to the GLTP for the standard representation of the unitary group. However, since any irreducible representation is an external degree of the standard representation [8], it is easy to verify that the result is valid for them too. Note also that the properties of the Grassmanian are also valid for the index $p = 1/2$, but this is not necessary, since the loop continuity is already needed in the GLTP formulation itself.

Now we can easily calculate the Fredholm index of the GLTP directly in terms of the GTO.

Proposition 3. If G is a simply connected classical simple compact Lie group, then for $f \in L_pG$ with $p > 1/2$ we have $\mathrm{ind}\ T_f^r = 0$.

Proof. As is known, the Fredholm index is a homotopic invariant and divides the components of the set $\mathcal{F}(H)$ of Fredholm operators in H. On the other hand, the homotopic class of a given loop-coefficient in the fundamental group $\pi_1(G)$ is also a homotopic invariant of the situation. Therefore from the condition of the coefficient group being simply connected we immediately conclude that each GTO T_f^r lies in the component of the unit of a semi-group of Fredholm operators and thereby $\mathrm{ind}\ T_f^r = 0$.

We shall now proceed to multiply connected groups. The most interesting case of the unitary group is $G = U(n)$. As it was already noted, it is enough to consider only irreducible representations r, since the index is obviously additive with respect to the sums of representations giving the GLTP. From the same arguments it is clear that we can consider only base irreducible representations which are given by external degrees $S^{\wedge k}$ of the standard representation S on $\mathbb{C}^n$ [8].

Proposition 4. Let $G = U(n)$ and $r = S^{\wedge t}$, where $1 \leq t \leq n$. Then for any loop $f \in L_pG$, where $p > 1/2$, we have

$$\mathrm{ind}\ T_f^r = \binom{n-1}{t-1} \sum_{i=1}^{n} k_i(f) . \qquad (9)$$

Proof. We shall use the existence of the GBF for the loop-coefficient and substitute (7) in the conjugation condition (3). After that we can introduce new unknowns $Y_0 = (r(f_+))^{-1}(x_0)$, $Y = r(f_-)(x_\infty)$

which will be holomorphic in the required domains simultaneously with x_0, x_∞, since $f_\pm$ do not violate the holomorphy. The new unknowns must satisfy the conjugation condition for the GLTP with respect to the narrowing of the representation r by the maximum torus T^n. Since representations of the tori are always expanded in the sum of one-**dimensional** ones and the Fredholm index is additive, it only remains to calculate the indices of all arisen one-dimensional LTPs. This can easily be done by means of the classical formula in terms of the loop-coefficient index which, by definition, coincides with the partial index (the only one in our case) [1], [17]. The latter, in its turn, can be easily expressed through the coefficient ranks of the initial problem. Formula (9) is obtained from these observations by means of a simple combinatorial analysis.

Remark 3. It is understood that in terms of holomorphic vector fiberings our calculations reduce to the calculation of the Chern class of the external degree of a given fibering, but the concepts discussed above are preferable because they also allow us to obtain results on a structure of the GLTP kernel.

Porism. The dimension of the GLTP kernel P_f^r is equal to the sum of positive terms in the expression obtained by substituting the partial indices of the loop-coefficient in the elementary symmetrical sum $S(w(r))$ of the higher weight $w(r)$ of the representation r [8].

Indeed, from the proof of Proposition 4 it is evident that the kernel, too, can be expanded in the direct sum of kernels of one-dimensional LTPs and then the desired result follows from the definitions and the classical formula for the dimension of the LTP kernel [1].

Remark 4. Note that a given loop can be ascribed to different enveloping loop groups, which is encountered, e.g., in the case of the embedding of the considered groups like $U(n) \subset O(2n)$. Ranks may depend on "ascription" and there arises a problem of describing all possible relations between them. Such a problem has already been encountered earlier in the form of results on factorization of matrix-functions of special classes [1]. Note that a complete description of all possible changes remains yet unknown.

For the orthogonal group we can use Atiyah-Zinger's $\mathbb{Z}_2$ - index [10] and obtain a similar result.

Proposition 5. For $G = O(n)$ the $\mathbb{Z}_2$ - index of the GLTP $P_f^{s\wedge t}$ coincides with the parity of expression (9).

5. Fredholm Structures on Loop Groups.

We shall begin our consideration with the following commutative diagram. Let E be a complex Banach space, L(E) be an algebra of bounded linear operators in the topology of the operator norm, C(E) be a closed bilateral ideal of compact operators, GL(E) be a group of units, i.e. a group of all continuously invertible operators and let $\mathcal{F}_n(E)$ be a set of Fredholm operators of the index n. We also introduce the Kalkin algebra Q(E) = L(E)/C(E), whose group of units will be denoted by GQ and its component of the unit by G_0Q. All these objects can now be combined into a diagram whose commutativity expresses the main concepts of the linear Fredholm theory

$$\begin{array}{ccccccc} GL(E) & \longrightarrow & \mathcal{F}_0(E) & \longrightarrow & \mathcal{F}(E) & \longrightarrow & L(E) \\ \downarrow p & & \downarrow q & & \downarrow q & & \downarrow q \\ GL(E)/GC(E) & \overset{j}{\text{---}} & G_0Q & \longrightarrow & GQ & \longrightarrow & Q(E) \end{array} \qquad (10)$$

Recall that p is a multiplicative factorization with respect to the so-called Fredholm subgroup $GC(E) = GL(E) \cap \{I + C(E)\}$, consisting of Riesz operators [4] , whereas the lower left arrow is the homeomorphism j.

Following [4] we shall use the term "Fredholm structure on the manifold M" to call al atlas (U_i, g_i) of the required smoothness class on M such that all differentials $D(g_i g_j^{-1})(x)$ lie in GC(E), where E is a model space for M. In other words, it is required to construct the reduction of the structural group GL(E) of the tangential fibering TM to the Fredholm subgroup GC(E) [12] . According to [12] such reductions exist but of course it is not easy to give them effectively. That is why we shall now describe an alternative way for introducing a Fredholm structure suitable for our purpose.

First we recall that by virtue of Theorem 2.2 from [12] $\mathcal{F}$-structures can be given by $\mathcal{F}_0$-mappings of the required smoothness class into the model space, i.e. in our case by infinitely differentiable or analytic mappings of the index zero into LA. The second step consists in that instead of these mappings we can take smooth mappings into the space $\mathcal{F}_0(E)$ provided only that the tangent bundle of the manifold M is trivial, which is frequently observed in infinite-dimensional Banach manifolds. Indeed, according to [12] to give an $\mathcal{F}$-structure we need only to know the homotopic class from $[\mathcal{F}_0(M,E)]$ ·, whereas Proposition 2.4 establishes the bijection $[F_0(M,E)] \cong [M, F_0(E)]$ between $[F_0(M,E)]$ and $[M, F_0(E)]$. It is not difficult to describe the result of the combination of these two constructions in a more explicit form directly from (10).

Proposition 6. A smooth mapping $F : M \longrightarrow F_0(E)$, where E is the model of the Banach manifold M with the contractible group GL(E), gives a smooth Fredholm structure on M.

Indeed, the simplest thing is to justify this as follows: Let $t : M \times E \longrightarrow TM$ be the trivialization of the tangential fibering and (V_j, h_j) be a maximum atlas of the manifold M. We select in (V_j, h_j) a subatlas (U_i, g_i) assuming that for any x the differential $D(g_i \circ t^{-1})(x)$ lies in GC(E). It is not difficult to verify that precisely this subatlas gives the Fredholm structure corresponding to the composition of the above constructions.

Recalling that the GLTP for a simply connected group provides us exactly with a family of Fredholm operators of the index zero and that by Proposition 3 this family is determined by the correspondence $T^r : LG \longrightarrow F_0(H_+)$, $f \longmapsto T^r_f$, we see that to apply Proposition 6 it remains to verify only two technical conditions.

Proposition 7. The family of GTOs is analytic.

The proof is obtained by direct verification since we need only to ascertain that the multiplication operator M^r_f corresponds to the symbol f , which is nearly obvious.

As to the triviality of the tangential fibering, it can be obtained in the simplest way by choosing a group of loops of adequate regularity. E.g., if we take L_pG, then according to [5] we shall obtain simply a Hilbert manifold and the desired result follows from Kuiper's theorem [4] . However in reality we can do almost with any natural choice of a regularity class (L_pG with $p > 1/2$, Hölder loops, differentiable loops), since the general results from [23] guarantee the fulfilment of Kuiper's theorem for the corresponding model spaces consisting of mappings of the mentioned regularity class into the Lie algebra G. For this reason we continue to assume that we consider L_pG with $p > 1/2$ and this will be enough till the end of the paper.

Now it is not difficult for us to consider loop groups together with the Birkhoff stratification in the Fredholm context.

Theorem 2. Let G be a classical simple compact Lie group and let r be a linear representation of G. Then on the group L_pG with $p > 1/2$ there arises a Fredholm structure F^r induced by r.

In the case of a simply connected group the proof is immediately obtained from the preceding arguments. Namely, we must take as F^r a structure corresponding by virtue of the construction of Proposition 6 to the mapping $T^r : LG \longrightarrow F_0(H_+)$. For a multiply connected group we may first introduce the required structure on a component of the loop group unit and then distribute it over other components the more so that they are isomorphic. Note that the structure F^r on L_pG is infinitely smooth.

The latter circumstance allows us to define the notion of a smooth submanifold of L_pG with respect to the Fredholm structure F^r.

Theorem 4. For any representation r of the classical simple compact Lie group G and for any $K \in \mathbb{Z}^d$ the Birkhoff stratum B_K is a contractible analytic submanifold of the group L_pG with $p > 1/2$ with respect to the Fredholm structure F^r.

Indeed, according to [5] , each B-stratum is a submanifold of the loop group L_pG as a complex manifold and in L_pG has finite codimension expressed by formula (8). In [12] it is shown that such submanifolds are Fredholm ones The contractibility is provided by Theorem 8.4.5 [5] .

It should be emphasized here that we do not mean the contractibility to the categories of Fredholm manifolds of which we might think because we have the defined notions of a Fredholm mapping, Fredholm homotopy, etc., but we mean the ordinary contractibility. As a matter of fact, B-strata are not contractible in the Fredholm sense, as follows, in particular, from §6.

Remark 5. By virtue of the above reasoning about admissible classes of loop regularity the results remain valid for Hölder loops, too, which form the basis of the classical considerations [1] , [17] , [7]. Therefore they provide answers to the geometric questions on B-strata from [7] and [9] . Moreover, it becomes clear that the formula for a **homotopy type of a B-stratum** given in [9] , p.301, is not correct.

Now we can gain a better insight into the GLTP geometry. For this we first fix some topological invariants whose existence is due to L_pG and B_K being Fredholmian.

Corollary 3. For any G and r , as above, Chern classes $Ch_i(F^r)$ [4] are defined as homology classes in $H^*(L_pG)$ with $p > 1/2$.

Corollary 4. The embedding of each stratum B_K in L_pG with $p > 1/2$ defines the fundamental cohomological class $[B_K]$ in $H^*(L_pG)$ [22] .

6. Cohomological Invariants of Fredholm structures.

The results of the preceding section give rise to a number of natural questions about a more exact topological description of the obtained Fredholm objects. For example, it would be desirable to clarify whether structures F^r differ for various representations and to characterize the imbedding of the Birkhoff stratum B_K in LG in terms of partial indices k_i.

To accomplish this we should proceed from the classification theorems of the Fredholm structure theory [4], [12]. In particular, we conclude from these theorems that isomorphism classes are well defined by characteristic classes investigated in [22]. In our case it is enough to use the Chern classes from Corollary 3; which we can try to identify within the framework of the well-known description of loop group cohomologies by R.Bott [14]. The same also refers to fundamental classes $[B_K]$, since the knowledge of the latter would provide us with sufficiently complete information on the imbedding of B_K in LG. It is understood that the corresponding calculations are rather difficult and hence we have no chance to consider all interesting cases though the general principles are already clear. Accordingly, we shall briefly describe the approaches used and give one example to illustrate each invariant.

In calculating $ch_i(F^r)$ we shall follow D.Freed [13] who pointed out a very elegant, in our opinion, procedure for reducing the traces of some kernel operators, appearing from the loop group curvature in the sense of the classical Chern-Weyl theory, to calculations. Special mention should be made here of the maximum use of the smallness of GTO operators, which fits the general strategy of investigation of algebras of crypro-integral operators [24], as well as the new theorem on the index for Fredholm representations of Lie groups [25]. For simplicity, like in §3, we shall consider only adjoint-type group, which corresponds to GLTP for $r = ad$, and take $G = SU(n)$.

In fact we can do with any connected simply connected simple compact group of the adjoint type and the results will be similar to the results given below for the case $G = SU(n)$.

Thus we calculate Chern classes $ch_i(F^r_{SU})$ for $r = ad$, where $ad : SU(n) \longrightarrow su(n)$ is the adjoint representation. We can work with real coefficients, since LSU has no torsion [5]. To formulate the result we would like first to recall the description of $H^*(LSU)$. As is well-known, from the fact that LG is a group it follows that $H^*(LG; \mathbb{R})$ is the Hopf algebra with primitive generatrices given by elements from $\pi_*(LG)\otimes\mathbb{R}$. Furthermore, $\pi_n(LG) \cong \pi_{n+1}(G)$ and tensor products $\pi_{n+1}(G)\otimes\mathbb{R}$ are well-known [5]. Namely, to each considered group we associate certain odd numbers $2m_i - 1$ called indices of the group G, in the number equal to rank G, such that $\pi_{n+1}(G) \otimes \mathbb{R} \cong \mathbb{R}$ for $n + 1 = 2m_i - 1$ and otherwise is trivial. For simple groups $m_1 = 2$. To summarize, the real cohomology $H^*(LG; \mathbb{R})$ is a symmetrical algebra with generatrices Y_{2m_i-2} in dimension $2m_i - 2$, where m_i are indices of G. In particular for SU(n) the indices are equal to 2,3,...,n and thus we have

$$H^*(LSU(n); \mathbb{R}) = \mathbb{R}\left[y_2, y_4, \ldots, y_{2n-2}\right]. \tag{11}$$

Now we can describe the answer, in which instead of formulas for individual Chern classes we can give immediately, following [13] , a complete Chern class.

<u>Proposition 8</u>. A complete Chern class of the Fredholm structure F^{ad} on SU(n) is equal to

$$\exp(2n\left\{y_2 + y_6/3 + y_{10}/5 + \ldots + y_{4m-2}/(2m-1)\right\}), \tag{12}$$

where the generatrices y_{2k} are the same as in (11) and $m = [n/2]$.

To prove the proposition we have to verify that in the case when $r = ad$ our family of GTOs is homotopic to Kähler Toeplitz operators from [13] and then to apply Freed's index formula from [25] following the Chern-Weyl scheme in the sense of the work [13] .

It should be emphasized that the described calculation contains many nontrivial steps and requires a great number of preliminary data and references and, therefore, a separate exposition. Here we simply wish to show that it is possible to calculate the obtained invariants. We would also like to add that they allow us to prove that structures F^r do not coincide for various r and thereby to exhaust the first concentre of natural topological questions concerning the GLTP.

As to fundamental classes $[B_K]$, there are some subtleties here too. It is enough to say that these classes were initially defined in [9] using the theory of analytic Banach sets and the description of singularities of sets of operators of the fixed rank and corank. We have managed to bypass this here due to using "ready" results of the theory of Fredholm structures from [22] . However, to accomplish actual calculations we could invent nothing but the reduction to the mentioned subsets in $\mathcal{F}_0(H)$ as outlined in [9] . Note that here, too, we have a novelty connected with the fact that Bott's description of cohomologies $H^*(LG)$ used in [9] is also suitable for other classical simple groups [14] and therefore we have a much greater stock of examples of B-strata. Analyzing the calculations in [9], we can verify that the restrictions imposed here on dimension really belong to the rank of the group G and therefore for groups of rank 2 and 3 the calculations can be performed to the end. For simplicity, we shall limit our consideration to the structures of the form F^s, where s is the standard representation defined for any classical simple group [8].

Proposition 9. For groups of a rank not exceeding three, fundamental cohomological classes B_K are expressed in $H^*(LG)$ by the formulas from Proposition 5.3 and Theorem 6.1 from [9].

Note that this, in particular, leads to the Fredholm non-contractibility of B-strata though they are contractible in the usual sense.

7. The non-zero genus case.

As it was mentioned in the introduction, GLTP arises naturally on Riemann surfaces of arbitrary genus. Unfortunately, an attempt to connect it with holomorphic bundles leads to great difficulties because of the complication of their classification even in the case of torus [15]. That is why we have begun with the Riemann sphere case in order not to lose clarity as a result of the necessary reference to more involved results of complex analysis like the Riemann-Roch theorem [26]. And now we give a very brief description of the GLTP in the non-zero genus case.

Let $M = M^2_g = M_g$ be a compact Riemann surface of the genus g, and r, as above, a linear representation of a classical compact Lie group G in a vector-space V, $g \geq 1$.

In addition to the preliminaries of § 2 let us fix a distinguished point x_∞ on M and a local parameter z^{-1} around x_∞. In other words, z is a holomorphic mapping from a neighbourhood of x_∞ in some neighbourhood of ∞ in $\mathbb{P}$. We assume that $z(x_\infty) = \infty$ and z is an isomorphism on the domain $\{|z| > 1/2\}$. Then one may identify $\mathbb{T}$ with the circumference $S = \{|z| = 1\}$ around x_∞ on M. Let M_∞ denote the domain of M, where $|z| > 1$, and M_0 denote the complement of the closure of the former. In other words, $\overline{M}_0 \cap \overline{M}_\infty = S \cong \mathbb{T}$

Let $A(M_0, V)$ denote a subspace in $C(\overline{M}_0, V)$ consisting of mappings which are holomorphic in M_0 (and analogously for M_∞). Let again $\mathbb{H} = L^2(\mathbb{T}, V)$ polarized as $\mathbb{H} = H_+ \oplus H_-$, where H_+ is the usual Hardy space from § 2. The new point is that one can also associate with these data the subspace H_M consisting of boundary values of mappings from $A(M_0, V)$ which should enter in the formulation of the GLTP on M.

Finally fix a loop $f\in L_pG$ with $p>1/2$. Then by the complete analogy with §1 we formulate the GLTP $P^r_{M,f}$ (on M with the coefficient f with respect to the representation r) as a question about the existence and about the quantity of pairs $(X_o, X_\infty)\in A(M_o, V)\times A(M_\infty, V)$ with $X_\infty(\infty) = 0$ satisfying the following transition condition on S:

$$X_o = r(f)\, X_\infty . \tag{13}$$

There arises again the Riemann operator $R^r_{M,f}$ given by formula (4) but the analogues of GTO should be introduced in a slightly different way. For this purpose denote by $P_M : H \longrightarrow H_M$ the orthogonal projection and set

$$T^r_{M,f} = P_M M^r_f : H_M \longrightarrow H_M , \tag{14}$$

where M^r_f is the operator of multiplying by f in the representation r.

In this case one also has the connections between various descriptions of the GLTP established in §2, and we shall develop our considerations in the language of GTO. As a first step one has to establish the Fredholmness of the GLTP in the scheme of §2 but now it is necessary to use instead of the Grassmanian model of the loop group LG the so-called fundamental homogeneous space associated with the Riemann surface M as introduced in chapter 8 of the book [5] .

Theorem 5. If G is a classical simple Lie group then for any loop $f\in L_pG$ with $p>1/2$ the problem $P^r_{M,f}$ is Fredholm.

In fact, from the properties of the fundamental homogeneous space it follows that the subspace H_M together with the initial Hardy space forms a Fredholm pair [7] . As in §2, the index of this pair is finite and it is equal to the index of the problem. The desired conclusion follows because the projection of one element of a Fredholm pair on another is Fredholm.

These observations make it clear that in order to calculate the index one has only to calculate the virtual dimension of the subspace H_M. Recall that from the connection with principal bundles follows a cohomological interpretation of this dimension. Namely, after denoting by E_f the associated bundle with the fibre V one has:

$$\text{vir.dim } H_M = \dim H^0(M,E_f) - \dim H^1(M,E_f) - \dim V. \tag{15}$$

The equality (15) enables one to use the Riemann-Roch theorem (26) and it remains to compute the Chern class of the bundle corresponding to the given loop. As we have already seen it suffices to determine its partial indices $k_i(f)$. Finally, calculating the homologies of M with the coefficients in the associated bundle with the representation space as a fibre we obtain in the simply-connected case the following

Proposition 10. Let G be a connected simply connected classical simple Lie group. Then for any M and r as above one has the equality

$$\text{ind } T_M^r = -(g-1)\dim V.$$

In fact, we have only to refer to the Prop. 8.11.10 from [5] which provides the value of virtual dimension.

Let now G = U(n). We write $|K(f)|$ for the sum of partial indices.

Theorem 6. For $G = U(n)$, $M = M_g$, $r = S^{\wedge t}$ with $1 \leqslant t \leqslant n$ and $f \in L_pG$ with $p > 1/2$ one has the following formula

$$\text{ind } P^r_{M,f} = -\binom{n}{t}(g-1)+\binom{n-1}{t-1}|K(f)| \; . \qquad (16)$$

As a matter of fact, this is simply a transition of the Riemann-Roch theorem into the language of partial indices, and $|K(f)|$ may be also interpreted as the increment of the determinant of the loop [17].

We can continue following the scheme of the preceding sections verifying the smoothness of the GTO family and the possibility of construction of analogues of F^r structures.

Theorem 7. For any M,G,r as above there exists a smooth Fredholm structure F^r_M on L_pG with $p > 1/2$, and every Birkhoff stratum is a smooth Fredholm submanifold of the finite codimension with respect to any of the structures F^r_M.

Of course, we could proceed investigating the same questions but at present we do not have analogues of the results from §6 because the geometric picture of GLTP and holomorphic bundles is considerably more difficult now.

8. Concluding remarks.

Let us summarize. In fact, in the present paper we have dealt only with the one-dimensional holomorphic GLTP on Riemann surfaces, especially we have developed the Fredholm theory and constructed some specific structures on the coefficients group, which provide certain topological invariants of this infinite-dimensional manifold.

Considering now from this viewpoint the GRBVP from §1 one finds quite a number of analogous topics and constructions which are difficult to classify or compare with respect to actuality. That is why we restrict ourselves to rather isolated declarations having as a purpose a more complete description of GRBVP.

First of all, nothing prevents us from searching for solutions in the class of generalized analytic functions [19] , [27] , i.e. solutions of some natural elliptic system. This is also reasonable in the non-zero genus case because the recent monograph [27] on generalized analytic functions on Riemann surfaces provides all the necessary analytic techniques (Liouville theorem, Riemann-Roch theorem et al.). The essence of our declaration is that GBF is valid also in this framework and then the desired results follow along the lines of the present paper. We would also like to point out a connection between the modification just mentioned and K.Uhlenbeck's theory of harmonic mappings into a Lie group which is exposed already in the Russian edition of the book [5] .

Secondly, all the foregoing shows that a reasonable multi-dimensional theory of the GLTP (e.g. on a sphere $S^{n-1}\subset\mathbb{R}^n$) requires some analogues of holomorphic functions in $\mathbb{R}^n$. We declare that the so-called Clifford holomorphic functions (see [28]) are compatible here, i.e. the solutions should lie in the kernel of the Dirac operator (square root of Laplacian).

As is well known, the Dirac operator in $\mathbb{R}^n$ may be constructed if $n = 2^k$, so that for such values of n there exists a natural formulation of the GLTP for Clifford analytic functions. The form of the analogues of GBF is unclear here, and one has to proceed in a different way.

First note that it is again easy to solve the "problem of the jump" [1] because there is an analogue of the Plemelji-Sochocky formula [28] . Consequently, when the coefficient $f\in H_p(S^{n-1},G)$ is close to the identical one we may take its logarithm and solve the corresponding "problem of the jump" with the values in the Lie algebra, and then return to the group $H_p(S^{n-1},G)$ using the exponential mapping. This scheme was already used in [2] for solving non-linear GLTP and it may be made quite correct also in the case under consideration but unfortunately it works only for coefficients sufficiently close to one. In the case of general coefficients one could try to "deloop" them so that they should possess logarythms but the form of the corresponding monodromy factor is still unclear. Nevertheless, the problem is again Fredholm for Hölder coefficients as it follows from the theory of singular integral equations with Cauchy kernel [28] . Thus emerges an analogue of the theory developed in the preceding sections which requires further comprehension.

In connection with the outlined "logarythmical" method of constructing the factorisation we would like to point out a special problem of characterizing those classes of coefficient groups which permit factorisation on contours or, in general, on spheres. The most natural candidates seem to be the so-called exponential groups but this is only a hypothesis. One may consider this as an attempt to admit the non-compact groups of coefficients, and there arises a perspective of considering various infinite-dimensional operator Lie groups like the general linear group of the Hilbert space or the Fredholm group GC(H). Here we come to the connections with holomorphic Banach bundles [29] which is quite reasonable even in the case of a Riemann surface and Fredholm group GC($\mathbb{H}$) because from the existence of Fredholm determinants follows the non-triviality of the corresponding first Chern class [16] . Here we are also forced to restrict ourselves by stating the existing perspectives.

Finally, a few words should also be said about non-linear modifications of the GLTP. The simplest possibility within our framework is to permit arbitrary actions r of the group G on its homogeneous spaces, which has also some physical significance [5] , [18] . Two examples of this kind - the principal action on itself and the adjoint representation - were considered in [2] , and there were again essential the coefficient partial indices which enable one to establish the existence as well as the stability of associated holomorphic bundles arising as solutions of the non-linear GRP. At the same time other non-linear modifications are also available: e.g., the transmission condition (2) itself may be made non-linear by taking some polynomial of $r(f)$ instead of $r(f)$, which is evidently always reasonable, or one may impose some geometric condition of restricting values of the solution to some fixed family of curves as in the non-linear Hilbert problem from [30] .

We have no possibility to describe this in some detail here and we restrict ourselves by declaring that the "stability storage" of the RBVP with respect to the Fredholm theory is very high so that all the constructions outlined above lead to Fredholm operators, possibly non-linear. Thus, all the foregoing may be mnemonically formulated as a list of reasonable examples within the Fredholm structures theory arising in the connection with generalizations of the Riemann boundary value problem.

References

1. Gakhov F.D. Boundary value problems. Moscow, 1977 (in Russian).
2. Khimshiashvili G.N. On Riemann boundary value problem with values in a compact Lie group. - Trudy IPM TGU, v.3, No.1, 1988 (in Russian).
3. Khimshiashvili G.N. On Riemann-Hilbert problem for a compact Lie group. - DAN SSSR (Soviet Math. Doklady), 1990, v.310, No.5.
4. Eells J. Fredholm structures. In: Proc. Symp. Pure Math., vol.18, AMS, 1970.
5. Pressley A., Segal G. Loop groups. Clarendon Press, Oxford, 1988.
6. Boyarski B. Abstract problem of a linear conjugation and Fredholm pairs of subspaces. In: Boundary value problems. Tbilisi, 1979 (in Russian).
7. Bojarski B. Some analytical and geometrical aspects of the Riemann-Hilbert problem. In: Complex analysis. Berlin, 1983.
8. Adams J.F. Lectures on Lie groups. W.A.Benjamin Inc., N.Y. - Amsterdam, 1969.
9. Disney S. The exponents of loops on the complex general linear group. - Topology, 1973, V.12, No.4.
10. Atiyah M., Singer I. Index theory for skew-adjoint G Fredholm operators. - Publ. Math. IHES, 1970, V.37, 5-26.
11. Khimshiashvili G. Lie groups and transmission problems on Riemann surfaces. - Soobshch. Akad. Nauk Gruz. SSR, 1990, V.137, No.1.
12. Elworthy K., Tromba A. Differential structures and Fredholm maps on Banach manifolds. - Proc. Symp. Pure Math., 1970, V.15.
13. Freed D. The geometry of loop groups. - J. Diff. Geometry, 1988, V.28, No.3.
14. Bott R. The space of loops on a Lie group. - Mich. Math. J., 1958, V.5, No.1.
15. Palamodov V.P. Deformations of complex spaces. In: Encyclopaedia of Mathematical Sciences. Vol.10, Springer, 1989.
16. Laiterer Yu. Holomorphic vector bundles and the Oka-Grauert principle. In: Encyclopaedia of Mathematical Sciences. Vol.10. Springer, 1989.
17. Muskhelishvili N.I. Singular integral equations. Moscow, 1977 (in Russian).
18. Uhlenbeck K. Harmonic maps in a Lie group. - Preprint, Univ. Chicago, 1985.
19. Bitsadze A.V. Introduction to the theory of analytic functions. Moscow, 1974 (in Russian).
20. Onishchik A.L. Some notions and applications of the theory of non-abelian cohomologies. - Trudy Mosk. Mat. obshchestva. V.17, 1962 (in Russian)
21. Doi H. Nonlinear equations on a Lie group. - Hiroshima Math. J., 1987, V.17, 535-560.
22. Koshorke U. K-theory and characteristic classes of Fredholm

bundles. - Proc. Symp. Pure Math., 1970, V.15, 95-133.

23. Mityagin B.S. Homotopic structure of a linear group of a Banach space. - Uspekhi mat. nauk (Russian Math. Surveys), 1970, V.25, vyp.5.

24. Khimshiashvili G.N. To the theory of algebras of singular operators. - In: Trudy Tbil. matem. instituta, 1987 (in Russian).

25. Freed D. An index theorem for families of Fredholm operators. - Topology, 1988, V.27, No.3.

26. Forster O. Riemannian Surfaces. Springer, 1977.

27. Rodin Yu. Generalized analytic functions on Riemann surfaces. - Lect. Notes in Math., 1987, V.1288.

28. Brackx F., Delanghe R., Sommen F. Clifford Analysis. Pitman, 1982.

29. Zaĭdenberg M., Kreĭn S., Kuchment P., Pankov A. Banach bundles and linear operators. - Uspekhi mat. nauk (Russian Math. Surveys), 1975, v.30, No.5.

30. Shnirel'man A.S. Nonlinear Hilbert problem and degree of quasiregular mapping. - Mat. sbornik (Soviet Math. Sbornik), 1972, v.89, No.3.

STATIONARY SOLUTIONS OF NONLINEAR STOCHASTIC EQUATIONS

A.S.Mishchenko
Department of Mechanics and Mathematics
Moscow State University
119899, Moscow, USSR

The water balance of a closed reservoir is usually described by the differential equation

$$\frac{dw}{dt} = v(t) + \alpha(t) F(w) , \qquad (1)$$

where w is a full volume of water in a reservoir, v is the rate of the water's influx which comprises a river run-off, the surface and underground influxes; the second summand of the right-hand side of equation (1) expresses the value of the rate of water's evaporation from the water surface of a reservoir minus the rate of the moisture's ingress in the form of precipitation. To be more exact, the second summand should take account of the evaporation not only from the smooth water surface but also from the parts of the banks which can be washed down by the closed reservoir. Conditionally the second summand is presented in the form of the product

$$\alpha(t) F(w) , \qquad (2)$$

where α is the evaporation coefficient reduced to the unit of the square of the smooth water surface under the given external conditions of evaporation, i.e. the characteristics of the environment such as air humidity, rate fields of the movement of atmosphere and the characteristics of the inflow of the thermal energy in the evaporation region. Thus the second co-factor in (2) is a reduced square of the smooth water surface, whose value depends on the reservoir's level, i.e. on the value of w. The values of $F(w)$ represent a far more complicated dependence than the dependence of the geometric square of the smooth water surface on the volume, since it takes account of the change of evaporation rate depending on the configuration of the reservoir's bottom, the distribution of temperature in different water layers, the structure of the distribution of moisture in the washed down parts of banks etc., under invariable external conditions.

In practical applications the function $F(w)$ is essentially nonlinear. Note that in the papers, dealing with the study of the water balance of the closed reservoir, the Caspian Sea, in particular, (see, for example [1]), on the basis of probability processes it was groundlessly assumed that the evaporation coefficient does not depend on the configuration of a reservoir, in particular, on the value of w. As a result of such an assumption a linear approximation of the dependence of the square of the smooth water surface on the volume was chosen as the function $F(w)$. However, uncomplicated calculations with respect to the actual values of the level of a closed reservoir and the general run-off over past years under the averaging when w is of a constant value show that the dependence $F(w)$ is not even monotonous and its deviation from the best linear approximation in the case of the Caspian Sea is comparable by the order with the general volume of the annual river run-off. The above arguments make it possible to consider the problem of the behaviour of the function $w(t)$, when the dependence $F(w)$ is a nonlinear nonmonotonous function. The obtained qualitative results are interesting for the refinement of the mathematical model, describing the level regime of the closed reservoirs.

Thus, we consider an ordinary differential equation (1), in which the functions $v(t)$ and $\alpha(t)$ are stationary random processes and F is essentially nonlinear with respect to argument w. It is assumed that both the processes $v(t)$ and $\alpha(t)$ are given by one smooth flow g_t on some compact smooth manifold x in the form of the total space of the smooth locally trivial bundle over the circle with fibre Y. It is assumed that the flow g_t preserves a certain smooth measure on manifold x and it is ergodic. Then the functions $v(t)$ and $\alpha(t)$ have the following form

$$v(t) = v(g_t(x)), \qquad \alpha(t) = \alpha(g_t(x)), \qquad x \quad X \tag{3}$$

Let, additionally, the function $\alpha(t)$ have the form

$$\alpha(x) = -1 + E(x),$$

so that the inequality

$$|E(x)| \leq E_0 < 0$$

holds and let the functions $v(x)$ and $\alpha(t)$ be smooth and the function $F(w)$ have the form

$$F(w) = w(w + w_-)(w - w_+), \quad w_-, w_+ \quad 0. \tag{4}$$

<u>Theorem 1</u>. Equation (1) has a bounded stationary solution, measurable with respect to variable x and smooth of class C^1 with respect to variable t. Equation (1) has no more than three such solutions, which differ from each other on the set of a positive measure.

Under some additional assumptions any solution asymptotically tends to a linear combination of stationary solutions of the form

$$w(x,t) = \chi_1(x) w_1(g_t(x)) + \chi_2(x) w_2(g_t(x)) + \chi_3(x) w_3(g_t(x))$$

for $t \to \infty$. The theorem can be extended onto the case of stationary processes and an arbitrary degree of polynomial F by a simple modification of the construction.

1. Existence of stationary solutions.

Consider equation (1), where $F(w)$ is a function of the form (4) and the functions $v(t)$ and $E(t)$ have the form (3). It means that x denotes a variable point of a probability space X, on which acts a one-parameter group g_t of the transformations, preserving the measure:

$$g_t : X \longrightarrow X$$

Moreover, further we shall assume that the dynamic system g_t is an ergodic system of the following special form: space X is isomorphic to the Cartesian product

$$X = Y \times I,$$

where I is a unitary semi-interval of real numbers, Y is another probability space and the group g_t effects a uniform movement of the points along the parameter $t \in I$ with a spasmodic transition from the upper base (for $t = 1$) onto the lower (for $t = 0$) with the help of some ergodic automorphism

$$\varphi : Y \longrightarrow Y,$$

preserving the measure on space Y. More exactly, if

$$x = (y,s) \in X = Y \times I, \; y \in Y, \; s \in I, \; 0 \leq s < 1,$$

then

$$g_t(y,s) = \left(\varphi^{([t+s])}(y), \{t+s\} \right), \qquad (1.2)$$

where $[t+s]$ and $\{t+s\}$ are an integer and fractional part of numbers $t+s$, respectively.

Theorem 1.1 There exists such function $w(x) \in L^1(X)$ that the function

$$w(x,t) = w(g_t(x)) \qquad (1.3)$$

satisfies equation (1).

Proof. Consider a space $Z = Y \times R^1$. It is obvious that space Z covers space X by an epimorphism

$$\pi : Z \longrightarrow X.$$

Any function f on space X is lifted onto space Z in the form of composition $\hat{f} = f \circ \pi$, which is a function of two variables and which satisfies the condition

$$\hat{f}(y,t+n) = \hat{f}(\varphi^n(y),t). \qquad (1.4)$$

Conversely, any function, satisfying condition (1.4), is covered by some function from space X. In particular, functions $v(x)$ and $E(x)$ are lifted onto space Z, where there arise functions $\hat{v}(y,t)$ and $\hat{E}(y,t)$, satisfying condition (1.4). Then equation (1) is reduced to the following equation onspace Z:

$$\frac{dw}{dt}(y,t) = \hat{v}(y,t) + (-1+\hat{E}(y,t))w(w + w_-)(w - w_+). \qquad (1.5)$$

Consider the solution to equation (1.5) with initial conditions

$$\hat{w}(y,\ 0) = \hat{w}_0. \qquad (1.6)$$

Let a number $\hat{w}_0$ be sufficiently large. Then for any Y and sufficiently large t the expression in the right-hand side of (1.5) can only be negative. This means that the solution $\hat{w}(y,t)$ satisfies the following strict inequality:

$$\hat{w}(y,\ 1) < \hat{w}_0,\quad y \in Y. \tag{1.6}$$

From condition (1.4) it follows that the functions

$$\hat{w}_n(y,t) = \hat{w}(\varphi^{-n}(y),\ t+n) \tag{1.7}$$

are also solutions to equation (1.5). The initial value of the solution $\hat{w}_1(y,0)$ satisfies the inequality

$$\hat{w}_1(y,0) < \hat{w}(\varphi^{-1}(y),1) < \hat{w}_0 = \hat{w}(y,0). \tag{1.8}$$

Consequently, for other values of parameter $t > 0$ there takes place the inequality

$$\hat{w}_1(y,t) < \hat{w}(y,t), \tag{1.9}$$

i.e.

$$\hat{w}(\varphi^{-1}(y),t+1) < \hat{w}(y,t). \tag{1.10}$$

Consequently,

$$\hat{w}(\varphi^{-n}(y),t+n) < \hat{w}(\varphi^{-n+1}(y),t+n-1). \tag{1.11}$$

In other words,

$$\hat{w}_n(y,t) < \hat{w}_{n-1}(y,t). \tag{1.12}$$

Replacing the initial condition (1.6) by the condition

$$\tilde{w}(y,0) = \tilde{w}_0, \tag{1.13}$$

where $\tilde{w}_0$ is a sufficiently large negative number, we obtain solution $\tilde{w}(y,t)$ to equation (1.5), satisfying the inequality

$$\tilde{w}(y,t) > \tilde{w}_0 \tag{1.14}$$

and the inequalities

$$\tilde{w}(y,t) < \tilde{w}_n(y,t). \tag{1.15}$$

Thus, the sequence of functions $\hat{w}_n(y,t)$ monotonously decreases and it is bounded from below. Consequently, there exists a limit function

$$w(y,t) = \lim_{n\to\infty} \hat{w}_n(y,t). \tag{1.16}$$

Let us show that the function $w(y,t)$ is a smooth function with respect to parameter t, satisfying condition (1.4).

First of all for every fixed $y \in Y$ the functions $\hat{w}_n(y,t)$ of variable t satisfy an integral equation

$$\hat{w}_n(y,t) = \hat{w}_n(y,0) + \int_0^t \hat{v}(y,s)ds +$$

$$+ \int_0^t \left(-1+\hat{E}(y,s)\right) \hat{w}_n(y,s)\left(\hat{w}_n(y,s) + w_-\right)\left(\hat{w}_n(y,s) - w_+\right) ds. \tag{1.17}$$

From the pointwise convergence (1.16) follows a uniform convergence on a measurable set $E \subseteq [0,t]$, whose measure is arbitrarily close to the measure of a segment $[0,t]$. Therefore for any $\varepsilon > 0$ there exists such a set $E \subseteq [0,t]$ that $\mu(E) > t-\varepsilon$ and such a number N, that for $n > N$ there takes place the inequality

$$\left| \hat{w}_n(y,s) - w(y,s) \right| < \varepsilon$$

for all $s \in E$. Then

$$\hat{w}_n(y,t') - \hat{w}_{n'}(y,t') =$$

$$= \hat{w}_n(y,0) - \hat{w}_{n'}(y,0) +$$

$$+ \int_0^{t'} \left(-1+\hat{E}(y,s)\right) \left[F\left(\hat{w}_n(y,s)\right) - F\left(\hat{w}_{n'}(y,s)\right)\right] ds, \tag{1.18}$$

where $F(w) = w(w + w_-)(w - w_+)$. Estimate the integral in the right-hand side of (1.18), dividing the domain of integration into two sets $E' = E \cap [0,t']$ and $[0,t'] \setminus E'$. For $n,n' > N$ we obtain

$$\left| \int_0^{t'} \left(-1+\hat{E}(y,s)\right) \left[F\left(\hat{w}_n(y,s)\right) - F\left(\hat{w}_{n'}(y,s)\right)\right] ds \right| \leq$$

$$\leq \; tC_1 \varepsilon + C_2 \varepsilon . \qquad (1.19)$$

Here C_1 is a constant, for which the inequality

$$\left| \left(-1+\hat{E}(y,s)\right)\left(F(w_1) - F(w_2)\right)\right| \leq C_1 \left| w_1 - w_2 \right|$$

is fulfilled for all y,s and such w_1, w_2 that $|w_i| \leq \hat{w}_0$, $i=1,2$. The constant C_2 is chosen in such a way that the inequality

$$2\left| \left(-1+\hat{E}(y,s)\right) F(w) \right| \leq C_2$$

is fulfilled for all y,s and such w that $|w| \leq \hat{w}_0$. Constants C_1 and C_2 do hot depend on the value of t. Thus, the estimate (1.19) and inequality (1.18) give us a uniform convergence of the sequence $\hat{w}_n(y,t)$ for a fixed y on a segment $[0,t]$. Then the limit function satisfies equation (1.17) and, consequently, it is a smooth function of class C^1 with respect to variable t. Therefore for a fixed y function $w(y,t)$ satisfies equation (1.5).

Function $w(y,t)$ satisfies inequality (1.15). From (1.7) it follows that

$$\hat{w}_{n+1}(y,t) = \hat{w}_n(\varphi^{-1}(y),t+1).$$

Consequently,

$$w(y,t) < \hat{w}_n(\varphi^{-1}(y),t+1) < \hat{w}_n(y,t). \qquad (1.20)$$

From inequality (1.20) it follows that

$$\lim_{n\to\infty} \hat{w}_n(\varphi^{-1}(y),t+1) = w(y,t),$$

i.e.

$$w(\varphi^{-1}(y),\; t+1) = w(y,t). \qquad (1.21)$$

Comparing (1.21) with (1.4), we obtain the covering of function $w(y,t)$ by a certain function on space X.

2. Finiteness of the number of stationary solutions.

As it is shown in Section 1, there exists at least one stationary solution to equation (1.5). Let

$$w_1(x,t) = w(g_t(x)), \quad w_2(x,t) = w_2(g_t(x))$$

be two such solutions. Assume that functions $w_1(x)$, $w_2(x)$ differ from each other on the set of a positive measure. It is obvious that if for some point x X the inequality

$$w_1(x) < w_2(x)$$

is satisfied, then also for any t the inequality

$$w_1(g_t(x)) < w_2(g_t(x))$$

is fulfilled. Consequently, on the set of a complete measure holds the inequality

$$w_1(x) < w_2(x). \tag{2.1}$$

Let $w_1, w_2, \ldots, w_n$ be a finite number of stationary solutions to equation (1.5) and on the set of a complete measure hold the inequalities

$$w_1(x) < w_2(x) < \ldots < w_n(x). \tag{2.2}$$

Put

$$w_k(x) = w_1(x) + \alpha_k(x)$$

and, respectively,

$$w_k(x,t) = w_1(x,t) + \alpha_k(x,t), \tag{2.3}$$

$$\alpha_k(x,t) > 0, \quad K=2,\ldots n.$$

Since all functions (2.3) satisfy equation (1.5), then

$$\frac{d\alpha_k}{dt} = \left(-1+ \hat{E}(x,t)\right) \times$$

$$\times\alpha_k\left(3w_1^2 + 3w_1\alpha_k + (w_+ - w_-)(2w_1 + \alpha_k) - w_-w_+ + \alpha_k^2\right). \tag{2.4}$$

Similarly for $k \geq 3$ we have

$$w_k(x,t) = w_2(x,t) + \alpha_k(x,t) - \alpha_2(x,t).$$

Therefore

$$\frac{d}{dt}\left(\alpha_k(x,t) - \alpha_2(x,t)\right) = \left(-1+\hat{E}(x,t)\right)\left(\alpha_k(x,t) - \alpha_2(x,t)\right)\times$$

$$\times\left(3w_1^2 + 3w_1\alpha_k + 3w_1\alpha_2 + \alpha_k\alpha_2 + \alpha_k^2 + \alpha_2^2 + \right.$$

$$\left. + 2(w_+ - w_-)w_1 + (w_+ - w_-)(\alpha_k - \alpha_2) - w_-w_+\right) . \tag{2.5}$$

Subtracting equality (2.4) from equality (2.5) for $k=2$, $k=k$ we obtain

$$\frac{\frac{d}{dt}(\alpha_k - \alpha_2)}{\alpha_k - \alpha_2} - \frac{\frac{d}{dt}\alpha_k}{\alpha_k} - \frac{\frac{d}{dt}\alpha_2}{\alpha_2} = \left(-1 + \hat{E}(x,t)\right)\times$$

$$\times\left(-3w_1^2 + \alpha_k\alpha_2 - 2(w_+ - w_-)w_1 + w_+w_-\right) . \tag{2.6}$$

Then

$$\frac{\alpha_k - \alpha_2}{\alpha_k\alpha_2}\frac{\alpha_k^0\alpha_2^0}{\alpha_k^0 - \alpha_2^0} =$$

$$= \exp\left(\int_0^t \left(-1+\hat{E}(x,t)\right)\left(-3w_1^2 + \alpha_k\alpha_2 - 2(w_+ - w_-)w_1 + w_+w_-\right) dt\right). \tag{2.7}$$

Since the functions

$$\left(-1+\hat{E}(x,t)\right)\left(-3w_1^2 - 2(w_+ - w_-)w_1 + w_+w_-\right)$$

and

$$\left(-1 + E(x,t)\right)\alpha_k\alpha_2$$

are bounded measurable functions on manifold X and the action of g_t

is ergodic, then

$$A = \int_X \left(-1+E(x)\right)\left(-3w_1^2 - 2(w_+-w_-)w_1 + w_+w_-\right) d\mu(x) =$$

$$= \lim_{t\to\infty} \frac{1}{t} \int_0^t \left(-1+E(g_t(x))\right)\left(-3w_1^2(g_t(x)) - 2(w_+-w_-)w_1(g_t(x)) + w_+w_-\right) dt,$$

$$B_k = \int_X \left(-1+E(x)\right)\alpha_k(x)\,\alpha_2(x)\,d\mu(x) =$$

$$= \lim_{t\to\infty} \frac{1}{t} \int_0^t \left(-1+E(g_t(x))\right)\alpha_k(g_t(x))\,\alpha_2(g_t(x))dt \ .$$

In other words,

$$\int_0^t \left(-1+E(g_t(x))\right)\left(-3w_1^2(g_t(x)) - 2(w_+-w_-)w_1(g_t(x)) + w_+w_-\right) dt =$$

$$= t\left(A + \delta(t)\right), \quad \delta(t) \longrightarrow 0. \tag{2.8}$$

$$\int_0^t \left(-1+E(g_t(x))\right)\alpha_k(g_t(x))\,\alpha_2(g_t(x))dt =$$

$$= t\left(B_k + \varepsilon_k(t)\right), \quad \varepsilon_k(t) \rightarrow 0. \tag{2.9}$$

Then

$$\frac{\alpha_k(g_t(x)) - \alpha_2(g_t(x))}{\alpha_k(g_t(x))\alpha_2(g_t(x))} \; \frac{\alpha_k(x) \quad \alpha_2(x)}{\alpha_k(x) - \alpha_2(x)} =$$

$$= e^{t(A + B_k + \delta(t) + \varepsilon(t))} \ . \tag{2.10}$$

From inequalities (2.2) it follows that on a certain set E of a positive measure simultaneously hold the inequalities

$$\alpha_k(x) \geqslant \eta > 0,$$

$$\alpha_{k_1}(x) - \alpha_{k_2}(x) \geqslant \eta > 0, \quad k_1 > k_2, \quad x \in E \tag{2.11}$$

Then there exists such a sequence t_n, $\lim_{n\to\infty} t_n = \infty$ that $g_{t_n}(x) \in E$ for an appropriate $x \in E$. This means that the sequence of left-hand parts of (2.10) for $t = t_n$ is bounded both from above and below. Therefore the sequence

$$t_n(A + B_k + \delta(t_n) + \mathcal{E}(t_n)$$

is bounded. Therefore for all $k \geq 3$ hold the inequalities

$$A + B_k = 0$$

i.e.

$$B_k = -A. \qquad (2.12)$$

On the other hand,

$$B_4 = \int_X \Big(-1+E(x)\Big)\alpha_4(x)\alpha_2(x)d\mu(x) =$$

$$= \int_X \Big(-1+E(x)\Big)\Big(\alpha_k(x) - \alpha_3(x)\Big)\alpha_2(x)d\mu(x) + \int_X \Big(-1+E(x)\Big)\alpha_3(x)\alpha_2(x)d\mu(x) \leq$$

$$\leq -(1-q)\eta^2\mu(E) + B_3 < B_3, \qquad (2.13)$$

which contradicts (2.12).

Thus we prove that $n \leq 3$.

3. Asymptotics of solutions under three stationary solutions.

For simplicity let us assume that equation (1.5) has the following form:

$$\dot{w} = v - (w^3 - w). \qquad (3.1)$$

Let $w_1 \leq w_3 \leq w_2$ be three stationary solutions. Put

$$w_2 = w_1 + \alpha_1, \quad w_3 = w_1 + \alpha_2, \quad w = w_1 + \alpha, \ 0 \leq \alpha \leq \alpha_2 \leq \alpha_1 .$$

Then

$$\frac{\dot{\alpha}_1}{\alpha_1} = 1 - 3w_1^2 - w_1\alpha_1 - \alpha_1^2 ,$$

$$\frac{\dot{\alpha}_2}{\alpha_2} = 1 - 3w_1^2 - w_1\alpha_2 - \alpha_2^2$$

$$\frac{\dot{\alpha}}{\alpha} = 1 - 3w_1^2 - w_1\alpha - \alpha^2 .$$

On the other hand,

$$\dot{w}_3 - \dot{w}_2 = -(w_3^3 - w_2^3) + (w_3 - w_2) ,$$

or

$$\frac{\dot{\alpha}_2 - \dot{\alpha}_1}{\alpha_2 - \alpha_1} = 1 - 3w_1^2 - 3w_1\alpha_2 - 3w_1\alpha_1 - (\alpha_1^2 + \alpha_1\alpha_2 + \alpha_2^2),$$

and analogously,

$$\frac{\dot{\alpha}_2 - \dot{\alpha}}{\alpha_2 - \alpha} = 1 - 3w_1^2 - 3w_1\alpha_2 - 3w_1\alpha - (\alpha^2 + \alpha\alpha_2 + \alpha_2^2).$$

Compose the following two expressions

$$\frac{\dot{\alpha}_1}{\alpha_1} + \frac{\dot{\alpha}_2}{\alpha_2} + \frac{\dot{\alpha}_2 - \dot{\alpha}_1}{\alpha_2 - \alpha_1} = 1 - 3w_1^2 - 3\alpha_1\alpha_2 , \tag{3.2}$$

$$\frac{\dot{\alpha}}{\alpha} + \frac{\dot{\alpha}_2}{\alpha_2} + \frac{\dot{\alpha}_2 - \dot{\alpha}}{\alpha_2 - \alpha} = 1 - 3w_1^2 - 3\alpha\alpha_2 . \tag{3.3}$$

Integrating these expressions, we obtain

$$\frac{\alpha_1\alpha_2}{\alpha_1 - \alpha_2} = \frac{\alpha_1^0\alpha_2^0}{\alpha_1^0 - \alpha_2^0} e^{\int_0^t \left[(1-3w_1^2) + \alpha_1\alpha_2\right] dt} , \tag{3.4}$$

$$\frac{\alpha\alpha_2}{\alpha - \alpha_2} = \frac{\alpha^0\alpha_2^0}{\alpha^0 - \alpha_2^0} e^{\int_0^t \left[(1-3w_1^2) + \alpha\alpha_2\right] dt} . \tag{3.5}$$

Since the functions α_1 and α_2 are stationary, then there exists such point $x_0 \in X$ that the functions $\alpha_1(x_0, t)$, $\alpha_2(x_0, t)$, $\alpha_1(x_0,t) - \alpha_2(x_0,t)$ are bounded from above and from below by positive constants for an infinite sequence of values t_n for $t_n \to \infty$. This means that integrals

$$\int_0^{t_n}\left[(1-3w_1^2)+\alpha_1\alpha_2\right]dt$$

are jointly bounded. On the other hand,

$$\int_0^{t_n}\left[(1-3w_1^2)\right]dt = At_n + t_n o(1)\ ,$$

$$\int_0^{t_n}\alpha_1\alpha_2 dt = Bt_n + t_n o(1).$$

So, $A + B = 0$. Further

$$\int_0^{t_n}\alpha_1\alpha_2 dt = \int_0^{t_n}(\alpha_1-\alpha_2)\alpha_2 dt + \int_0^{t_n}\alpha_2^2 dt = B_1 t_n + B_2 t_n + t_n o(1),$$

and $B_1 > 0$, $B_2 > 0$. Therefore $B_1 + B_2 = -A$. Finally, since $\alpha \leq \alpha_2$, then

$$\int_0^{t_n}\alpha\alpha_2 dt \leq B_2 t_n + t_n o(1).$$

Thus

$$\int_0^{t_n}\left[(1-3w_1^2)+\alpha\alpha_2\right]dt \leq At_n + B_2 t_n + t_n o(1) =$$

$$= -B_1 t_n + t_n\ o(1) \longrightarrow -\infty$$

Consequently,

$$\left|\frac{\alpha\alpha_2}{\alpha-\alpha_2}\right| \leq \left|\frac{\alpha^0\alpha_2^0}{\alpha^0-\alpha_2^0}\right| e^{-Ct} \qquad (3.6)$$

for some positive constant C, starting from some value t_n, and both C and t_n do not depend on the choice of function α.

4. Asymptotics of solutions under two stationary solutions.

We shall make use of the notations from the previous section, the only difference being the fact that function w_3 is not assumed to be stationary. Dividing (3.4) by (3.5), we obtain

$$\frac{\alpha_2(\alpha_1-\alpha)}{\alpha(\alpha_1-\alpha_2)} = \frac{\alpha_{20}(\alpha_{10}-\alpha_0)}{\alpha_0(\alpha_{10}-\alpha_{20})}\, e^{\int_0^t \alpha_1(\alpha_2-\alpha)dt} \tag{4.1}$$

Since $\alpha_1(\alpha_2-\alpha) \geq 0$, then the function

$$\int_0^t \alpha_1(\alpha_2-\alpha)dt$$

is a monotonously increasing function. If

$$0 < \lim_{t\to\infty} \int_0^t \alpha_1(\alpha_2-\alpha)dt < +\infty$$

then

$$\lim_{t\to\infty} \alpha_1(\alpha_2-\alpha) = 0 \quad .$$

And

$$\lim_{t\to\infty} \frac{\alpha_2(\alpha_1-\alpha)}{(\alpha_1-\alpha_2)} = \frac{\alpha_{20}(\alpha_{10}-\alpha_0)}{\alpha_0(\alpha_{10}-\alpha_{20})}\, A, \tag{4.2}$$

where

$$A = e^{\lim_{t\to\infty} \int_0^t \alpha_1(\alpha_2-\alpha)dt} > 1.$$

Assume that $\alpha_1 \geq \varepsilon > 0$ for all t. This condition, obviously, takes place for a continuous function α_1. In this case

$$\lim_{t\to\infty} (\alpha_2-\alpha) = 0$$

Consequently, either $\lim_{t\to\infty} (\alpha_1-\alpha) = 0$ or $\lim_{t\to\infty} \alpha = 0$. Indeed, if there exists such sequence t_n that $\lim_{n\to\infty} (\alpha_1-\alpha)(t_n) > 0$ and $\lim_{n\to\infty} \alpha(t_n) > 0$, then a lift-hand side of (4.2) is equal to unit, the right-hand side being greater than unit.

In the case when

$$\lim_{t\to\infty} \int_0^t \alpha_1(\alpha_2-\alpha)dt = 0 \;,$$

we have

$$\lim_{t\to\infty} \alpha(\alpha_1-\alpha_2) = 0 \quad .$$

Put $\alpha = \beta\alpha_1$. Then $0 \leq \beta \leq 1$. From (3.5) we obtain

$$\frac{\beta}{1-\beta} = \frac{\beta_0}{1-\beta_0}\frac{\alpha_{10}}{\alpha_1} e^{\int_0^t (1-3w_1^2)dt + \int_0^t \beta\alpha_1^2 dt} . \qquad (4.3)$$

Hence we have

$$\beta = \frac{\frac{\beta_0}{1-\beta_0}\frac{\alpha_{10}}{\alpha_1} e^{\int_0^t (1-3w_1^2)dt + \int_0^t \beta\alpha_1^2 dt}}{1 + \frac{\beta_0}{1-\beta_0}\frac{\alpha_{10}}{\alpha_1} e^{\int_0^t (1-3w_1^2)dt + \int_0^t \beta\alpha_1^2 dt}} .$$

Note that the condition

$$\frac{\dot{\alpha}_1}{\alpha_1} = 1 - 3w_1^2 - w_1\alpha_1 - \alpha_1^2$$

means that

$$(1 - 3w_1^2)d\mu(x) = \int (w_1\alpha_1 - \alpha_1^2)d\mu(x).$$

Analogously,

$$(1 - 3w_2^2)d\mu(x) = \int (-w_2\alpha_1 - \alpha_1^2)d\mu(x) =$$

$$= \int (-w_1\alpha_1)d\mu(x).$$

Then

$$\int (1-3w_1^2)d\mu(x) + \int (1-3w_2^2)d\mu(x) \pm \int (-\alpha_1^2)d\mu(x) < 0.$$

Let

$$\lim \frac{1}{t}\int_0^t \left[(1-3w_1^2) \right] dt = -A, \quad A>0. \qquad (4.5)$$

Then

$$\int_0^t \left[(1-3w_1^2) \right] dt = -At + t\varphi(t), \quad \lim \varphi(t) = 0.$$

Put

$$u(t) = \int_0^t \beta\alpha_1^2 dt - At. \qquad (4.6)$$

Then

$$\dot{u}(t) = \beta \alpha_1^2 - A,$$

i.e.

$$\dot{u} = \frac{\alpha_1^2 \frac{\beta_0}{1-\beta_0} \frac{\alpha_{10}}{\alpha_1} e^{t\varphi(t)+u}}{1 + \frac{\beta_0}{1-\beta_0} \frac{\alpha_{10}}{\alpha_1} e^{t\varphi(t)+u}} - A.$$

Assume that there exists such constant C, that $|t\varphi(t)| \lesssim C$ for all values of t. Then for $u \lesssim 0$ and a sufficiently small β_0 holds the estimate

$$\left| \frac{\alpha_{10}}{1-\beta_0} e^{t\varphi(t)+u} \right| \lesssim C_1 .$$

Then

$$\alpha_1^2 \frac{\frac{\beta_0 \alpha_{10}}{1-\beta_0} e^{t\varphi(t)+u}}{\alpha_1 + \frac{\beta_0 \alpha_{10}}{1-\beta_0} e^{t\varphi(t)+u}} \lesssim \alpha_1 \sqrt{\alpha_1 \beta_0 C_1} . \qquad (4.7)$$

Consequently, for a sufficiently small β_0 holds the estimate

$$\dot{u}(t) \lesssim -\frac{A}{2} ,$$

i.e.

$$u(t) \lesssim -\frac{A}{2}t. \qquad (4.8)$$

Then

$$\alpha = \frac{\alpha_1 \frac{\beta_0 \alpha_{10}}{1-\beta_0} e^{t\varphi(t)+u(t)}}{\alpha_1 + \frac{\beta_0 \alpha_{10}}{1-\beta_0} e^{t\varphi(t)+u(t)}} \lesssim \left(\alpha_1 + \frac{\beta_0 \alpha_{10}}{1-\beta_0} e^{t\varphi(t)+u(t)}\right)^{\frac{1}{2}}$$

i.e.

$$\lim_{t\to\infty} \alpha(t) = 0 .$$

Instead of the inequality $|t\varphi(t)| \leq c$ it is sufficient that the inequality $t\varphi(t) \leq c$ should hold. Moreover, assume that the inequality $t\varphi(t) \leq \psi(t)$ is fulfilled for some function $\psi(t)$, for which $\psi(0) = 0$, $\psi'(t) \leq \varepsilon$ for some $\varepsilon < A$ and

$$\lim_{t\to\infty} \psi'(t) = 0. \tag{4.9}$$

Put

$$v(t) = u(t) + \psi(t).$$

Then

$$\dot{v}(t) = \frac{\alpha_1^2 \frac{\beta_0 \alpha_{10}}{1-\beta_0} e^{(t\varphi(t) - \psi(t)) + v(t)}}{\alpha_1 + \frac{\beta_0 \alpha_{10}}{1-\beta_0} e^{(t\varphi(t) - \psi(t)) + v(t)}} - A + \dot{\psi}(t) .$$

Since

$$\frac{\alpha_{10}}{1-\beta_0} e^{(t\varphi(t) - \psi(t))} \leq c_2 ,$$

then

$$\dot{v}(t) \leq \alpha_1 \sqrt{\alpha_1 \beta_0 c_2} - A + \varepsilon$$

for all $v \leq 0$ and sufficiently small β_0. Consequently, choosing smaller β_0, if necessary, we obtain

$$\dot{v}(t) \leq -c_3,$$

or

$$v(t) \leq -c_3 t, \qquad c_3 > 0. \tag{4.10}$$

Then

$$\alpha = \beta\alpha_1 = \frac{u'+A}{\alpha_1} = \frac{v'+A-\psi'(t)}{\alpha_1} =$$

$$= \frac{\alpha_1 \frac{\beta_0 \alpha_{10}}{1-\beta_0} e^{(t\varphi(t)-\psi(t))+v(t))}}{\alpha_1 + \frac{\beta_0 \alpha_{10}}{1-\beta_0} e^{(t\varphi(t)-\psi(t))+v(t))}} \leq$$

$$\leq \left(\alpha_1 \frac{\beta_0 \alpha_{10}}{1-\beta_0} e^{(t\varphi(t) - \psi(t))+v(t))}\right)^{\frac{1}{2}} . \tag{4.11}$$

Consequently,

$$\lim_{t\to\infty} \alpha(t) = 0 \quad .$$

Thus, it required to find such function $\psi(t)$ that

$$\psi'(t) \leq \mathcal{E} < A, \quad t\varphi(t) \leq \psi(t).$$

The fulfillment of condition (4.9) is not necessary.

Since $\lim_{t\to\infty} \varphi(t) = 0$, then there exists such value t_0, that for $t > t_0$ holds the inequality

$$|\varphi(t)| < \mathcal{E}.$$

Then

$$t\varphi(t) < \mathcal{E} t.$$

Let

$$C_4 = \max_{0 \leq t \leq t_0} t\varphi(t).$$

Then

$$t\varphi(t) \leq \mathcal{E} t + C_4.$$

So, one should put

$$\psi(t) = \mathcal{E} t + C.$$

This completes the proof of the existence of solution $\alpha(t)$ with condition (4.11).

From inequality (4.11) it follows that $\lim_{t\to\infty} \beta = 0$. Indeed,

$$\alpha_1 = \alpha_{10}\, e^{\int_0^t (1-3w_1^2 - 3w_1\alpha_1 - \alpha_1^2)dt} \quad . \qquad (4.12)$$

Since on the set of a positive measure the function $\alpha_1 > \mathcal{E} > 0$, then for an infinite sequence $t_n \to \infty$ holds the inequality

$$\alpha_1(t_n) > \mathcal{E} > 0.$$

Consequently, from (4.12) it follows that

$$\lim_{t\to\infty} \chi(t) = \lim_{t\to\infty} - \int_0^t (1-3w_1^2 - 1w_1\alpha_1 - \alpha_1^2)dt = 0 \quad .$$

Then the right side of (4.11) tends to zero when $t \to \infty$. Indeed, the square of the right side of (4.11) is equal to

$$\frac{\beta_0 \alpha^2_{10}}{1-\beta_0}\, e^{(t\varphi(t) - \psi(t))+v(t) - t\chi(t)}.$$

Taking account of (4.10), we obtain

$$\lim_{t\to\infty} \Big(t\varphi(t) - \psi(t)+v(t)-t\chi(t)\Big) = -\infty ,$$

Q.E.D.

Thus, from (4.11) it follows that

$$\lim_{t\to\infty} \beta = 0.$$

So, we showed that if condition (4.5) is fulfilled, there exists a solution $w \longrightarrow w_1$. If holds the condition

$$\lim \frac{1}{t} \int_0^t \Big[(1-3w_2^2) \Big] dt = -B, \quad B > 0, \qquad (4.13)$$

then there exists solution $w_3 \to w_2$ and, thus, the third, intermediate, stationary solution. In the case of three stationary solutions both conditions (4.5) and (4.13) are simultaneously fulfilled.

Consequently, if the equation has only two stationary solutions, then either

$$\lim \frac{1}{t} \int_0^t \Big[(1-3w_2^2) \Big] dt = 0,$$

or

$$\lim \frac{1}{t} \int_0^t \left[(1-3w_1^2) \right] dt = 0.$$

Reference

1. S.N.Kritskiĭ, D.V.Korenistov, D.Ya.Ratkovich. Fluctuations of the level of the Caspian Sea (calculation analysis and probability forecast). - Moscow, Nauka, 1975 (in Russian).

CONTINUATION OF SOLUTIONS TO ELLIPTIC EQUATIONS AND LOCALIZATION OF SINGULARITIES

B.Yu.Sternin
Moscow State University
Lenin Hills,
Moscow 119 899,
USSR

V.E.Shatalov
Moscow Institute of Electronic
Engineering
B.Vuzovskii per., 3/12
Moscow 109028, USSR

This paper presents a review of the present state of the theory of continuation of solutions to elliptic differential equations from the point of view of the theory of differential equations in complex domains. That is why we present here a detailed discussion only of that part of the theory which uses the complex-analytic continuation of solutions; the investigations using different methods are just mentioned. Certainly, the choice of topics for detailed discussion was greatly influenced by the scientific interests of the authors. The detailed discussion of other questions concerued with the problem can be found in the literature cited below.

We are grateful to Prof. Boris Shapiro, who turned our attention to paper [**KS 2**], and to Prof. Harold Shapiro for useful discussions during our visit to Royal Institute of Technology (Stockholm, Sweden, in autumn 1990).

1. Statement of the problem

Let

$$\hat{L} = \sum_{|\alpha|\leq m} a_\alpha(x) \left(\frac{\partial}{\partial x}\right)^\alpha \tag{1}$$

be an elliptic differential operator in a domain Ω in space $\mathbf{R}^n$. Let $D \subset \Omega$ be a subdomain of Ω and $u(x)$ be a solution to the equation

$$\hat{L} u = 0 \tag{2}$$

in D. The problem is to continue the solution $u(x)$ into $\Omega\backslash D$ and to locate the singularities of the continuation.

This statement of the problem is a classical one. We mention here the paper of Gustav Herglotz [**Hr 1**] where the problem of continuation of Newtonian potential into the domain occupied by gravitating masses is examined. Similar problems arise when considering inverse problems of scattering theory (see, e.g., [**Co 1**],[**Co 2**] for axially symmetric case, [**Sl 1**] where the limitation of axial symmetricity is removed or [**WBA 1**] where similar results for vector problems are considered). The continuation of solutions to the Helmholtz equation arises also when considering the Rayleigh hypothesis (see, e.g., [**AK 1**] and references therein, [**Mi 1**], [**Mi 2**], [**Mi 4**], [**Mi 5**]), and in some problems of computational mathematics (see [**Kü 1**], [**Kü 2**), and so on.

First of all we point out that the continuation of the solution to the equation (2) is unique, if any. This fact follows from the real analyticity of the solutions to the elliptic equations (see, for example, [**Lw 2**], [**Ha 1**]). Of course, we must require that the coefficients of the operator (1) be real-analytic functions in Ω.

Next, we can easily see that not every solution to the equation (2) given in a domain D can be continued as a solution to the same equation into a wider domain. To show this let us consider a 2π-periodic function $g(\varphi)$, such that it has m continuous derivatives but its m-th derivative is not differentiable on a dense set in $[0, 2\pi]$. Let $u(x)$ be the solution of the Dirichlet problem

$$\begin{cases} \Delta u = 0, \; x \in D_1 \\ u|_{\partial D_1} = g(\varphi) \end{cases}$$

in the unit disk $D_1 \subset \mathbf{R}^2$. It is clear that $u(x)$ cannot be continued into any wider domain. Otherwise the restriction $g(\varphi)$ of the function $u(x)$ to the boundary ∂D_1 would be an analytic function in some open subset of ∂D_1, which is impossible.

Finally, the real analyticity of solutions to elliptic equation (2) leads to occurence of multiple-valued solutions to this equation. The reason is just the same as in the theory of complex-analytic functions: the continuation along different paths in space $\mathbf{R}^n$ to one and the same point can give different solutions in a neighbourhood of this point. This is the case, for example, when the continuation of solution of exterior boundary value problem for an ellipse inside the ellipse is considered. In this case there exist two points where the solution ramifies; these two points are just the foci of the ellipse.

Thus, we have to describe the class of the equations (2) whose solutions admit continuation across the boundary of their domains, then to examine the maximal domain into which the continuation is possible and, finally, to locate the singularities (in general, ramifying points) of the continued solutions. From the above considerations it is evident that the maximal domains of continuation are in general coverings over $\mathbf{R}^n$ rather then domains in $\mathbf{R}^n$.

We shall consider here two variants of statement of the problem for which the continuation is possible.

The first variant is as follows. Let $u(x)$ be a solution to the equation

$$\hat{L} u = f \tag{3}$$

in the entire space $\mathbf{R}^n$ where f is a function vanishing in D and coinciding in $\mathbf{R}^n \backslash D$ with a real-analytic function (this function will be also denoted by f). Then $u(x)$ is a solution to the equation (2) in the domain D and the problem ie to continue $u(x)$ outside D *as a solution to the equation* (2). One of such problems, mentioned above, is the continuation of Newtonian potential into domain occupied by gravitating masses. The other variant of this problem arises in mathematical theory of electrostatics (see [**Je** 1], §204). The problem is to replace a given distribution of charges with another one in which charges are contained in a smaller domain without changing the field outside the domain containing the former distribution. This problem was named "balayage" by Poincaré; in fact the change is "swept" inside the domain in such a fashion that it does not alter the external field ("balayage inwards"). The detailed discussion of this problem will be presented below; see also the papers of D.Khavinson and H.Shapiro [**Kh 2**], [**KS 2**], [**Sh 1**].

The second variant is the problem of continuation outside the domain D of the solution to the equation (2) subject to linear conditions on the boundary ∂D of this domain. For example, for a second-order operator $\hat{L}$ such conditions can be given in the form

$$\sum a_j(x)\frac{\partial u}{\partial x_j} + b(x)\, u|_{\partial D} = v(x) \tag{4}$$

Such statement of the problem arises in problems of continuation of solutions of boundary value problems; see, for example [**AK 1**], [**Lw 1**], [**Mi 1 - 16**].

Certainly, all the objects involved in the problems must be analytic: the coefficients of the operator (1), the coefficients of the operator in the left-hand side of the condition (4), right-hand sides $f(x)$, $v(x)$ and the boundary ∂D of the domain D under consideration.

The most advanced results are obtained for the operator $\hat{L}$ of the form

$$\hat{L}u = \Delta u + \sum a_j(x)\frac{\partial u}{\partial x_j} + b(x)u \tag{5}$$

especially in the two-dimensional case. In what follows we assume that the operator $\hat{L}$ has the form (5) except for generalizations which will be mentioned specially.

2. Methods and results

In this section we present a brief survey of methods used for solving the problem of continuation and of results obtained by these methods. Methods presented in the literature upon the subject can be divided into six main groups.

1. Methods connected with the reflection principle. This method goes back to Schwarz's paper [**Sc 1**] (see also [**Da 1**], [**Sh 1**], [**KS 3**]). It is based on the fact that for any harmonic function $u(x)$ vanishing on the analytic arc Γ the relation

$$u(z) + u(R(z)) = 0 \tag{6}$$

holds; here $z \in \mathbf{C} \cong \mathbf{R}^2$, and the mapping R is defined as follows: for any $x_0 \in \Gamma$, Γ divides a neighbourhhod of x_0 into two disjoint parts. The mapping R interchanges these parts. The similar relation is valid for harmonic function $u(z)$ whose normal derivative vanishes on Γ.

The relation (6) can be used for continuation of a harmonic function from one side of Γ to another provided that this function (or its normal derivative) vanishes on Γ.

However, the reflection principle (6) cannot be generalized to higher dimensions as well as to operators more general than the Laplace operator. Indeed, even for the Laplace operator for $n \geq 3$, the following statement is valid (see **[KS 3]**):

Let $n \geq 3$, $(Tu)(x) = f(x)[u \circ R(x)]$ for a mapping $R: U \to U$, $R|_\Gamma = id$, $R \neq id$. Then, if for all functions $u(x)$ harmonic in U the function $(Tu)(x)$ is also harmonic in U, then (i) R is a conformal mapping, (ii) Γ is either a plane or a sphere, (iii) T is either a reflection in the plane or a Kelvin transformation.

If $n = 2$ *then even for the Helmholtz operator $\hat{L} = \Delta + \lambda^2$ the reflection principle is not valid* (except for the reflection w.r.t. the straight line). More exactly, the following statement is valid (**[KS 3]**).

Let U be a domain in $\mathbf{R}^2$, Γ be an analitic arc in U. If there exist two points $A, B \in U$ divided by Γ such that for any solution $u(x)$ of the problem $\hat{L}u = 0$ in U, $u|_\Gamma = 0$ the relation $u(A) = 0$ yields $u(B) = 0$ then Γ is a piece of a straight line.

An attempt to use the reflection principle for investigation of continuation problem for Helmholtz equation was made by V.F.Apeltzin in **[AK 1, Ch. III]**. Unfortunately, *the affirmation on the reflection principle in this paper is invalid as well as Theorem 4.1* (on synthesis of solution of the diffractional problem)*based on this principle.*

Thus, the possibilities of using the reflection principle are limited to the case of two-dimensional harmonic functions. That is why H.Lewy (**[Lw 1]**) considers (for $n = 2$ more general operators than $\hat{L} = \Delta$) the reflection of functions for the operator (5) rather than the reflection of points. Lewy's results will be discussed in details below; for a more detailed discussion on reflection principles see **[Ga 1]**, **[Kh 1]**, **[KS 3]**, **[Sh 1]**, **[Mi 14]**, **[Mi 15]** and references cited therein.

2. Methods connected with complex-analytic Cauchy problem. These methods are based on the explicit formulas for solutions of the Cauchy problem in complex domains. In two dimensions such formulas can be derived by Riemann's method of solving of Cauchy problem for the operators of the form (5); see **[Ve 1]**, **[He 1]**, **[Ga 2]**, **[Ga 3]**. In higher dimensions such formulas can be obtained for arbitrary operators with constant coefficients by means of the Fourier-Radon transform of complex-analytic functions worked out in the series of papers **[StS 2, StS 1 - 15]**.

A different techniques of investigation of the Cauchy problem based on globalization of Leray methods of uniformization (see **[GKL 1]**, **[Le 1]**) was developed by Gunnar Johnsson **[Jo 1]** *for the particular case of quadratic initial surfaces.*

Of course, before using the mentioned formulas, the problem under consideration is to be reduced to a complex-analytic Cauchy problem. For the problem of "balayage" such reduction was done in **[KS 2]**, **[Sh 1]**; the corresponding Cauchy problem has zero Cauchy data on the initial manifold. A more detailed discussion on this topic in presented below. For boundary value problems this method requires preliminary analytic continuation of the Cauchy data into the corresponding complex domain. The technique of continuation was developed by R.Millar **[Mi 1-16]** for two-dimensional case and for same specific types of boundaries in higher dimensions (a plane boundary and an axially-symmetric boundary in $\mathbf{R}^3$).

3. Methods connected with complex-analytic Goursat problem. These methods, developed only in two dimensions, are based on the explicit formulas carried out with the help of Riemann's method in [**Ve 1**], [**He 1**], [**Ga 2**], [**Ga 3**], [**Lw 1**]. Certainly, on the preliminary stage one has to continue the *Goursat data* for the solution to a complex domain. Such continuation can be carried out in the following way.

Firstly, *every* solution to the homogeneous equation $\hat{L}u = 0$ given in the real domain D (where $\hat{L}$ is an operator of the form (5)) can be continued to the complex domain called Vekua hull of the domain D (we emphasize that *no conditions of regularity of the solution $u(x)$ near the boundary ∂D of the domain D are necessary for such continuation*).

Secondly, further continuation is possible in the case when the function $u(x)$ is subject to "boundary conditions" of the form (4) with analytic functions $a_1(x)$, $a_2(x)$, $b(x)$, $v(x)$. The method of continuation based on the reduction of the problem to Volterra equations was developed by H.Lewy [**Lw 1**] (even for the case of nonliner "boundary conditions"). We give a more detailed presentation of this method below; note that for the operators of the type (5) *this method gives a complete solution of the problem of continuation of solutions of boundary value problems in two dimensions.* The singularities of the continuation correspond to the singularuties of $u(x)$ and the singularities of the so-called Schwarz function which is the equation of boundary Γ in "characteristical coordinates" (see below).

The notion of the Schwarz function, which is essentially due to Schwarz [**Sc 1**], was explicitly introduced by D.-A.Grave [**Gr 1**] (see also [**Da 1**]). This notion is connected also with focal properties of algebraic curves, with the notion of quadrature domains for harmonic functions, witn notions of Hilbert and Friedrichs operators, etc. We cannot discuss this topic in detail here; one can find a detailed discussion and complete references in the review [**Sh 1**] by Harold Shapiro.

We note here that the continuation of solutions of the equation $\hat{L}u = 0$ into the Vekua hull can be carried out for more general operators than (5) using the results of [**BS 1**], [**Ze 1**]. Such generalization was carried out in [**KS 1**]. We note that, in contrast to Vekua's method, this method does not present explicit formulas for continuation.

REMARK 1.

It is evident that both methods connected with the Cauchy problem and methods connected with the Goursat problems nesessarily require a continuation of the solution *into a complex domain* due to complexity of characteristics of elliptic operators.

4. The technique of "unitary solutions". This method is based on the idea that all singularities of solution of the Cauchy problem

$$\begin{cases} \hat{L}u(x) = f(x) \\ u(x)\,\text{has zero of order 2 on}\,\Gamma \end{cases}$$

for the Laplace operator $\hat{L}$ (or, more generally, for an operator $\hat{L}$ with constant coefficients) with an entire right-hand part $f(x)$ are included into the singularity set of the solution of this problem for $f = 1$ (the unitary solution or the Schwarz potential of the manifold Γ). This conjecture was stated by D.Khavinson and H.Shapiro (see [**Kh 2**], [**KS 2**], [**Sh 1**]) and was proved by them in some particular cases, such as in the case when Schwarz potential can be calculated explicitly ([**KS** 2]), when Γ is of special form:

$$\Gamma = \{z_1, z_2, z_3 | z_3 = \varphi(z_1 + i z_2)\}$$

([**Kh** 2]) and so on. So far no complete proof of this conjecture does exist.

5. The "inner" (functional-analytical and algebraic) methods which don't require the continuation of solutions into the complex domain. We cannot present here a detailed discussion of these methods, related to a priori estimates or corresponding algebraic theorems. Such methods were worked out in [**Co** 1], [**Co** 2], [**Sl** 1], [**WBA** 1], [**Sh** 2], [**Sh** 3], [**Wi** 6]. As a rule, these methods do not give exact location of the singularity set of the continued solution. For example, in [**Co** 1], [**Co** 2], [**Sl** 1], [**Mi** 6] only a convex hull of the set of singularities is calculated. We mention here a very interesting paper [**Sh** 2] by H.Shapiro where the question of existence of entire solutions for some problems is investigated.

6. Methods connected with integral representations and integral equations in $\mathbf{R}^n$ (see, for example [**AK** 1, **Ch. II, III**]). The results obtained by these methods are, as a rule, direct consequences of the corresponding results obtained with the help of the analytic continuation. For example, Theorem 2.1 of [**AK** 1, **Ch. III**] (V.F.Apeltzin) is a trivial consequence of H.Lewy's results.

However, there are certain integral representations which are of interest by themselves. An example of such representations is the representation of scattered field in terms of scattering diagram used by A.Kürktchan [**AK** 1, **Ch. II**], for localization of singularities of the continuation of scattered field into the domain occupied by a scattering body. Such representations allow to calculate, even in three dimensions, the convex hull of the set of singularities of continuation of a scattered field and have applications to inverse problems of the scattering theory.

In conclusion of this section we note that the main purpose of this review is to show the relationship between the continuation problems and the differential equations on complex-analytic manifolds. That is why very interesting questions concerning, for example, the theory of quadrature domains, the theory of Hilbert and Friedrichs operators and others are beyond detailed discussion here. For these questions, we refer the reader to the review [**Sh** 1] by Harold Shapiro.

3. Riemann method

Here we briefly recall the basic facts connected with applications of the Riemann method to the theory of elliptic equations with analytic coefficients. More detailed presentation of this method can be found in [**Ve** 1], [**He** 1], [**Ga** 2], [**Ga** 3]. In this and the following sections we consider only two-dimensional problems.

Let

$$\hat{L}u = \Delta u + a(x, y)\frac{\partial u}{\partial x} + b(x, y)\frac{\partial u}{\partial y} + c(x, y)u = f(x, y) \tag{8}$$

be an elliptic differential equation of second order with constant coefficients in the principal part. We suppose, for simplicity, that both the coefficients of the operator and the right-hand side of the equation (8) can be continued to entire functions of the complex variables x and

y (the equation (8) being originally given in a domain D of real plane $\mathbf{R}^2$). If the solution to (8) can be continued into a domain in $\mathbf{C}^2$, then the change of variables $z = x + iy$, $\zeta = x - iy$ reduces the equation (8) to the form

$$\hat{L}_{\mathrm{C}} U(z, \zeta) = \frac{\partial^2 u}{\partial z \partial \zeta} + a(z, \zeta)\frac{\partial U}{\partial z} + B(z, \zeta)\frac{\partial U}{\partial \zeta} + C(z, \zeta) U = F(z, \zeta) \tag{9}$$

which will be called *the complex form of the equation (8)*. Here

$$A(z, \zeta) = \frac{1}{4}\left[a\left(\frac{z+\zeta}{2}, \frac{z-\zeta}{2i}\right) + ib\left(\frac{z+\zeta}{2}, \frac{z-\zeta}{2i}\right)\right],$$

$$B(z, \zeta) = \frac{1}{4}\left[a\left(\frac{z+\zeta}{2}, \frac{z-\zeta}{2i}\right) - ib\left(\frac{z+\zeta}{2}, \frac{z-\zeta}{2i}\right)\right],$$

$$C(z, \zeta) = \frac{1}{4}\left(\frac{z+\zeta}{2}, \frac{z-\zeta}{2i}\right), \qquad F(z, \zeta) = \frac{1}{4} f\left(\frac{z+\zeta}{2}, \frac{z-\zeta}{2i}\right)$$

are entire functions of the variables (z, ζ).

The equation

$$\hat{L}^*_{\mathrm{C}} u = \frac{\partial^2 U}{\partial z \partial \zeta} - \frac{\partial}{\partial z}[A(z, \zeta)U] - \frac{\partial}{\partial \zeta}[B(z, \zeta)U] + C(z, \zeta)U = G(z, \zeta) \tag{10}$$

is called the conjugate equation for the equation (9). Note that the solutions of (8) can be obtained from the solutions of (9) via the formula $u(x, y) = U(z, \bar{z})$. We call the domain $V(D) = D \times \bar{D}$ in the space $\mathbf{C}^2_{z,\zeta}$ *a Vekua hull of the domain* D (a fundamental domain in Vekua's terminology; see [**Ve 1**]). Note that the all theory presented below is valid if A, B, C, F are holomorphic only in $D \times \bar{D}$.

DEFINITION 1.

The (unique) solution $G(z, \zeta, z_0, \zeta_0)$ of the Goursat problem

$$\begin{cases} L^*_{\mathrm{C}}(G) = 0, \\ G|_{z=z_0} = exp\left\{\int_{\zeta_0}^{\zeta} A(z_0, \eta)\, d\eta\right\}, \\ G|_{\zeta=\zeta_0} = exp\left\{\int_{z_0}^{z} B(\xi, \zeta_0)\, d\xi\right\}, \end{cases} \tag{11}$$

w.r.t. variables (z, ζ) is called *a Riemann's function* of the equation (8).

The following affirmation is valid ([**Ve 1**]).

THEOREM 1. *There exists a unique solution of the problem (11) which is an entire function w.r.t. its arguments. This solution satisfies the equation*

$$L_{\mathrm{C}}(G) = 0 \tag{12}$$

w.r.t. the arguments (z_0, ζ_0).

Using the obrious relation

$$2[V\,\hat{L}_{\mathrm{C}}(U) - U\,\hat{L}^{*}_{\mathrm{C}}(V)] = (U_z V - V_z U + 2BUV)_\zeta + (U_\zeta V - V_\zeta U + 2AUV)_z \qquad (13)$$

valid for any functions $U(z\,,\,\zeta)\,,\,V(z\zeta)$ holomorphic in the Vekua hull $D\times\bar{D}$, one can easily derive the following integral relation

$$\begin{aligned} U(z\,,\,\zeta) &= U(z_0\,,\,\zeta_0)G(z_0\,,\,\zeta_0\,,\,z\,,\,\zeta) \\ &+ \int_{z_0}^{z}\{U_z(\xi\,,\,\zeta_0) + B(\xi\,,\,\zeta_0)U(\xi\,,\,\zeta_0)\}G(\xi\,,\,\zeta_0\,,\,z\,,\,\zeta)d\xi \\ &+ \int_{\zeta_0}^{\zeta}\{U_\zeta(z_0\,,\,\eta) + A(z_0\,,\,\eta)U(z_0\,,\,\eta)\}G(z_0\,,\,\eta\,,\,z\,,\,\zeta)d\eta \\ &+ \int_{z_0}^{z}d\xi\int_{\zeta_0}^{\zeta}G(\xi\,,\,\eta_0\,,\,z\,,\,\zeta)\hat{L}_{\mathrm{C}}(U)(\xi\,,\,\eta)d\eta \end{aligned} \qquad (14)$$

where the subscripts denote derivatives w.r.t. the corresponding variables. With the help of the formula (14) one can easily obtain the solution of the Goursat problem for the equation (9).

There exists another integral relation adapted for solving Cauchy problems which is essentially due to H.Lewy ([**Lw** 1]). Let $\zeta = S(z)$ be an analytic in D function which has an inverse function $S^{-1}(\zeta)$ on $\bar{D}$. Let $\gamma_1[z\,,\,z_0]$ be a path from z to z_0 lying in the domain D, and let $\gamma_2[\zeta\,,\,\zeta_0]$ be a path from ζ to ζ_0 lying in the domain $\bar{D}$. Suppose that $\zeta = S(z_0)$ $\zeta_0 = S(z)$, so that $(z_0\,,\,\zeta)\in\Gamma\,,\,(z\,,\,\zeta_0)\in\Gamma$. Let $z = z(1-s)\,,\,\zeta = \zeta(s) = S(z(s))\,,\,s\in[0\,,\,1]$ be parametrizations of these paths. Denote by $\gamma[(z_0\,,\,\zeta)\,,\,(z\,,\,\zeta_0)]$ the path $z = z(s)\,,\,\zeta = \zeta(s)\,,\,s\in[0\,,\,1]$; it is evident that $\gamma[(z_0\,,\,\zeta)\,,\,(z\,,\,\zeta_0)]$ lies in the manifold $\Gamma = \{(z\,,\,\zeta)\,|\,\zeta = S(z)\}$. The (real) two-dimensional surface $\sigma = \{z = z(s')\,,\,\zeta = \zeta(s'\prime)\,,\,0 \le s' \le 1\,,\,s' \le s'\prime \le 1\}$ (see figure 1) has the boundary $\gamma + \gamma_1 - \gamma_2$.

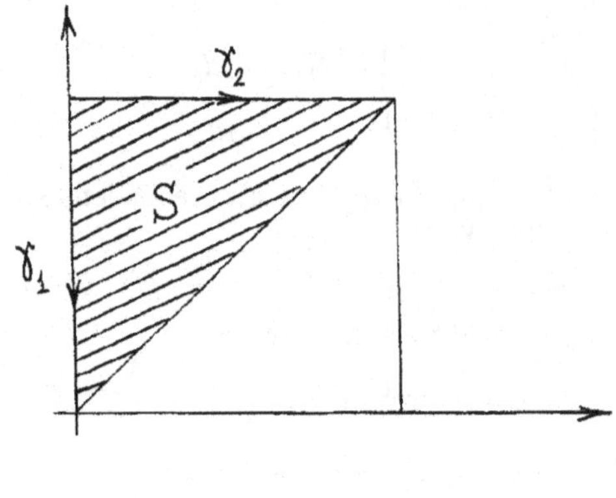

Fig. 1

Multiplying (13) by $dz \wedge d\zeta$ and integrating it over the surface σ (for $V = G(z, \zeta, z_0, \zeta_0)$ and any solution $U(z, \zeta)$ of the equation $\hat{L}_C U(z, \zeta) = 0$), we obtain the formula

$$\begin{aligned} U(z, \zeta) = \frac{1}{2}\Big\{ & U(S^{-1}(\zeta), \zeta) G(S^{-1}(\zeta), \zeta, z, \zeta) + U(z, S(z)) G(z, S(z), z, \zeta)) \\ & + \int_\gamma \Big[-\Big(\frac{\partial U}{\partial z_0}(z_0, \zeta_0) G(z_0, \zeta_0, z, \zeta) - \frac{\partial G}{\partial z_0}(z_0, \zeta_0, z, \zeta) U(z_0, \zeta_0) \\ & + 2B(z_0, \zeta_0) U(z_0, \zeta_0) G(z_0, \zeta_0, z, \zeta)\Big) dz_0 \\ & + \Big(\frac{\partial U}{\partial \zeta_0}(z_0, \zeta_0) G(z_0, \zeta_0, z, \zeta) - \frac{\partial G}{\partial \zeta_0}(z_0, \zeta_0, z, \zeta) U(z_0, \zeta_0) \\ & + 2A(z_0, \zeta_0) U(z_0, \zeta_0) G(z_0, \zeta_0, z, \zeta)\Big) d\zeta_0 \Big] \Big\} \end{aligned} \tag{15}$$

(where (z_0, ζ_0) is denoted by (z, ζ) and vice versa). This formula allows us to calculate the solution of Cauchy problem for the operator (9) with initial data on Γ, since all expressions in the right-hand side of the formula (15) use the values of the function $U(z, \zeta)$ and its first derivatives only at points of Γ.

The formula (14) gives a form of general solution to the equation (9) in the Vekua hull $D \times \bar{D}$:

$$\begin{aligned} U(z, \zeta) = {} & A G(z_0, \zeta_0, z, \zeta) + \int_{z_0}^{z} G(\xi, \zeta_0, z, \zeta) \Phi(\xi) \, d\xi \\ & + \int_{\zeta_0}^{\zeta} G(z_0, \eta, z, \zeta) \Psi(\eta) \, d\eta \end{aligned} \tag{16}$$

where A is an arbitrary constant and $\Phi(z)$, $\Psi(\zeta)$ are arbitrary functions holomorphic in D and $\bar{D}$, respectively. As we shall see below, this formula gives also the general solution to the equation (8) for $z = x + iy$, $\zeta = x - iy$, $(x, y) \in D$, i.e. any solution $u(x, y)$ to the equation (8) regular in D can be represented in the form (16).

Using the formula (16) as a starting point, we can prove the following affirmation ([Ve 1]).

THEOREM 2. *There exist such an entire function $\tilde{\Omega}(z, \zeta, z_0, \zeta_0)$ that the function*

$$\Omega(z, \zeta, z_0, \zeta_0) = \frac{1}{2\pi} G(z_0, \zeta_0, z, \zeta) \ln[(z - z_0)(\zeta - \zeta_0)] + \tilde{\Omega}(z, \zeta, z_0, \zeta_0) \tag{17}$$

is a solution to the equation $\hat{L}_C \Omega = 0$ w.r.t. the variables (z, ζ) and to the equation $\hat{L}_C^* \Omega = 0$ w.r.t. the variables (z_0, ζ_0).

It is evident that the function

$$\omega(x, y, x_0, y_0) = \Omega(x + iy, x - iy, x_0 + iy_0, x_0 - iy_0) \tag{18}$$

is a fundamental solution of the equation (8). Note here that, in spite of the fact that Ω is a multiple-valued function, the function ω is a single-valued function defined on the entire plane.

4. Continuation method based on the solution of the Goursat problem

1. Continuation into the Vekua hull. As it was mentioned above (see sect. 2) this method is based on the fact (cf. Vekua [**Ve 1**]) that each solution of the problem (8) regular in a domain D can be analytically continued into the Vekua hull $V(D)$ of D (we recall that $(x, y) \in V(D)$ iff $(z, \zeta) = (x+iy, x-iy) \in D \times \bar{D})$. This fact can be easily derived from the Green's formula using the fundamental solution (18). More precisely, let $T \subset D$ be a simple connected subdomain of D with smooth boundary (we suppose that $\partial T \subset D$). We have

$$\begin{aligned} u(x, y) = \int_{\partial T} \{ [u(x_0, y_0) \frac{\partial}{\partial x_0} \omega(x, y, x_0, y_0) \\ - \omega(x, y, x_0, y_0) [\frac{\partial u}{\partial x_0}(x_0, y_0) - a(x_0, y_0) u(x_0, y_0)]] \, dy_0 \\ - [u(x_0, y_0) \frac{\partial}{\partial y_0} \omega(x, y, x_0, y_0) \\ - \omega(x, y, x_0, y_0) [\frac{\partial u}{\partial y_0}(x_0, y_0) - b(x_0, y_0) u(x_0, y_0)]] \, dx_0 \} \end{aligned} \tag{19}$$

Substituting $z = x+iy$, $\zeta = x-iy$ into the latter formula and using the analytic continuation (17) of the function (18), we obtain the analytic continuation $U(z, \zeta)$ of the function $u(x, y)$ which

i) is analytic in $T \times \bar{T}$ since the singularities of Ω lie at $z = z_0$ or $\zeta = \zeta_0$ and in the formula (19) we have $z_0 \in \partial T$, $\zeta_0 \in \partial T$;

ii) coincides with the function $u(x, y)$ for real values of (x, y), i.e. $u(x, y) = U(x + iy, x - iy)$.

As T is an arbitrary subdomain of D, we obtain the analytic continuation of u into the Vekua huul $V(D)$. Note that if D is not simply connected, the continuation can be a multiple-valued function; the type of nonuniqueness is investigated in [**Ve 1**].

We point out that due to results of F.John [**Jh 1**], [**Jh 2**], these results can be generalized to higher dimensions. More precisely, let $\hat{L}$ be an elliptic differential operator in $\mathbf{R}^n$ with analytic coefficients (we suppose that the coefficients can be continued to entire functions in $\mathbf{C}^n$) and with constant coefficients in the principal part. Then, due to [**Jh 1**] (see also [**Jh 2**], **Ch. III**) there exists a fundamental solution $\omega(x, x_0)$, $x = (x^1, \dots, x^n)$, $x_0 = (x_0^1, \dots, x_0^n)$ to the operator $\hat{L}$ which is analytic for all values of $x \in \mathbf{C}^n x_0 \in \mathbf{C}^n$, $x \notin K_{x_0}$, K_{x_0} being the characteristic cone of the operator L with the vertex x_0.

The characteristic cone K_{x_0} is defined as a union of characteristic (with respect to the operator $\hat{L}$) curves with origin at x_0; as the coefficients of the principal part of L are constants, these characteristic curves are straight lines, K_{x_0} thus being a cone in $\mathbf{C}^n$ with the vertex x_0. Since the operator $\hat{L}$ is elliptic, the cone K_{x_0} for $x_0 \in \mathbf{R}^n$ does not intersect the real space $\mathbf{R}^n$ at points different from x_0.

We denote by $\hat{L}^*$ the *conjugate operator*

$$\hat{L}^* u = \sum_{|\alpha|\leq m} (-1)^{|\alpha|} \left(\frac{\partial}{\partial x}\right)^{\alpha} [a_\alpha(x) u] \quad \text{if} \quad \hat{L} u = \sum_{|\alpha|\leq m} a_\alpha(x) \left(\frac{\partial}{\partial x}\right)^{\alpha} u \ .$$

We use Green's formula (see **[LM 1]**)

$$\int_D [v \hat{L} u - u \hat{L}^* v] \, dx = \int_\Gamma \sum_{j=0}^{m-1} \hat{B}_j u \cdot \hat{C}_j v \, ds$$

where $\hat{B}_j$, $\hat{C}_j$ are differential operators of order $\leq m-1$ and ds is the volume element on the boundary Γ of the domain D.

Using Green's formula for arbitrary solution $u(x)$ to the equation $\hat{L} u = 0$ and for fundamental solution $v(x) = \omega(x, x_0)$ to the equation $\hat{L}^* v = 0$ we obtain the analogue of the formula (19):

$$u(x_0) = \int_\Gamma \sum_{j=0}^{m-1} \hat{B}_j u(x) \hat{C}_j \omega^*(x, x_0) \, ds_x \tag{19'}$$

We note that since the operator $\hat{L}^*$ has the same principal part as the operator $\hat{L}$, the function $\omega^*(x, x_0)$ has singularities only at points of the characteristical cone K_{x_0}.

We denote by $V(D)$ the domain in the complex space $\mathbf{C}^n$ which contains D and is bounded by the envelope of set of characteristical cones $\{K_{x_0}, x_0 \in \Gamma\}$. The formula (19') shows that *any solution $u(x)$ of the equation $\hat{L} u = 0$ which is regular in D can be continued into the domain $V(D)$*. This domain we call *the Vekua hull of the domain D* in n-dimensional space.

2. Continuation across the boundary. Witn the constructed continuation at hand, we can prove the existence of the analytic continuation across the boundary of the domain of solutions the homogeneous equation (8) (i.e. for $f(x, y) = 0$) which satisfy the condition

$$\alpha(x, y) u_x + \beta(x, y) u_y + \gamma(x, y) u|_\Gamma = v(x, y) \tag{20}$$

provided that the boundary Γ of the domain D is analytic and the coefficients α, β, γ and the function v are entire functions of their variables, α and β are real-valued at real values of x, y.

Suppose, for the sake of simplicity, that the equation of Γ has the form

$$P(x, y) = 0 \tag{21}$$

where $P(x, y)$ is a polynomial with real coefficients, grad $P \neq 0$ on Γ. Let $u(x, y)$ be a solution of the homogeneous equation (8) which is regular up to the boundary $\Gamma = \partial D$ and satisfies the condition (20). Suppose that one of two conditions is valid:

a) $\alpha^2(x, y) + \beta^2(x, y) \neq 0$ on Γ,

b) $\alpha(x, y) \equiv \beta(x, y) \equiv 0$, $\gamma(x, y) \neq 0$ on Γ.

As u is regular in D up to the boundary, it can be continued as the function $U(z, \zeta)$ of variables (z, ζ) to the Vekua hull $D \times \bar{D}$ of the domain D. Hence, due to the formula (14)

$U(z, \zeta)$ can be represented in a form (16).

Let

$$\zeta = S(z) \tag{22}$$

be a solution of the equation (21) of Γ in characteristic coordinates z, ζ. This means, that $S(\zeta)$ is a solution of the equation

$$P\left(\frac{z+\zeta}{2}, \frac{z-\zeta}{2i}\right) = 0 \tag{23}$$

As $\operatorname{grad}_{x,y} P(x, y) \neq 0$ it can be easily shown that there exists a unique holomorphic solution of the equation (23) at least in a neighbourhood of the real part of Γ. The following affirmation is essentially due to H.A.Schwarz; for the proof see [**Da 1**], [**Sh 1**].

PROPOSITION 1. *Under the above assumptions there exists a neighbourhood $V \subset \mathbf{R}^2 = \mathbf{C}$ of the real part of Γ such that the function $\overline{S(z)}$, $z = x + iy$ maps V into itself. The mapping $z \to \overline{S(z)} = R(z)$ is anti-conformal, $R(z) = z$ for $z \in \Gamma$ and $R(z)$ transposes parts V_1, V_2 of the neighbourhood V into which Γ divides V. Besides, R is an involution.*

The function $S(z)$ is called the Schwarz function of Γ, and the mapping $R(z)$ is called the corresponding Schwarz reflection (see section 2).

Let us rewrite the condition (20) in the form

$$\alpha^*(z, \zeta)\frac{\partial U}{\partial z} + \beta^*(z, \zeta)\frac{\partial U}{\partial \zeta} + \gamma^*(z, \zeta)\, U|_\Gamma = v(z, \zeta) \tag{24}$$

The equality (24) is valid on the real part of Γ (i.e. for $z = x + iy$, $\zeta = \bar{z}$).

Differentiating the formula (16) w.r.t. z and ζ and substistuting the results in (24) we obtain the relation

$$\Phi(z) \cdot \tilde{\alpha}(z, \zeta) + \Psi(\zeta) \cdot \tilde{\beta}(z, \zeta) + \int_{z_0}^{z} \Phi(\xi)\, K_1(\xi, z, \zeta)\, d\xi$$
$$+ \int_{\zeta_0}^{\zeta} \Psi(\eta)\, K_2(\eta, z, \zeta)\, d\eta = \tilde{v}(z, \zeta) \tag{25}$$

which holds on the real part of Γ. Here

$$\tilde{\alpha}(z, \zeta) = \alpha^*(z, \zeta)\, G(z, \zeta_0, z, \zeta), \qquad \tilde{\beta}(z, \zeta) = \beta^*(z, \zeta)\, G(z_0, \zeta, z, \zeta); \tag{26}$$

$$K_1(\xi, z, \zeta) = \alpha^*(z, \zeta)\frac{\partial G}{\partial z}(\xi, \zeta_0, z, \zeta) + \beta^*(z, \zeta)\frac{\partial G}{\partial \zeta}(\xi, \zeta_0, z, \zeta)$$
$$+ \gamma^*(z, \zeta) G(\xi, \zeta_0, z, \zeta); \tag{27}$$
$$K_2(\eta, z, \zeta) = \alpha^*(z, \zeta)\frac{\partial G}{\partial z}(z_0, \eta, z, \zeta) + \beta^*(z, \zeta)\frac{\partial G}{\partial \zeta}(z_0, \eta, z, \zeta)$$
$$+ \gamma^*(z, \zeta) G(z_0, \eta, z, \zeta); \tag{28}$$
$$\tilde{v}(z, \zeta) = v(z, \zeta) - A\left\{\alpha^*(z, \zeta)\frac{\partial G}{\partial z}(z_0, \zeta_0, z, \zeta) + \beta^*(z, \zeta)\frac{\partial G}{\partial \zeta}(z_0, \zeta_0, z, \zeta)\right.$$
$$\left. + \gamma(z, \zeta) G(z_0, \zeta_0, z, \zeta)\right\} \tag{29}$$

are the entire functions of the variables (z, ζ). Note here that since

$$\alpha^* = \alpha\left(\frac{z+\zeta}{2}, \frac{z-\zeta}{2i}\right) + i\beta\left(\frac{z+\zeta}{2}, \frac{z-\zeta}{2i}\right),$$
$$\beta^* = \alpha\left(\frac{z+\zeta}{2}, \frac{z-\zeta}{2i}\right) - i\beta\left(\frac{z+\zeta}{2}, \frac{z-\zeta}{2i}\right),$$
$$\gamma^* = \gamma\left(\frac{z+\zeta}{2}, \frac{z-\zeta}{2i}\right)$$

and due to conditions (11) the functions $\tilde{\alpha}$, $\tilde{\beta}$ do not vanish on the real part of Γ if $\alpha^2+\beta^2 \neq 0$ on Γ. Hence, under the condition a) above, the relation (26) determines the function $\Phi(z)$ on the real part of Γ uniquely for given $\Psi(\zeta)$ and vice versa. Similarly, if the condition b) is valid, the kernels K_1, K_2 do not vanish for $z = \xi$, $\zeta = \eta$ respectively. Hence, in this case the function Φ is uniquely determined by (25) for given Ψ and vice versa.

Suppose that $T \subset \mathbf{R}^2 \backslash D$ is such a domain that $\partial T \cup T \cup D$ contains a real part of Γ, $S(\zeta)$ is regular for $\zeta \in \bar{T}$, $S(z)$ is regular for $z \in T$, values of $S^{-1}(\zeta)$ for $\zeta \in \bar{T}$ lye in D, values of $S(z)$ for $z \in T$ lye in $\bar{D}$ and one of the conditions

a^*) $\alpha^*(z, S(z)) \neq 0$ and $\beta^*(S^{-1}(\zeta), \zeta) \neq 0$,

b^*) $\alpha^* \equiv 0$, $\beta^* \equiv 0$, $\gamma^* \equiv 1$

is valid. The existence of such a domain follows from Proposition 1 and conditions a) and b) for the functions α, β, γ.

The continuation of the function $U(z, \zeta)$ will be constructed with the help of the formula (16). To do so, we have to continue the functions $\Phi(z)$, $\Psi(\zeta)$ to the domains T, $\bar{T}$ respectively. This continuation can be carried out with the help of equation (25) which is continued to complex domain. More exactly, if we substitute $z = S^{-1}(\zeta)$, $\zeta \in \bar{T}$ in (25) we obtain the

Volterra integral equation for the function $\Psi(\zeta)$:

$$\Psi(\zeta)\,\tilde{\beta}(S^{-1}(\zeta),\zeta)+\int_{\zeta_0}^{\zeta}\Psi(\eta)\,K_2(\eta,S^{-1}(\zeta),\zeta)\,d\eta$$
$$=(S^{-1}(\zeta),\zeta)-\Phi(S^{-1}(\zeta))\,\tilde{\alpha}(S^{-1}(\zeta),\zeta)$$
$$-\int_{z_0}^{S^{-1}(\zeta)}\Phi(\xi)\,K_1(\xi,S^{-1}(\zeta),\zeta)\,d\xi \tag{30}$$

We suppose that (z_0,ζ_0) lies on the real part of Γ. Then the part from z_0 to $S^{-1}(\zeta)$ can be chosen in the domain D and, hence, the right-hand part of (30) is a known holomorphic function in $\tilde{T}$. We define $\Psi(\zeta)$ in $\tilde{T}$ as a unique solution to this Volterra integral equation. As the equation (30) for $\zeta\in\overline{\partial D}$ is reduced to (25), the solution $\Psi(\zeta)$ of (30) is a continuation of the function $\Psi(\zeta)$ which is previously determined in D.

Similarly, the function $\Phi(z)$ can be continued into the domain T as the solution to the equation

$$\Phi(z)\,\tilde{\alpha}(z,S(z))+\int_{z_0}^{z}\Phi(\xi)\,K_1(\xi,z,S(z))\,d\xi$$
$$=v(z,S(z))-\Psi(S(z))\,\tilde{\beta}(z,S(z))$$
$$-\int_{\zeta_0}^{S(z)}\Psi(\eta)\,K_2(\eta,z,S(z))\,d\eta \tag{31}$$

Thus, we have proved that the solution $u(x,y)$ can be continued into the domain $D\cup\partial T\cup T$.

Evidently, while constructing the continuation one must take into account the possibility of multiple-valuedness of the continued function (see the preliminary discussion of this possibility in section 1). The character of this multiple-valuedness can be investigated by methods described in [Ve 1].

The further continuation can be constructed step-by-step with the help of the following affirmation, which is proved by the direct generalization of the above considerations.

THEOREM 3. *Let $X\to\mathbf{C}\backslash Y$ be a covering over $\mathbf{C}$, where $Y\subset\mathbf{C}\backslash D$ with marked lifting $\tilde{D}$ of the domain D. Let T be a domain in X such that there exist analytic continuations of the functions Φ, Ψ into T, $\tilde{T}$ respectively. Let T_1 be a domain in X such that:*

i) $T_1\cup T$ is a connected domain;

ii) $S(z)$ and $S^{-1}(\zeta)$ are regular on T_1, $\tilde{T}_1$ respectively;

iii) the values of $S(z)$ lie in $\tilde{T}$ for $z\in T_1$ and the values of $S^{-1}(\zeta)$ lie in T for $\zeta\in\tilde{T}_1$;

iv) one of the conditions a^*), b^*) is valid on T_1.

Then there exists an analytic continuation of the functions Φ, Ψ into the domains $T\cup T_1$, $\tilde{T}\cup\tilde{T}_1$ respectively and, hence, the solution $u(x,y)$ can be continued into the domain $T\cup T_1$.

It is evident that the obstacles for continuation of solution are defined by zeroes of the functions $\alpha^*(z,S(z))$, $\beta^*(S^{-1}(\zeta),\zeta)$ and by the singular points of the functions S and S^{-1}.

The latter form just the characteristical conoid of the initial surface Γ (considered in the complex space $\mathbf{C}^2$) originated from both finite and infinite points of this surface.

The relation (15) leads us to the formula for the continued solution. For example, for the Dirichlet conditions

$$u(x\,,\,y)|_\Gamma = 0 \tag{32}$$

we obtain

$$U(z\,,\,\zeta) = \frac{1}{2}\int_\gamma G(z_0\,,\,\zeta_0\,,\,z\,,\,\zeta)\left[\frac{\partial u}{\partial \zeta_0}(z_0\,,\,\zeta_0)\,d\zeta_0 - \frac{\partial u}{\partial z_0}(z_0\,,\,\zeta_0)\,dz_0\right], \tag{33}$$

where γ is any path $z_0 = z_0(s)\,,\,\zeta_0 = S(z_0(s))$ joining the points $(z\,,\,S(z))$ and $(S^{-1}(\zeta),\,\zeta)$.

Note that the proof of the Theorem 3, given by H.Lewy [**Lw 1**] for the case of plane boundary, requires no essentially changes in the general case. A similar result for the Helmholtz equation is formulated in [**KS 3**].

3. Discussion on the reflection principle. We shall discuss the reflection principle w.r.t. the Dirichlet conditions (32). For the Laplace equation the Riemann's function is identically equal to 1 and, hence, the relation (33) becomes

$$U(z\,,\,\zeta) = \frac{1}{2}\int_\gamma \left[\frac{\partial u}{\partial \zeta_0}(z_0\,,\,\zeta_0)\,d\zeta_0 - \frac{\partial u}{\partial z_0}(z_0\,,\,\zeta_0)\,dz_0\right],$$

where γ is a path in Γ joining points $(z\,,\,S(z))$ and $S^{-1}(\zeta),\,\zeta)$. The values of the solution on the real plane are equal to

$$u(x\,,\,y) = U(z\,,\,\bar{z}) = \frac{1}{2}\int_\gamma \left[\frac{\partial u}{\partial \zeta_0}(z_0\,,\,\zeta_0)\,d\zeta_0 - \frac{\partial u}{\partial z_0}(z_0\,,\,\zeta_0)\,dz_0\right], \tag{34}$$

where γ joins the points $(z\,,\,S(z))$ and $S^{-1}(\bar{z}),\,\bar{z})$. Let us calculate the value of $U(z\,,\,\bar{z})$ at the point $R(z) = \overline{S(z)}$ (see Proposition 1). Evidently, $U(R(z),\,\overline{R(z)})$ is given by the right-hand part of the formula (34) with γ joining the points $\overline{S(z)},\,S(\overline{S(z)}))$ and $S^{-1}(S(z)),\,S(z))$. It is easy to prove the relations

$$\overline{S(z)} = S^{-1}(\bar{z}), \qquad S(\overline{S(z)}) = \bar{z}, \qquad S^{-1}(S(z)) = z$$

Thus, we have

$$U(R(z),\,\overline{R(z)}) = -U(z\,,\,\bar{z}) \tag{35}$$

This is the *Schwarz reflection principle* for the Laplace equation. For the Neumann's conditions a similar formula (with the opposite sign w.r.t. (35)) can be obtained.

However, as it was mentioned before, even for the case of Helmholtz equation the reflection principle fails. The cause of this faulure is, evidently, the dependence of the Riemann's function $G(z_0\,,\,\zeta_0\,,\,z\,,\,\zeta)$ on the arguments $z\,,\,\zeta$.

There is a very elegant interpretation of the "reflected point" $R(z)$ given by E.Study [**St 1**]. Namely, let us draw two characteristical lines l_1 and l_2 through a point $(z\,,\,\bar{z})$ close to Γ. Consider the points $A\,,\,B$ of intersection of the lines l_1 and l_2 with Γ which are close to the point $(z, \bar{z})$. If we draw through A the characteristic line l_3 different from l_1 and through B the characteristic line l_4 different from l_2, then the intersection of lines l_3 and l_4 gives the point $(R(z), \overline{R(z)})$.

5. Continuation method based on the Cauchy problem

1. Solution of the problem of balayage. We discuss here the case of an arbitrary elliptic operator $\hat{L}$ with constant coefficients. For simplicity we suppose that the full symbol $L(p)$ of $\hat{L}$ does not vanish for real values of p. Let $D \subset \mathbf{R}^n$ be a domain bounded with an algebraic surface Γ with the equation of the type (21); we suppose the polynomial $P(x)$ involved in (21) to the irreducible. For any function $f(x) = f(x^1, \dots, x^n)$ we denote by $f_D(x)$ the function which coincides with f in D and vanishes in $\mathbf{R}^n \backslash D$. We require $f(x^1, \dots, x^n)$ may be continued to an entire function in $\mathbf{C}^n$.

Denote by U^f the unique solution to the equation

$$\hat{L} U^f = f_D(x) \tag{36}$$

in $\mathbf{R}^n$. For any distribution $w(x)$ with support in D we denote by U^w the the solution to the equation

$$\hat{L} U^w = w(x) \tag{37}$$

It is evident that the functions U^f and U^w satisfy the homogeneous equation $\hat{L} u = 0$ outside the domain D. The problem is as follows:

Find a distribution $w(x)$ with minimal support such that $U^f = U^w$ outside the domain D.

We present the reduction of this problem to a Cauchy problem in the space $\mathbf{R}^n$ which is essentially due to H.Shapiro (see [**Sh 1**]). Consider the difference $u = U^f - U^w$. As supp $w \subset D$, we have $\hat{L} u = f_D(x)$ in some neighbourhood T of the boundary Γ of the domain D. Since $f_D(x) \in L_2(\mathbf{R}^n)$, we obtain that $u(x) \in H^m(T)$ where m is an order of the operator $\hat{L}$. Due to the embedding theorem all the derivatives of u up to the order $m-1$ are continuous on Γ. On the other hand, $u(x) \equiv 0$ in $\mathbf{R}^n \backslash D$ and, hence, all the derivatives up to the order $m-1$ vanish on Γ. Thus, in $D \backslash \operatorname{supp} w$ the function $u(x)$ satisfies the Cauchy problem

$$\begin{cases} \hat{L} u = f \text{ in } D \backslash \operatorname{supp} w, \\ u \text{ has a zero of order } m \text{ on } \Gamma \end{cases} \tag{38}$$

Inversely, if u is a solution of the Cauchy problèm (38) in $D \backslash K$ for some compact $K \Subset D$, then there exists a distribution w with its support in arbitrary small neighbourhood of K, such that $U^f = U^w$ outside the domain D. For construction of such distribution we consider a neighbourhood T' of K, $T' \subset D$ and denote by U^w the function

$$U^w = U^f - u_{D \backslash T'}$$

Evidently, $w = \hat{L} U^w$ and supp $w \subset \partial T'$. Thus, for solving the problem of balayage it is sufficient to continue the solution $u(x)$ of the problem (38) into the maximal such subdomain of the domain D (note that the solution of (38) exists in a neighbourhood of Γ due to

Holmgren theorem). The singularities of continued solution will determine the support of the distribution w.

A solution of the problem (38) may be obtained as the restriction to a real space $\mathbf{R}^n$ of a solution $u(x)$ of the corresponding *complex-analytic* problem

$$\begin{cases} \hat{L}u = f \text{ in } \mathbf{C}^n, \\ u \text{ has a zero of order } m \text{ on } \Gamma \end{cases} \tag{39}$$

where by Γ we mean an algebraic set $P(x) = 0$, $x \in \mathbf{C}^n$. The solution of the problems of the type (39) was given in the series of works [**StS 1 - 15**] with the help of the Laplace-Radon transform. The explicit solution of the problem (39) can be represented in an integral form

$$u(x) = (n-1)! \left(\frac{i}{2\pi}\right)^{n-1} \int_{\tilde{h}(x)} \operatorname{Res} \frac{\check{u}(p)\,\omega(p)}{(p\cdot x)^n} \tag{40}$$

where Res is the Leray residue ([**Le 2**]),

$$p = (p_0, p_1, \dots, p_n),\ \omega(p) = \sum_{j=0}^{n} (-1)^j p_j dp_0 \wedge \dots \wedge dp_{j-1} \wedge dp_{j+1} \wedge \dots \wedge dp_n$$

is the Leray form, $p\cdot x = p_0 + p_1 x^1 + \dots + p_n x^n$, $h(x)$ is a relative homology class which will be described below and $\check{u}(p)$ is the solution of the Cauchy problem

$$\begin{cases} L\left(p' \frac{\partial}{\partial p_0}\right) \check{u}(p) = \check{f}(p), \\ \check{u}(p) \text{ has a zero of order } m \text{ on } \tilde{\Gamma} \end{cases} \tag{41}$$

for an ordinary differential operator $L\left(p' \frac{\partial}{\partial p_0}\right)$, $p' = (p_1, \dots, p_n)$. Here $\tilde{\Gamma}$ is a Legendre transform of the surface Γ, i.e. the closure of the set of such points p that the plane

$$L_p = \{x \mid p\cdot x = 0\} \tag{42}$$

is tangent to Γ at some its regular point and the function $\check{f}(p)$ is defined as an integral

$$\check{f}(p) = \int_{h(p)} \operatorname{Res} \frac{f(x)\,dx}{p\cdot x}, \tag{43}$$

where $h(p)$ is a relative homology class. The descriptions of classes $h(p)$, $\tilde{h}(x)$ is quite analogous; we carry out the description of the class $h(p)$. Let x_0 be such point of Γ that the tangent plane L_{p_0} to Γ at this point is tangent to Γ quadratically (we suppose that such point exists). Then for values of p sufficiently close to p^0, $p \notin \tilde{\Gamma}$ the intersection of the plane L_p with Γ is biholomorphic equivalent to complex quadrics in some neighbourhood of x^0. For such values of p we denote by $h(p)$ the relative vanishing class

$$h(p) \in H_{n-1}(L_p, L_p \cap \Gamma)$$

of this quadrics (see [**Le 2**]). With the help of Thom's theorem (see [**Ph 1**]) this class can be continued to all points p exept for the points of some homogeneous analytic set. Due to Thom's theorem this set can be described as a set of such points p for which the projective plane $L_p^* = \{(x^0 : x^1 : \cdots : x^n) | p_0 x^0 + \cdots + p_n x^n = 0\}$ is tangent to some stratum of the set $\Gamma^* \cup \infty$, where $\Gamma^* \subset \mathbf{CP}^n$ is a "compactification" of Γ, i.e. $\Gamma^* \cap \{x^0 = 1\} = \Gamma$ and ∞ is the infinity section of $\mathbf{CP}^n$, i.e. $\infty = \{x^0 = 0\}$. The reason of appearance of ∞ in $\Gamma^* \cup \infty$ is that it is *a singularity set of the function* $f(x)$ *in* $\mathbf{CP}^n$. The above description of the singularity set of the function (43) allows us to calculate these singularities explicitly. The same remark applies also to the calculation of singularities of the integral (40) provided that singularities of the function $\check{u}(p)$ are known.

The definition of the class

$$\tilde{h}(x) \in H_{n-1}(\tilde{L}_x, \tilde{L}_x \cap \tilde{\Gamma}),$$
$$\tilde{L}_x = \{(p_0 : p_1 : \cdots : p_n) | p_0 + p_1 x^1 + \cdots + p_n x^n = 0\}$$

is just the same as the definition of the class $h(p)$; we mention only that due to homogenuity of the integrand this integral is taken over the homology class *in the projective space.*

REMARK 2.

While calculating singularities in the described way one can obtain sham singularities, i.e. such singularities which don't actually occur. To remove sham singularities one can use the following criteria.

1. *The solution has no singularities on the manifold* Γ *provided that the right-hand part* $f(x)$ *of (39) is not singular on* Γ.

2. *All singularities of solution can either be characteristical ones or coincide with singularities of* $f(x)$.

We denote by X the set of singularities of solution of the problem (39) calculated by the method described above. Let X' be the intersection of a real part of X with $\bar{D}$. Under the above assumptions $X' \cap \partial D = \varnothing$. Consider a system Z of cuts with the boundary X' such that the solution (40) of the problem (39) possesses an univalent brauch $\tilde{u}(x)$ on $D \backslash Z$. Set $\tilde{u}(x) = 0$ outside D. Evidently, $\tilde{u}(x) \in C^1(\mathbf{R}^n \backslash Z)$ and $\Delta \tilde{u}(x) = f(x)$ in $D \backslash Z$. Let T be an arbitrary small tubular neighbourhood of X' in $\mathbf{R}^n$. We redefine $\tilde{u}$ in T, setting $\tilde{u}(x) = 0$ for $x \in T$. Denote

$$U = U^f - \tilde{u}, \qquad w = \hat{L} U \tag{44}$$

(the operator $\hat{L}$ is applied in the sense of distribution theory). Then $U^w = U$ and U^w coincides with U^f outside $\bar{D}$. We see that the support supp w of the distribution w is contained in $\partial(T \cup Z)$ since $\hat{L} U = 0$ outside this set.

REMARK 3.

If the function $\tilde{u}(x)$ given on the set $D \backslash Z$ *admits regularization up to a distribution on* D, we can aviod the cut-off procedure on the tubular neighbourhood T of the set X', thus ensuring the equality supp $w = z$. However, this is generally not the case for arbitrary $f(x)$.

For example, it can be shown that the solution $u(x)$ of the Cauchy problem

$$\begin{cases} \frac{\partial^2 u}{\partial x^2} + \frac{\partial^2 u}{\partial y^2} = e^{x+iy}, \\ u \text{ has a zero of order 2 on } \Gamma = \{(x, y) | x^2 + y^2 = 1\} \end{cases}$$

does not admit regularization up to a distribution on $D = \{x^2+y^2 \leq 1\}$ since it has an essential singularity at the origin. Thus, the distribution w constructed above may be considered (in some sense) as a solution of the problem of continuation with minimal support.

REMARK 4.

The above construction shows that the distribution w with minimal support *with respect to inclusion* does not exist in general. For example, let D be an ellipse and $\hat{L}$ be a Laplace operator. Then "the minimal support of the swept charge" consists of two foci-centered circles of *arbitrary* small radii together with an *arbitrary* curve in D connecting these circles.

REMARK 5.

We note that it is not nesessary for $f(x)$ to be an entire function. For the described technique to be applicable it is sufficient for the right-hand part $f(x)$ of (38) to be continuable up to the function $f(x)$ which is regular in $\mathbf{C}^n \backslash Y$ for an analytic set Y. We suppose, of course, that Y does not intersect $\bar{D}$.

2. Continuation of solutions of boundary value problems. As it was mentioned in the section 2, this continuation is based on a preliminary continuation of the Cauchy data in the complex domain. We demonstrate here this method for boundary value problems for the Laplace operator in $\mathbf{R}^2$ (see [Ml 8]).

If $u(x, y)$ is a harmonic function regular up to the (analytic) boundary ∂D of a plane domain D, then, due to Green's formula, it satisfies the integral relation inside the domain D:

$$2\pi u(x, y) = \int_{-l}^{l} \left[u(s) \frac{\partial}{\partial \nu} \ln \frac{1}{R} - v(s) \ln \frac{1}{R} \right] \partial s, \tag{45}$$

where $R = \sqrt{(x - \xi(s))^2 + (y - \eta(s))^2}$; $x = \xi(s)$, $y = \eta(s)$ are the equations of the boundary, s being the arc length; $u(s)$, $v(s)$ are the rescriptions of the functions $u(x, y)$, $\frac{\partial u}{\partial \nu}(x, y)$ to the boundary, ν being the outer normal vector to ∂D. The functions $\xi(s)$, $\eta(s)$ are, due to our assumptions, real analytic functions of s. If the point (x, y) tends to a boundary point corresponding to the parameter $t \in [-l, l]$, the relation (45) reduces, as it is well-known, to the following integral equation

$$\pi u(t) = \int_{-l}^{v} (s) \ln R \, ds - \int_{-l}^{l} u(s) R^{-1} \frac{\partial R}{\partial \nu} ds \tag{46}$$

Our next goal is to construct the continuation of the equation (46) into to the complex domain. Due to periodicity it is sufficient to continue this equation to such complex values of t, that $-l \leq \mathrm{Re}\, t \leq l$. To do so, we define the function

$$r = r(s, t) = \{[\xi(s) - \xi(t)]^2 + [\eta(s) - \eta(t)]^2\}^{\frac{1}{2}} = |r| e^{i\psi}, \qquad t \in \mathbf{C}, s \in \mathbf{R} \tag{47}$$

where the choice of the argument ψ is fixed by the condition $\psi = 0$ for $t = t_0 \in [-l, l]$, $-l < t_0 < s < l$. The continuation of the equation (46) to the domain $-l \le \mathrm{Re}\, t \le l$, $\mathrm{Im}\, t \gtrless 0$ is

$$\pi \left[u(t) \mp i \int_{-l}^{t} v(s)\, ds \right] = \int_{-l}^{l} v(s) \ln r\, ds - \int_{-l}^{l} u(s) r^{-1} \frac{\partial r}{\partial \nu}(s, t)\, ds \tag{48}$$

where the upper sign corresponds to the continuation into the upper half-plane and the lower sign corresponds to that into the lower half-plane. We note here that (for *fixed* function $u(x, y)$) both integrals in the right-hand side of (48) are *known* analytic functions in the considered domain. In the integrands of these integrals $s \in [-l, l]$ is *real*, the continuation of the kernel $\ln r(s, t)$ of the first integral is given by the formula (47) and the continuation of the kernel $r^{-1} \frac{\partial r}{\partial \nu}(s, t)$ of the second integral can be carried out directly since

$$\frac{1}{r} \frac{\partial r}{\partial \nu}(s, t) = \frac{(\xi(s) - \xi(t))\, \eta'(s) - (\eta(s) - \eta(t))\, \xi'}{(\xi(s) - \xi(t))^2 + (\eta(s) - \eta(t))^2} \tag{49}$$

and $\xi(s), \eta(s)$ are analytic functions of their argument s. We note that the kernel (49) has singularity at $t = s$.

Now, let $u(x, y)$ be a solution of Dirichlet problem for the Laplace equation, i.e. $u(x, y)$ is a harmonic function which satisfies the Dirichlet's condition

$$u(x, y)|_\Gamma = v_0(x, y)|_\Gamma$$

on the boundary Γ of a domain D, where $v(x, y)$ is an entire function. Denoting by $u_0(t)$ the restriction $u_0(t)(x, y)|_\Gamma$, we obtain the equation for $U(t)$

$$\pm i \pi \int_{-l}^{t} v(s)\, ds = \int_{-l}^{l} v(s) \ln r\, ds - \int_{-l}^{l} u_0(s) r^{-1} \frac{\partial r}{\partial \nu}\, ds - u_0(t) = F(t)$$

with the known function $F(t)$. Solving this equation w.r.t. $v(t)$ we reduce the problem of continuation to the solution of complex Cauchy problem. This solution can be obtained either with the help of Riemann's method (see section 3) or with the help of Laplace-Radon transform (see this section).

Similarly, the equation (48) allows to calculate the function $u(t)$ provided that the continuation of the function $v(t)$ is given. This reduces the Neumann problem to the complex Cauchy problem. Finally, for the boundary condition of the form

$$\frac{\partial u}{\partial n} + \alpha(x, y)\, u|_\Gamma = u_0$$

the equation (48) can be reduced to Volterra equation for the unknown function $u(t)$. Thus, for two-dimensional problems this method allows to calculate the singularities of continuated solutions in an explicit manner. This method can be used for the Helmholtz equation as well.

This method can be generalized for the axisymmetric boundary value problems in three-dimensional space $\mathbf{R}^3$ (see [**Mi 10**], [**Mi 12**]).

To conclude this section we mention an interesting paper of R.Millar ([**Mi 9**]) where was carried out the analytic continuation of Cauchy data in three-dimensional space for the case of plane boundary. This continuation was obtained with the help of method of E.E.Levi ([Lv 1]). We note also that we consider this method as far from being completed.

References

[AK1]. V.F.Apeltzin, A.G.Kürktchan, *Analytic Propeties of Wave Fields*, MGU, Moscow, 1990.

[BS1]. J.Bony, P.Shapira, *Existence et prolongement des solution holomorphes des equations aux dérivés partielles*, Indent.Math. **17** (1972), 95 – 105.

[Co1]. D.Colton, *On the Inverse Scattering Problems for Axially Symmetric Solutions of the Helmholtz Equation*, Quart. J. Math. Oxford (2) **22** (1971), 125 – 130.

[Co2]. D.Colton, *Integral Operators and Inverse Problems in Scattering Theory*, Function Theoretic Method for Partial Differential Equations, Darmstadt, 1976, Springer Verlag Berlin/Heidelberg/New York, 1976, pp. 17 – 28.

[CH1]. R.Courant, D.Hilbert, *Methoden der mathematischen Phisik*, vol. 2, Berlin, 1937; *Methods of Mathematical Phisics Interscience*, New York, 1962.

[Da1]. Ph.Davis, *The Schwarz Functions and its Applications*, Carus Mathematical Monographs, MAA, 1979.

[Ga1]. P.Garabedian, *Partial Differential Equations with More Than Two Independant Variables in the Complex Domain*, J.Math. Mech. **9** no. 2 (1960), 241 – 271.

[Ga2]. P.Garabedian, *Lectures on Function Theory and PDE*, Rice University Studies **49** no. 4 (1963).

[Ga3]. P.Garabedian, *Partial differential equations*, John Wiley and Sons Inc., New York - London - Sidney, 1964.

[Gr1]. D.-A.Grave, *Sur le problème de Dirichlet*, Assoc. Francaise pour l'Avancement des Sciences **24** (1895), 111 – 136, Completes Rendus (Bordeaux).

[Ha1]. J.Hadamard, *Le problème de Cauchy et les équations aux dérivées partielles linéaires hyperboliques*, Paris, 1932.

[HLT1]. Y.Hamada, J.Leray, A.Takeuchi, *Prolongements analytiques de la solution du problème de Cauchy linéaire*, J.Math. Pures Appl. **64** (1985), 257 – 319.

[He1]. P.Henrici, *A Survey of I.N.Vekua's Theory of Elliptic Partial Differential Equations with Analytic Coefficeints*, Z. Anegew. Math. Phys. **8** (1957), 169 – 203.

[Hr1]. G.Herglotz, *Über die analytishe Forsetzung des Potentials ins Innere der anziehenden Massen*, Gekrönte Preisschr. der Jablonwskischen Gesellsch zu Leipzig, 1914, pp. 56.

[Je1]. J.Jeans, *The Mathematicial Theory of Electricity and Magnetism*, fifth ed., Reprinted 1966 by Cambrige Univ. Press, 1925.

[Jh1]. F.John, *The Fundamental Solution of Linear Elliptic Differential Equations with Analytic Coefficients*, Comm. pure and appl. Math. **2** (1950), 213 – 304.

[Jh2]. F.John, *Plane Waves and Spherical Means Applied to Partial Differential Equations*, Interscience Publishers Inc., New York, Interscience Publishers, Ltd., London, 1955.

[Jo1]. G.Johnsson, *Global Existence Results for Linear Analytic Partial Differential Equations*, research report TRITA - MAT 1989-12, Royal Institute of Technology, Stockholm, 1989.

[Kh1]. D.Khavinson, *On Reflection of Harmonic Functions in Surfaces of Revolution*, Preprint.

[Kh2]. D.Khavinson, *Singularities of Harmonic Functions in* $\mathbf{C}^n$, Proceedings of Symposia of Pure Applied Math., (to be published).

[KS1]. D.Khavinson, H.S.Shapiro, *The Vekua Hull of a Plane Domain*, Research report TRITA-MAT 1989 - 20, Royal Institute of Technology, Stockholm, 1989.

[KS2]. D.Khavinson, H.S.Shapiro, *The Schwartz Potential in* $\mathbf{R}^n$ *and Cauchy's Problem for the Laplace Equation*, Preprint, 1988, research report TRITA-MAT-1989-36,Royal Institute of Technology, Stockholm, 1989.

[KS3]. D.Khavinson, H.S.Shapiro, *Remarks on the Reflection Principle for Harmonic Functions*, Research report TRITA - MAT - 1989 - 13, Royal Institute of Technology, Stockholm, 1989, Journal d'analyse mathematique **54** (1990), 60 – 76.

[Kü1]. A.G.Kürktchan, *On a Method of Auxiliary Currents and Sources in Wave Differential Problems*, Radiotekhnika i Elektronika **29** no. 1 (1984), 2129 – 2139.

[Kü2]. A.G.Kürktchan, *Representation of Diffractional Fields by Wave potentials and Method of Auxiliary Currents in Problem on Diffraction of Electromagnetic Wavws*, Radiotekhnika i Elektronika **31** no. 1 (1986), 20 – 27.

[Le1]. J.Leray, *Unifoamisation de la solution du problème linéaire analytique près de la variété qui porte les donnes' de Cauchy*, Bull Soc. Math. France **85** (1957), 389 – 429.

[Le2]. J.Leray, *Le calcul différentiel et intégral sur une variété analytique complexe (Problème de Cauchy, III)*, Bull. Soc. Math. France **87, fasc.II** (1959), 81 – 180.

[Lv1]. E.E.Levi, *Sulle equazioni lineari totalmente ellittiche alle derivate parziali*, Rend. Circ. Mat. Palermo **24** (1907), 275 – 317.

[Lw1]. H.Lewy, *On the Reflection Laws of Second Order Differential Equations in Two Independant Variables*, Bull. Amer.Math.Soc. **65** (1959), 37 – 58.

[Lw2]. H.Lewy, *Neuer Bewies des analytishen Charakters der Loesungen elliptisher Differentialgleichungen*, Math.Ann. **101** (1929), 609 – 619.

[LM1]. J.-L.Lions, E.Magenes, *Problèmes aux limites non homogènes et applications*, vol. 1, Dunod, Paris, 1968.

[Mi1]. R.F.Millar, *On the Rayleigh Assumption in Scattering by a Periodic Surface*, Proceedings of the Cambridge philosophical society **65** no. 3 (1969), 773 – 792.

[Mi2]. R.F.Millar, *Rayleigh Hypothesis in Scattering Problems*, Electron Letters **5** (1969), 416 – 418.

[Mi3]. R.F.Millar, *The Location of Singularities of Two-dimensional Harmonic Functions*, *I*, Theory, SIAM J.Math.Anal. **1** (1970), 333 – 344; *II, Applications*, 345 – 353.

[Mi4]. R.F.Millar, *On the Rayleigh Assumption in Scattering by a Periodic Surface*, Proceedings of the Cambridge Philosophical Society **69** no. 1 (1971), 217 – 225.

[Mi5]. R.F.Millar, *Singularities of Solutions to Linear Second Order Analytic Elliptic Equations in Two Independant Variables, I.* Applicable Anal. **1** (1971), 101 – 121.

[Mi6]. R.F.Millar, *Singularities of Two-dimensional Exterior Solutions of the Helmholtz Equation*, Proceedings of the Cambridge philosophical Society **69** no. 1 (1971), 175 – 188.

[Mi7]. R.F.Millar, *Singularities of Solutions to Linear Second Order Analytic Equations in Two Independant Variables, II*, Applicable Anal. **2** (1973), 301 – 320.

[Mi8]. R.F.Millar, *The singularities of solutions to analytic elliptic boundary value problems*, Lecture Notes in Math no. 561 (1976), Function Theoretic Methods for Partial Differential Equations, Darmstadt, 1976, Springer-Verlag, Berlin/Heidelberg/New York.

[Mi9]. R.F.Millar, *Singularities of Solutions to Exterior Analytic Boundary Value Problems for Helmholtz Equation in Three Independant Variables, I. The plain boundary*, SIAM J. on Math. Anal. **7** no. 1 (1976), 131 – 156.

[Mi10]. R.F.Millar, *Singularities of Solutions to Exterior Analytic Boundary Value Problems for Helmholtz Equation in Three Independant Variables, II. The axisymmetric boundary*, SIAM J. Math. Anal. **10** no. 4 (1979), 682 – 694.

[Mi11]. R.F.Millar, *The Analytic Continuation of Solutions to Elliptic Boundary Value Problems in Two Independant Variables*, J.Math. Anal. and Appl. **76** no. 2 (1980), 498 – 515.

[Mi12]. R.F.Millar, *The Analytic Continuation of Solutions of the Generalized Axially Symmetric Helmholtz Equations*, Archive for Rat. Mech.Anal **81** no. 4 (1983), 349 – 372.

[Mi13]. R.F.Millar, *The Analytic Cauchy Problem for Fourth Order Elliptic Equations in Two Independent Variables*, SIAM J.Math. Anal. **15** no. 5 (1984), 964 – 978.

[Mi14]. R.F.Millar, *Integral Representations for Solutions to Some Differential Equations That Arise in Wave Theory*, Wave phenomena: Modern theory and applications (Toronto, 1983), North-Holland Math. Stud. 97, North-Holland Amsterdam - New York, 1984, pp. 391 – 408.

[Mi15]. R.F.Millar, *Singularities and the Rayleigh Hypothesis for Solutions to the Helmholtz Equation*, IMA J.Appl. Math. **37** no. 2 (1986), 155 – 171.

[Mi16]. R.F.Millar, *Application of the Schwarz Function to Boundary Problems for Laplace's Equation*, MAth. Methods Appl. Sci. **10** no. 1 (1988), 67 – 86.

[My1]. M.Miyake, *Global and Local Goursat Problems in a Class of Holomorphic or Partially Holomorphic Functions*, J. Diff Eq. **39** (1981), 445 – 463.

[Ph1]. F.Pham, *Introduction à l'étude topologique des singularits de Landau*, Mém. Sci. Math. fasc. 164, Gauthier-Villars, Paris, 1967.

[Sc1]. H.A.Schwarz, *Ueber die Integration der partiellen Differentialglerchung* $\frac{\partial^2 u}{\partial x^2} + \frac{\partial^2 u}{\partial y^2} = 0$ *unter vorgeschriebonen Grenz- und Unstetigkeitsbedingungen*, Monatsber. der Königl Akad. der Wiss. zu Berlin (1870), 767 – 795.

[Sh1]. H.S.Shapiro, *The Schwarz Function and its Generalization to Higer Dimensions*, University of Arkansas-Fayetteville Lecture Notes (1987), John Wiley.

[Sh2]. H.S.Shapiro, *An Algebraic Theorem of E.Fisher and the Holomorphic Goursat Problem*, Bull. London Math. Soc. **21** (1989), 513 – 537.

[Sh3]. H.S.Shapiro, *Global Geometric Aspects of Cauchy's Problem for the Laplace operator*, Research report TRITA - MAT - 1989 - 37, Royal Institute of Technology, Stockholm, 1989.

[Sl1]. B.D.Sleeman, *The Three-dimensional Inverse Scattering Problem for the Helmholtz Equation*, Proc. Cambridge Phil. Soc. **73** (1973), 477 – 488.

[StS1]. B.Yu.Sternin, V.E.Shatalov, *Characteristic Cauchy Problem on a Complex Analytic Manifold*, Lecture Notes in

Math. **1108** (1984).

[StS2]. B.Yu.Sternin, V.E.Shatalov, *Notes on a Problem of Balayage*, Preprint, Moscow, 1990, pp. 23.

[StS3]. B.Yu.Sternin, V.E.Shatalov, *The Laplace-Radon Transformation in Complex Analysis and some its Applications*, (to be published), Soviet Mathem. (1991).

[StS4]. B.Yu.Sternin, V.E.Shatalov, *On an Integral Transform of Complex-analytic Functions*, Doklady Akad. Nauk SSSR **280** no. 3 (1985), 553 – 556.

[StS5]. B.Yu.Sternin, V.E.Shatalov, *On an Integral Transform of Complex-analytic Functions*, Izvestiya Akad. Nauk SSSR (ser. matem.) **5** (1986), 1054 – 1076.

[StS6]. B.Yu.Sternin, V.E.Shatalov, *Singularities of Differential Equations on Complex Manifolds*, Doklady Akad. Nauk SSSR **292** no. 5 (1987), 1058 – 1062.

[StS7]. B.Yu.Sternin, V.E.Shatalov, *Singularities of Solutions of Differential Equations on Complex Manifolds (Characteristic Case)*, Geometry and Theory of Singularities in Non-linear Equations, VGU, Voronezh, 1987.

[StS8]. B.Yu.Sternin, V.E,Shatalov, *Laplace-Radon Integral Operators and Singularities of Solutions of Differential Equations on Complex Manifolds*, Global Analysis and Mathematical Phisics, VGU, Voronezh, 1987.

[StS9]. B.Yu.Sternin, V.E.Shatalov, *Asymptotic Expansions of Solutions of Differential Equations on Complex Manifolds*, Doklady Akad. Nauk SSSR **295** no. 1 (1987), 38 – 41.

[StS10]. B.Yu.Sternin, V.E.Shatalov, *On an Integral Representation and Connected Transform of Complex Analytic Functions*, Doklady Akad. Nauk SSSR **298** no. 1 (1988), 44 – 48.

[StS11]. B.Yu.Sternin, V.E.Shatalov, *Laplace-Radon Integral Transform and its Applications to Non- Homogeneous Cauchy Problem in Complex space*, Differentzialnye Uravneniya **24** no. 1 (1988), 167 – 174.

[StS12]. B.Yu.Sternin, V.E.Shatalov, *Differential Equations on Complex-analytic Manifolds and Maslov's Canonical Operstor*, Uspekhi Matematicheskikh Nauk **43(3)** no. 261 (1988), 99 – 124.

[StS13]. B.Yu.Sternin, V.E.Shatalov, *Maslov's Method in Analytic Theory of Differential Equations with partial derivatives*, (to appear), Nauka, Moscow, 1991.

[StS14]. B.Yu.Sternin, V.E.Shatalov, *Singularities of solutions of Differential Equations on Complex Analytic Manifolds and "Balayage" Problem*, (to appear), Proceedings of 3d School "Geometry ana Analysis" (1991), KGU, Kemerovo.

[StS15]. B.Yu.Sternin , V.E.Shatalov, *On Fourier-Maslov Transform in Space of Multiple-valued Analytic Functions*, Matematicheskiye Zametki, Moscow.

[St1]. E.Study, *Einige elementare Bemerkungen über den Prozess der analytischen Fortsetzung*, Math Ann. **63** (1907), 239 – 245.

[Ve1]. I.N.Vekua, *New Methods for Solving Elliptic Equations*, North-Holland Publishing Co. (translated from Russian), Amsterdam, 1967.

[WBA1]. V.H.Weston, J.J.Bowman, E.Ar, *On the electromagnetic inverse scattering problems*, Arch. Rational Mech. Anal. (1968/69), 199 – 213.

[Ze1]. M.Zerner, *Domaine d'holomorphie des fonctions vérifiant une équation aux dérivées partielles*, C. R. Acad. Sci. Paris **272** (1971), 1646 – 1648.

PROPERNESS OF NONLINEAR ELLIPTIC DIFFERENTIAL OPERATORS IN HÖLDER SPACES

V.G.Zvyagin and V.T.Dmitrienko
Department of Mathematics
Voronezh State University
394693, Voronezh, USSR

This article continues the study of properness of nonlinear maps in spaces of Holder functions $C^{k,\alpha}(\bar{\Omega})$ induced by boundary-value problems for nonlinear elliptic equations (see [1,2]).

We remind that a map $F: W \subset E_0 \longrightarrow E_1$ from a set W of Banach space E_0 into Banach space E_1 is called proper if for any compact $K \subset E_1$ the set $F^{-1}(K)$ is a compact in W. In [1] some statements about the properness of Fredholm maps, acting in pairs of Banach spaces, are obtained. These results make it possible to establish the properness of maps induced by nonlinear elliptic boundary-value problems in spaces $\Lambda^{k,\alpha}(\bar{\Omega})$, where $\Lambda^{k,\alpha}(\bar{\Omega})$ is the closure of $C^{\infty}(\bar{\Omega})$ in the norm space $C^{k,\alpha}(\bar{\Omega})$ (see [1]). In [3] a simple way of establishing the properness for the maps induced by quasilinear elliptic boundary-value problems in spaces $C^{k,\alpha}(\bar{\Omega})$ is described. In [2,4] one can find the methods for proving the properness of the maps induced by nonlinear elliptic boundary-value problems of a general type both in Sobolev and Holder spaces. The main idea of these two methods is to reduce the problem (by using differentiation) first to quasilinear and then to linear cases with the help of a priori estimates of Schauder type for linear elliptic problem. However for differentiation one needs to have smooth functions in the domain of values of the nonlinear map and smooth initial data of the problem.

In this paper the differentiation is replaced by an application of a certain analogue of a differential-difference operator and so the initial problem is reduced to a quasilinear problem with this operator. Then an a priori estimate of Schauder type for differential-difference operators, obtained by the second author is used. This makes it possible to weaken the smoothness requirement on the functions in the domain of values of the nonlinear map as well as on the functions which determine the equations and boundary conditions.

The paper consists of three sections. In the first one the necessary notations and facts from the theory of linear elliptic equations are imtroduced. In the second section certain analogue of linear differential-difference operator of elliptic type is introduced; a priori estimates of Schauder type for such operator are established. The third section contains the main statement on the convergence of solutions of general boundary problem for nonlinear elliptic equations and (as a consequence) the statement about properness of the induced map. Let us formulate the basic results of the work.

We shall consider nonlinear elliptic equation

$$F[u](x) = F(x,u(x), Du(x),\dots, D^{2m}u(x)) = 0, \ x \in \bar{\Omega} \qquad (I)$$

and general nonlinear boundary conditions

$$G_j[u](x) = G_j(x,u(x), Du(x),\dots, D^{m_j}u(x)) = 0, \ x \in \partial\Omega \qquad (2)$$

where Ω is a bounded domain in R^n with the boundary $\partial\Omega$ of class $C^{k,\alpha}$, $k \geq 2m$, $m_j < 2m$, $j = I,\dots,m$.

Theorem of convergence. Let $\{u_s\}$, a sequence of functions from $C^{k,\alpha}(\bar{\Omega})$, be such, that

$$u_s \xrightarrow[S \to \infty]{C^k(\bar{\Omega})} u_o, \quad F[u_s] \xrightarrow[S \to \infty]{C^{k-2m,\alpha}(\bar{\Omega})} F[u_o], \qquad (3)$$

$$G_j[u_s] \xrightarrow[S \to \infty]{C^{k-m_j,\alpha}(\partial\Omega)} G_j[u_o], \ j = I,\dots, m. \qquad (4)$$

If functions F and G_j, $j=I,\dots,m$, satisfy conditions $A_I - A_4$ (see section 3) in some neighbourhood of

$$\Gamma_{u_o} = \{(x,u_o(x), Du_o(x),\dots, D^{2m}u_o(x)), \ x \in \bar{\Omega}\}$$

then u_o is the function of class $C^{k,\alpha}(\bar{\Omega})$ and $u_s \xrightarrow[S \to \infty]{C^{k,\alpha}(\bar{\Omega})} u_o$

Here $\longrightarrow$ means the convergence in the norm of space indicated over the pointer.

The consequence of theorem of convergence is the statement about the properness of mapping induced by the problem (I)-(2).

Theorem. Let D be a closed bounded subset of functions from

$C^{k,\alpha}(\bar{\Omega})$ and

$$\Gamma_D = \left\{ (x, u(x), Du(x), \ldots, D^{2m}u(x)),\ x \in \bar{\Omega},\ u \in D \right\}.$$

If functions F and G_j, j = I,..., m satisfy condition $A_I)$-$A_4)$ in some neighbourhood for Γ_D, then the map

$$\mathbb{F} = (F, G_I, \ldots, G_m) : D \longrightarrow C^{k-2m,\alpha}(\bar{\Omega}) \times \prod_{j=I}^{m} C^{k-mj,\alpha}(\partial\Omega)$$

$$u \longmapsto \mathbb{F}[u] = (F[u], \ldots, G_m[u]) \tag{5}$$

is proper on D.

I. Notations and necessary statements

Let Ω be a bounded domain in R^n with boundary $\partial\Omega$ of class $C^{k,\alpha}$, where $o < \alpha < I$; k,m,n are the integers and $n \geqslant 2$, $m \geqslant I$, $k \geqslant 2m$. Denote by M (q) the number of multi-indices $\beta = (\beta_I, \ldots, \beta_n)$, with nonnegative integer coordinates β_i and $|\beta_i| \leqslant q$ where $|\beta| = \beta_I + \beta_2 + \ldots + \beta_n$. For arbitrary multi-index β and function u: $\bar{\Omega} \longrightarrow R$ let us introduce:

$$D^{\beta}u = \left(\frac{\partial}{\partial x_I}\right)^{\beta_I} \ldots \left(\frac{\partial}{\partial x_n}\right)^{\beta_n} u, \quad D^s u = \left\{ D^{\beta}u : |\beta| = s \right\},$$

where s is an integer nonnegative number. Define

$$|u|_o = \max_{x \in \bar{\Omega}} |u(x)|, \quad |u|_s = \sum_{|\beta| \leqslant s} |D^{\beta}u|_o,$$

$$\langle u \rangle_{s,\alpha} = \sum_{|\beta| = s} \sup_{\substack{x, \tilde{x} \in \bar{\Omega} \\ x \neq \tilde{x}}} \frac{|D^{\beta}u(x) - D^{\beta}u(\tilde{x})|}{|x - \tilde{x}|^{\alpha}}$$

here $|x - \tilde{x}|$ is a distance from x to $\tilde{x}$. Notations $C^k(\bar{\Omega})$ and $C^{k,\alpha}(\bar{\Omega})$ are standard. Let us assume that in $C^k(\bar{\Omega})$ and $C^{k,\alpha}(\bar{\Omega})$ the norms $|u|_k$ and $|u|_{k,\alpha} = |u|_k + \langle u \rangle_{k,\alpha}$ are introduced.

Now we formulate the Schauder's estimates in Holder space for elliptic operators

$$\mathbb{L}\, u(x) = \sum_{|\beta|=2m} a_\beta \, D^\beta u(x)$$

with constant coefficients (see [5]).

<u>Theorem I</u>. Let L be an elliptic operator and u be a function of class $C^{k,\alpha}$ with compact support in R^n. Then we have the estimate

$$\langle u \rangle_{k,\alpha} \leq C \cdot \langle \mathbb{L}u \rangle_{k-2m,\alpha} \tag{I.I}$$

where constant C depends only on α,k,m and a characteristic constant E.

Let u(x) be a function with support in semi-space $R^n_+ = \{ x=(x_I, \ldots x_{n-I}, x_n): x_n \geq 0 \}$. Let us assume that we have a set of boundary operators

$$\mathbb{B}_j u(x) = \sum_{|\gamma|=m_j} b_{j\gamma} \, D^\gamma u(x), \quad x \in R^n_+, \; x_n = 0, \; j = I,2,\ldots, m$$

with constant coefficients.

<u>Theorem 2</u>. Let $\mathbb{L}$ be an elliptic operator and operators $\mathbb{B}_j$, j=I, ...,m, satisfy a complimentary condition relative to $\mathbb{L}$. Then for any function u of class $C^{k,\alpha}$ with compact support in R^n_+ the following estimate is true

$$\langle u \rangle_{k,\alpha} \leq C \left(\langle \mathbb{L}u \rangle_{k-2m,\alpha} + \sum_{j=I}^{m} \langle \mathbb{B}_j u \rangle_{k-mj,\alpha} \right), \tag{I.2}$$

where constant C depends only on k, α ,m and a characteristic constant E.

These estimates will play an important role in § 2 during the establishment of Schauder's estimates for differential-difference operators. The following interpolation inequality in Holder space will also be used (see [6]).

<u>Lemma I</u>. For any positive ε there exists constant C, depending only on ε ,k,n, α and domain $\Omega \subset R^n$, such that

$$|u|_1 \leq \varepsilon \cdot \langle u \rangle_{k,\alpha} + C \cdot |u|_0 \tag{I.3}$$

for all functions $u \in C^{k,\alpha}(\bar{\Omega})$ and $o < l \leq k$.

2. Estimates of Schauder type in spaces $C^{k,\alpha}(\bar{\Omega})$ for differential-difference operators

In this section we shall obtain estimates of Schuder type in spaces $C^{k,\alpha}(\bar{\Omega})$ for differential-difference operators

$$Lu(x,\tilde{x}) = \sum_{|\beta|=2m} a_\beta(x,\tilde{x}) \frac{D^\beta u(x) - D^\beta u(\tilde{x})}{|x-\tilde{x}|^\alpha} \text{------} x,\tilde{x} \in \bar{\Omega},\ x \neq x \qquad (2.I)$$

which are uniformly elliptical on $\bar{\Omega}$ and for boundary operators

$$B_j u(x,\tilde{x}) = \sum_{|\gamma|=m_j} b_{j\gamma}(x,\tilde{x}) \frac{D^\gamma u(x) - D^\gamma u(\tilde{x})}{|x-\tilde{x}|^\alpha} \qquad x,\tilde{x} \in \partial\Omega,\ x \neq \tilde{x},\ j=I,\ldots,m \qquad (2.2)$$

satisfying the complimentary condition relative to operator L.

Operator L (2.I) will be called uniformly elliptical on $\bar{\Omega}$ if the following operator is uniformly elliptical:

$$\mathbb{L}\, u(x) = \sum_{|\beta|=2m} a_\beta(x,x) D^\beta u(x).$$

We shall say that operators B_j, $j=I,\ldots,m$ satisfy a complimentary condition relative to operator $\mathbb{L}$ if the corresponding operators $\mathbb{B}_j = \sum_{|\gamma|=m_j} b_{j\gamma}(x,x) D^\gamma$, $j=I,2,\ldots,m$ on $\partial\Omega$ satisfy a complimentary condition relative to operator $\mathbb{L}$ (see [5]).

Estimates of Schauder type will be established by standard arguments. At first the estimates are established for operators with constant coefficients defined on the functions with compact supports. Further, by means of methods of localization and "freezing" of coefficients the estimates for operators with variable coefficients are derived.

We shall present the main notations and statements, connected with the use of difference operators.

<u>2.I. Difference operators</u>. Operators $\mathbb{L}$ and B_j along with differential operators contain the difference operator D_α

$$D_\alpha v(x,\tilde{x}) = \frac{v(x)-v(\tilde{x})}{|x-\tilde{x}|^\alpha}, \qquad x,\tilde{x} \in \bar{\Omega},\ x \neq \tilde{x},$$

which is defined on functions $v : \bar{\Omega} \longrightarrow R^s$, $s \geq I$.

We shall present the simplest properties of operator D_α.

Lemma 2. Let $v, w : \bar{\Omega} \longrightarrow R^s$ be arbitrary functions. Then

$$I)\ D_\alpha (v+w) = D_\alpha (v) + D_\alpha (w);$$

$$2)\ D_\alpha (\lambda \cdot v) = \lambda \cdot D_\alpha v \text{ for any number } \lambda \in R;$$

$$3) \text{ if } s=I\ D_\alpha (v \cdot w)(x,\tilde{x}) = v(x) D_\alpha w(x,\tilde{x}) + w(\tilde{x}) D_\alpha v(x,\tilde{x}) = w(x) D_\alpha v(x,\tilde{x}) + v(\tilde{x}) D_\alpha w(x,\tilde{x});$$

$$4) \text{ if } s=I \text{ and } w(x) \neq 0 \text{ for all } x \in \bar{\Omega};$$

$$D_\alpha (v/w)(x,\tilde{x}) = D_\alpha v(x,\tilde{x})/w(x) - [v(\tilde{x})/(w(x)w(\tilde{x}))] \cdot D_\alpha w(x,x).$$

Let $D_\alpha^* v$ denote an extension (possibly multivalued) of the function $D_\alpha v(x,\tilde{x})$, determined on $\bar{\Omega} \times \bar{\Omega} \setminus d$ where $d = \{(x,x) : x \in \bar{\Omega}\}$, on $\bar{\Omega} \times \bar{\Omega}$. For a function v define

$$D_\alpha^* v(x,\tilde{x}) = D_\alpha v(x,\tilde{x}) \text{ for } (x,\tilde{x}) \in \bar{\Omega} \times \bar{\Omega} \setminus d$$

$$D^*_\alpha v(x,x) = \Big\{ z \in R^s : z = \lim_{j \to \infty} D_\alpha v(x_j, \tilde{x}_j) \text{ for some sequence}$$

$$(x_j, \tilde{x}_j) \in (\bar{\Omega} \times \bar{\Omega}) \setminus d \text{ such that } (x_j, \tilde{x}_j) \xrightarrow[j \to \infty]{} (x,x) \Big\}$$

Lemma 3. Let $v : \bar{\Omega} \longrightarrow R^s$ and $w : v(\bar{\Omega}) \longrightarrow R^p$, $p \geq I$, then

$$D_{\alpha \cdot \beta} (w \cdot v)(x,\tilde{x}) = D_\beta^* w (v(x), v(\tilde{x})) \cdot |D_\alpha v(x,\tilde{x})|^\beta \tag{2.3}$$

for $0 < \alpha \leq I$, $0 < \beta \leq I$.

Using the statement of lemma 3, it is easy to obtain the rule of change of variables.

Lemma 4. Let $u : \bar{\Omega} \longrightarrow R$, $y : \bar{\Omega} \longrightarrow G$ and y be a bijective mapping. Then

$$D_\alpha (u \cdot y^{-I})(y(x), y(\tilde{x})) = D_\alpha u(x,\tilde{x}) \cdot |D_I y(x,\tilde{x})|^{-\alpha}. \tag{2.4}$$

The statement of lemma follows from equality (2.3) and

$$/D_I y^{-I}(y(x),y(\tilde{x}))/ = /D_I y\,(x,\tilde{x})/^{-I}.$$

Let us note that spaces $C^{k,\alpha}(\overline{\Omega})$ can be considered as a set of functions from $C^k(\overline{\Omega})$ for which the functions $D_\alpha\,(D^\beta u)\,(x,\tilde{x})$ are uniformly bounded on $(\overline{\Omega} \times \overline{\Omega}) \smallsetminus d$ for all $\beta : |\beta| = k$. The application of operator D_α leads to the study of the functions $w: (\overline{\Omega} \times \overline{\Omega}) \smallsetminus d \longrightarrow R$. For such functions $\frac{\tilde{\partial}}{\partial x_i} w$ denotes the derivative of w in direction $e_i \cdot e_i \in R^n \times R^n$, where $e_I, e_2, \ldots, e_n$ are the basis in Euclidean space R^n and $\tilde{D}^\beta w$ denotes the following expression

$$\tilde{D}^\beta w = (\frac{\tilde{\partial}}{\partial x_i})^{\beta_I} \cdots \cdot (\frac{\tilde{\partial}}{\partial x_n})^{\beta_n} w$$

for multi-index β .

Let $U \subset (\overline{\Omega} \times \overline{\Omega}) \smallsetminus d$ be closed in topology $(\overline{\Omega} \times \overline{\Omega}) \smallsetminus d$ and $\tilde{C}^k(U)$ denote the set of functions $w : U \longrightarrow R$, having continuous and bounded derivatives $\tilde{D}^\beta w$ on U up to order k inclusively. Let us denote

$$\langle\!\langle w \rangle\!\rangle_{o(U)} = \sup_{(x,\tilde{x}) \in U} |w\,(x,\tilde{x})|, \quad \langle\!\langle w \rangle\!\rangle_{k(U)} = \sum_{|\beta|=k} \langle\!\langle \tilde{D}^\beta w \rangle\!\rangle_{o(U)}.$$

Symbol (U) is omitted if $U = (\overline{\Omega} \times \overline{\Omega}) \smallsetminus d$ or it is clear from the context which areas are under consideration.

    <u>Lemma 5</u>. If norm $\| w \|_k = \sum_{i=o}^{k} \langle\!\langle w \rangle\!\rangle_i$ is introduced in

$C^k (\overline{\Omega} \times \overline{\Omega} \smallsetminus d)$, then operator

$$D_\alpha : C^{k,\alpha}(\overline{\Omega}) \longrightarrow \tilde{C}^k((\overline{\Omega} \times \overline{\Omega}) \smallsetminus d)$$

is continuous.

    <u>Proof</u>. For k=o the statement is obvious. Assume that $u \in C^{I,\alpha}(\overline{\Omega})$, then derivatives $\frac{\partial u}{\partial x_i}$ are continuous on $\overline{\Omega}$ and values

$$\sup_{(x,\tilde{x}) \in (\overline{\Omega} \times \overline{\Omega}) \smallsetminus d} \left| \frac{\partial u(x)}{\partial x_i} - \frac{\partial u(\tilde{x})}{\partial x_i} \right| \cdot |x - \tilde{x}|^\alpha = \langle\!\langle D_\alpha \frac{\partial u}{\partial x_i} \rangle\!\rangle_o$$

are bounded. We shall show that derivatives $\frac{\tilde{\partial}}{\partial x_i} (D_\alpha u)$ are well-defined and continuous for all $i = I, 2, \ldots, n$.

We have

$$\frac{\tilde{\partial}}{\partial x_i}(D_\alpha u)(x,\tilde{x}) = \lim_{\varepsilon\to 0}\frac{D_\alpha u(x+\varepsilon\cdot e_i,\tilde{x}+\varepsilon\cdot e_i)-D_\alpha u(x,\tilde{x})}{2} =$$

$$= \lim_{\varepsilon\to 0}\frac{I}{\varepsilon\sqrt{2}}\left(\frac{u(x+\varepsilon\cdot e_i)-u(\tilde{x}+\varepsilon\cdot e_i)}{/(x+\varepsilon\cdot e_i)-(\tilde{x}+\varepsilon\cdot e_i)/^{\alpha}} - \frac{u(x)-u(\tilde{x})}{/x-\tilde{x}/^{\alpha}}\right) =$$

$$= \lim_{\varepsilon\to 0}\frac{I}{\sqrt{2}/x-\tilde{x}/^{\alpha}}\left(\frac{u(x+\varepsilon\cdot e_i)-u(x)}{\varepsilon} - \frac{u(\tilde{x}+\varepsilon\cdot e_i)-u(x)}{\varepsilon}\right) =$$

$$= \frac{I}{\sqrt{2}}\left(\frac{\partial}{\partial x_i}u(x) - \frac{\partial}{\partial x_i}u(\tilde{x})\right)\Big/ /x-\tilde{x}/^{\alpha} = \frac{I}{\sqrt{2}}D_\alpha\left(\frac{\partial u}{\partial x_i}\right)(x,\tilde{x}).$$

From equality

$$\frac{\tilde{\partial}}{\partial x_i}(D_\alpha u) = \frac{I}{\sqrt{2}}D_\alpha\left(\frac{\partial u}{\partial x_i}\right) \qquad (2.5)$$

follows the existence of $\frac{\tilde{\partial}}{\partial x_i}(D_\alpha u)$ and continuity on $(\bar{\Omega}\times\bar{\Omega})\setminus d$ as well as

$$\langle\langle D_\alpha U\rangle\rangle_I = \frac{I}{\sqrt{2}}\langle U\rangle_{I,\alpha}\,.$$

Successive application of formula (2.5) to function $u\in C^{k,\alpha}(\bar{\Omega})$ gives us

$$\tilde{D}^{\beta}(D_\alpha u) = 2^{-\frac{|\beta|}{2}}D_\alpha(D^{\beta}u), \qquad (2.6)$$

$$\|D_\alpha u\|_k = \sum_{i=0}^{k}2^{-\frac{i}{2}}\langle u\rangle_{i,\alpha}\,,\quad \langle\langle D_\alpha u\rangle\rangle_i = 2^{-\frac{i}{2}}\langle u\rangle_{i,\alpha} \qquad (2.7)$$

which completes the proof in the case of arbitrary **k**.

2.2. Operators with constant coefficients. Let us consider operators L and B_j, j=I,...,m with constant coefficients. Estimates of Schauder type are a simple consequence of theorems I and 2 for such operators.

Theorem 3. For an elliptic operator L and arbitrary function $u\in C^{k,\alpha}(R^n)$ with compact support holds the following estimate

$$\langle u\rangle_{k,\alpha} \leq C\cdot\langle\langle Lu\rangle\rangle_{k-2m}, \qquad (2.8)$$

where constant C depends only on α, k,m and a characteristic constant E.

Theorem 4. Let operators $B_j, j=I,\dots,m$ satisfy the complimentary condition relative to elliptic operator L.

Then for any function $u \in C^{k,\alpha}(R^n_+)$ with compact support in R^n_+ the following estimate is true

$$\langle u \rangle_{k,\alpha} \leq C(\langle\langle Lu \rangle\rangle_{k-2m} + \sum_{j=I}^{m} \langle\langle B_j u \rangle\rangle_{k-mj}), \qquad (2.9)$$

where constant C depends only on α, k, m and a characteristic constant E.

For proof it is enough to note that

$$Lu = D_\alpha(Lu), \qquad B_j u = D_\alpha(B_j u), \ j=I,\dots,m$$

and due to (2.7)

$$\langle Lu \rangle_{k-2m,\alpha} = 2^{\frac{k-2m}{2}} \cdot \langle\langle Lu \rangle\rangle_{k-2m}, \langle B_j u \rangle_{k-m_j,\alpha} = 2^{\frac{k-mj}{2}} \langle\langle B_j u \rangle\rangle_{k-mj}$$

for $j=I,\dots,m$. Substituting the obtained expressions for $\langle Lu \rangle_{k-2m,\alpha}$, $\langle B_j u \rangle_{k-m_j,\alpha}$ into estimates (I.I),(I,2) of theorems I and 2 we obtain the estimates (2.8),(2.9).

2.3. Operators with variable coefficients. Let us consider operators L, B_j, $j=I,\dots,m$ with variable coefficients. We shall obtain estimates of Schauder type in spaces $C^{k,\alpha}(\bar{\Omega})$ in the following theorem.

Theorem 5. Let $\bar{\Omega}$ be a closed bounded domain in R^n with the boundary $\partial\Omega$ of class $C^{k,\alpha}$, $k \geq 2m$; a_β be functions of class $\tilde{C}^{k-2m}((\bar{\Omega} \times \bar{\Omega}) \setminus d)$, which are continuous on $\bar{\Omega} \times \bar{\Omega}$ for all β : $|\beta| = 2m$; let $b_{j,\gamma}$ be functions of class $\tilde{C}^{k-m_j}((\partial\Omega \times \partial\Omega) \setminus d)$, which are continuous on $\partial\Omega \times \partial\Omega$ for all γ : $|\gamma| = m_j$, $j=I,2,\dots,m$. If operator L satisfies the condition of uniform ellipticity on $\bar{\Omega}$ and operators B_j, $j=I,2,\dots,m$, satisfy the complimentary condition relative to L on $\partial\Omega$, then for any function $u \in C^{k,\alpha}(\bar{\Omega})$ the following estimate is true

$$|u|_{k,\alpha} \leq C(\langle\langle Lu \rangle\rangle_{k-2m} + \sum_{j=I}^{m} \langle\langle B_j u \rangle\rangle_{k-mj} + |u|_0) \qquad (2.I0)$$

Constant C can be chosen as a constant for any family of operators $(L, B_I, \dots, B_m)$ for which the characteristic constants and values A=

$\sum_{|\beta|=2m} \|a_\beta\|_{k-2m,B} = \sum_{j=I}^{m} \sum_{|\gamma|=m_j} \|b_{j\gamma}\|_{k-m_j}$ are bounded and families of coefficients a_β, $|\beta| = 2m$, $b_{j\gamma}$, $j=I,\dots,m$, $|\gamma| = m_j$ are equicontinuous in some neighbourhood d.

<u>Proof</u>. The standard methods of transition to functions with compact support and "freezing" of operators coefficients are used for the proof.

Let us construct the finite cover of the domain $\bar{\Omega}$ by disks. For every interior point $x \in \Omega$ we select a disk $B_R(x)$ with the centre x and radius R which is smaller than a half of the distance from x to $\partial\Omega$. For the point $x \in \partial\Omega$ a disk $B_R(x)$ is selected in such a way that there is a bijective mapping T_x of class $C^{k,\alpha}$ from $\bar{\Omega} \cap B_{2R}(x)$ to $R^n_+ = \{ x=(x',x_n) : x' \in R^{n-I}, x_n \in R^+, x_n \geq 0 \}$ which satisfies the conditions $T_x(x) = 0$ and $T_x(\partial\Omega \cap B_{2R}(x)) \subset R^{n-I} = \{ x = (x',x_n) : x_n = 0 \}$. Let $\{ B_{R_i}(x_i) \}_{i=I}^{N}$ denote a finite subcover $\bar{\Omega}$. Let $\bar{R} = \max_{I \leq i \leq N} R_i$ and C_T be an estimate of norms in space $C^{k,\alpha}$ for maps T_{x_i} and inverse maps for all $x_i \in \partial\Omega$.

Let us represent function u in the form of $u = \sum_{i=I}^{N} w_i u$ where $\{ w_i \}_{i=I}^{N}$ is an arbitrary smooth partition of unity subordinate to the above cover. We shall prove the estimate (2.I0) for every summand $w_i u$.

Let $x_i \in \partial\Omega$. We introduce new variables $y=T_i x$ and notations

$$\Omega_i = \bar{\Omega} \cap B_{2R_i}(x_i), \qquad \partial\Omega_i = \partial\Omega \cap B_{2R_i}(x_i).$$

$$D_i = T_i(\Omega_i) \subset R^n_+, \qquad D_i = T_i(\partial\Omega_i) \subset R^{n-I}.$$

The functions w_i and u on Ω_i are transformed into functions

$$w(y) = w_i(T_i^{-I}(y)), \qquad v(y) = u(T_i^{-I}(y))$$

on D_i. Using the properties of operator D_α it is easy to check that

$$|w_i u|_{l,\alpha,\Omega_i} \leq C\, |w\, v|_{l,\alpha,D_i} \qquad 0 \leq l \leq k. \quad (2.II)$$

Here and further C denotes the constants depending on C_T, k, α, radius R and probably on values E,A,B, which characterise the operators L, B_j.

After the introduction of variables and multiplication by $/D_I T_i(x,\tilde{x})/^{-\alpha}$ the operators $L, B_j, j=I,2,\dots,m$, are transformed into operators $\tilde{L}, \tilde{B}_j, j=I,2,\dots,m$, where

$$\tilde{L}(w\cdot v)(y,\tilde{y}) = \sum_{|\beta|\le 2m} [a'_\beta(y,y) D_\alpha D^\beta (w\cdot v)(y,\tilde{y}) + \tilde{a}_\beta (y,\tilde{y}) D^\beta (w\cdot v)(y)] ,$$

$$y,\tilde{y} \in D_i, \; y \neq \tilde{y};$$

$$\tilde{B}_j(w\cdot v)(y,\tilde{y}) = \sum_{|\gamma|\le m_j} [b'_{j\gamma}(y,\tilde{y}) D_\alpha D^\gamma (w\cdot v)(y,\tilde{y}) + \tilde{b}_{j\gamma}(y,\tilde{y}) D^\gamma (w\cdot v)(y)],$$

$$y,\tilde{y} \in \partial D_i, \; y \neq \tilde{y}.$$

For any multi-index β : $|\beta| = 2m$

$$a'_\beta(y,\tilde{y}) = \sum_{|\beta'|=2m} a_{\beta'}(T_i^{-I}(y), T_i^{-I}(\tilde{y})) \cdot \sum_{|\Gamma|=\beta'} \Big(\prod_{\substack{l=I\\ s=I}}^{n} \Big(\frac{\partial y_l (T_i^{-I}(\tilde{y}))}{\partial x_s} \Big)^{\beta^l_s} \Big)$$

where $\Gamma = (\beta^I, \beta^2, \dots, \beta^n)$, β^l are multi-indices, $l=I,2,\dots,n$ $|\Gamma| = \beta^I + \beta^2 + \dots + \beta^n$; and y_l is the l-th coordinate of the map T_i. Analogously the coefficients $b'_{j\gamma}$ for any $j=I,2,\dots,m$ and multi-indices γ, $|\gamma| = m_j$, are expressed:

$$b'_{j\gamma}(y,\tilde{y}) = \sum_{|\gamma'|=m_j} b_{j\gamma'}(T_i^{-I}(y), T_i^{-I}(\tilde{y})) \sum_{|\Gamma|=\gamma'} \Big(\prod_{\substack{l=I\\ s=I}}^{n} \Big(\frac{\partial y_l (T_i^{-I}(\tilde{y}))}{\partial x_s} \Big)^{\gamma^l_s} \Big)$$

where $\Gamma = (\gamma^I, \gamma^2, \dots, \gamma^n)$, γ^l are multi-indices, $l=I,2,\dots,n$. Expressions of the remaining coefficients have a more complicated (for writing) form.

Using the interconnection of coefficients of operators L and $\tilde{L}$, B_j and $\tilde{B}_j$, we obtain the following estimates:

$$\tilde{A} \equiv \sum_{|\beta|\le 2m} (\langle\langle a'_\beta \rangle\rangle_{k-2m, D_i} + \langle\langle a'_\beta \rangle\rangle_{k-2m, D_i}) \le C_I A$$

$$\tilde{B} \equiv \sum_{j=I}^{m} \sum_{|\gamma|\le m_j} (\langle\langle b'_{j\gamma} \rangle\rangle_{k-m_j, \partial D_i} + \langle\langle b'_{j,\gamma} \rangle\rangle_{k-mj, \partial D_i}) \le C_I B \quad (2.I2)$$

$$\langle\langle a'_\beta - a'_\beta(0,0) \rangle\rangle_{0,D_i} \le C_I (\max_{|\beta'|=2m} \langle\langle a_{\beta'} - a_{\beta'}(x_i,x_i) \rangle\rangle_{0,\Omega_i} + R_i)$$

for $\beta : |\beta|=2m$.

$$\langle\!\langle b'_{j\gamma} - b'_{j\gamma}(0,0) \rangle\!\rangle_{0,\partial D_i} \le C_I(\max_{|\gamma'|=m_j} \langle\!\langle b_{j\gamma'} - b_{j\gamma'}(x_i,x_i) \rangle\!\rangle_{0,\partial\Omega_i} + R_i)$$

for $j=1,2,\ldots,m$ and $\gamma : |\gamma| = m_j$, where the constant C_I depends on m,k,n,α and C_T.

Let $\tilde{L}', \tilde{B}'_j$ denote the main parts of operators $\tilde{L}$ and $\tilde{B}_j$, respectively

$$\tilde{L}'(wv) = \sum_{|\beta|=2m} a'_\beta D_\alpha D^\beta (w\cdot v), \quad B'_j(w\cdot v) = \sum_{|\gamma|=m_j} b'_{j\gamma} D_\alpha D^\gamma (w\cdot v).$$

Non-degenerate change of variables conserves the condition of uniform ellipticity for $\tilde{L}'$ on D_i and complimentary condition of operators $\tilde{B}'_j, j=1,2,\ldots,m$ relative to $\tilde{L}'$ on ∂D_i. From the structure of operators and estimates (2.12), it is easy to establish the inequalities

$$\langle\!\langle \tilde{L}'(w\cdot v) \rangle\!\rangle_{k-2m,D_i} \le C_2(\langle\!\langle \tilde{L}(w\cdot v) \rangle\!\rangle_{k-2m,D_i} + |w\cdot v|_{k,D_i}),$$

$$\langle\!\langle \tilde{B}'_j(w\cdot v) \rangle\!\rangle_{k-mj,\partial D_i} \le C_2(\langle\!\langle \tilde{B}_j(w\cdot v) \rangle\!\rangle_{k-mj,\partial D_i} + |w\cdot v|_{k,\partial D_i}).$$

By virtue of the obvious estimates

$$\langle\!\langle \tilde{L}(w\cdot v) \rangle\!\rangle_{k-2m,D_i} \le C_3(\langle\!\langle L(w_i\cdot u) \rangle\!\rangle_{k-2m,\Omega_i} + |w_i u|_{k,\Omega_i}),$$

$$\langle\!\langle \tilde{B}_j(w\cdot v) \rangle\!\rangle_{k-m_j,\partial D_i} \le C_3(\langle\!\langle B_j(w_i u) \rangle\!\rangle_{k-m_j,\partial\Omega_i} + |w_i u|_{k,\partial\Omega_i}),$$

$$|w\cdot v|_{k,D_i} \le C_4 \cdot |w_i\cdot u|_{k,\Omega_i}, \quad |w\cdot v|_{k,\partial D_i} \le C_4 \cdot |w_i\cdot u|_{k,\partial\Omega_i}$$

we obtain

$$\langle\!\langle \tilde{L}'(w\cdot v) \rangle\!\rangle_{k-2m,D_i} \le C_5(\langle\!\langle L(w_i u) \rangle\!\rangle_{k-2m,\Omega_i} + |w_i u|_{k,\Omega_i}) \tag{2.13}$$

$$\langle\!\langle \tilde{B}'_j(w\cdot v) \rangle\!\rangle_{k-m_j,\partial D_i} \le C_5(\langle\!\langle B_j(w_i u) \rangle\!\rangle_{k-m_j,\partial\Omega_i} + |w_i u|_{k,\partial\Omega_i})$$

for j=I,2,...,m.

Let us "freeze" the coefficients of operators in zero. From estimate (2.9) for operators with constant coefficients

$$\widetilde{L}'_o = \sum_{|\beta|=2m} a'_\beta\ (0,0)\ D_\alpha D^\beta\ , \quad \widetilde{B}'_{j,o} = \sum_{|\gamma|=m_j} b'_{j\gamma}\,(0,0) D_\alpha D^\gamma, \quad j=I,\dots,m$$

we have

$$\langle w\cdot v\rangle_{k,\alpha,D_i} \leqslant C_6(\langle\!\langle \widetilde{L}'_o(w\cdot v)\rangle\!\rangle_{k-2m,D_i} + \sum_{j=I}^{m} \langle\!\langle \widetilde{B}'_{j,o}(w\cdot v)\rangle\!\rangle_{k-m_j,\partial D_i}) \tag{2.I4}$$

Then from the representation

$$\widetilde{L}'_o(w\cdot v)(y,\widetilde{y}) = \widetilde{L}'(w\cdot v)(y,\widetilde{y}) + \sum_{|\beta|=2m} (a'_\beta\ (0,0)*a'_\beta\ (y,\widetilde{y}))D_\alpha\ D^\beta\cdot$$

$$(w\cdot v)\ (y,\widetilde{y}),\ y,\widetilde{y} \in D_i,\ y \neq \widetilde{y}.$$

$$\widetilde{B}'_{j,o}(w\cdot v)(y,\widetilde{y}) = \widetilde{B}'_j(w\cdot v)(y,\widetilde{y}) + \sum_{|\gamma|=m_j} (b'_{j\gamma}\ (0,0)-b'_{j\gamma}\ (y,\widetilde{y}))\cdot$$

$$D_\alpha D^\gamma (w\cdot v)(y,\widetilde{y}),\ y,\widetilde{y} \in \partial D_i,\ y \neq \widetilde{y},$$

j=I,2,...,m, we obtain the estimate

$$\langle\!\langle \widetilde{L}'_o\ (w\cdot v)\ \rangle\!\rangle_{k-2m,D_i} = \langle\!\langle \widetilde{L}'(w\cdot v)\ \rangle\!\rangle_{k-2m,D_i} +$$

$$+C_7(\max_{|\beta|=2m} \langle\!\langle a'_\beta - a'_\beta\ (0,0)\rangle\!\rangle_{0,D_i} \cdot\langle w\cdot v\rangle_{k,\alpha,D_i} + \widetilde{A}\cdot|w\cdot v|_{k-I,\alpha,D_i})$$

Substituting these estimates into (2.I4), we obtain

$$\langle w\cdot v\rangle_{k,\alpha,D_i} \leqslant C_8(\langle\!\langle \widetilde{L}'(w\cdot v)\rangle\!\rangle_{k-2m,D_i} + \sum_{j=I}^{m} \langle\!\langle \widetilde{B}'_j\ (w\cdot v)\rangle\!\rangle_{k-m_j,\partial D_i} \tag{2.I5}$$

$$+ (\max_{|\beta|=2m} \langle\!\langle a'_\beta - a'_\beta\ (0,0)\rangle\!\rangle_{0,D_i} + \max_{j=I,\dots,m}\ \max_{|\gamma|=m_j} \langle\!\langle b'_{j\gamma} - b'_{j\gamma}\ (0,0)\rangle\!\rangle_{0,\partial D_i})\cdot$$

$$\cdot\langle w\cdot v\rangle_{k,\alpha,D_i} + |w\cdot v|_{k,D_i}).$$

Constant C_8 doest not depend on a diameter $2\bar{R}$ of a chosen cover, so we choose $\bar{R}$ in such a way that

$$C_I C_8 \Big(\max_{|\beta|=2m} \langle\!\langle a_\beta - a_\beta(x_i,x_i)\rangle\!\rangle_{0,\Omega_i} + \max_{j=I,\dots,m}\max_{|\gamma|=m_j} \langle\!\langle b_{j\gamma} - b_{j\gamma}(x_i,x_i)\rangle\!\rangle_{0,\partial\Omega_i} + \bar{R}\Big) < \frac{I}{2}.$$

Then (2.I5) can be transformed to

$$\langle w\cdot v\rangle_{k,\alpha,D_i} \le C_9\Big(\langle\!\langle \tilde{L}'(w\cdot v)\rangle\!\rangle_{k-2m,D_i} + \sum_{j=I}^{m} \langle\!\langle \tilde{B}_j'(w\cdot v)\rangle\!\rangle_{k-m_j,\partial D_i} + |w\cdot v|_{k,D_i}\Big)$$

The standard application of interpolation inequalities gives us

$$|w\cdot v|_{k,\alpha,D_i} \le C_{I0}\Big(\langle\!\langle \tilde{L}'(w\cdot v)\rangle\!\rangle_{k-2m,D_i} + \sum_{j=I}^{m} \langle\!\langle B_j'(w\cdot v)\rangle\!\rangle_{k-m_j,\partial D_i} + |w\cdot v|_{0,D_i}\Big).$$

By applying estimates (2.II),(2.I3) and interpolation inequality (I.3) we obtain the required inequality for function $w_i u$:

$$|w_i u|_{k,\alpha,\Omega_i} \le C_{II}\Big(\langle\!\langle L(w_i u)\rangle\!\rangle_{k-2m,\Omega_i} + \sum_{j=I}^{m} \langle\!\langle B_j(w_i u)\rangle\!\rangle_{k-m_j,\partial\Omega_i} + |w_i u|_{0,\Omega_i}\Big) \tag{2.I6}$$

It is easy to see that the inequality is also true when Ω_i is replaced by $\bar{\Omega}$.

In the case when x_i is an interior point of domain Ω the analogous arguments imply the estimate

$$|w_i u|_{k,\alpha,B_{2R_i}(x_i)} \le C_{I2}\Big(\langle\!\langle L(w_i u)\rangle\!\rangle_{k-2m,B_{2R_i}(x_i)} + |w_i u|_{0,B_{2R_i}(x_i)}\Big) \tag{2.I7}$$

The transition to new variables is not required here and in the proof estimate (2.9) is replaced by estimate (2.8).

Summing up the inequalities (2.I6) and (2.I7) for all i=I,..., N with the replacement of Ω_i and $B_{2R_i}(x_i)$ by $\bar{\Omega}$ and applying the estimates

$$|u|_{k,\alpha,\bar{\Omega}} \leq \sum_{i=I}^{N} |w_i u|_{k,\alpha,\bar{\Omega}}, |w_i u|_{0,\bar{\Omega}} \leq |u|_{0,\bar{\Omega}},$$

$$\langle\langle L(w_i u)\rangle\rangle_{k-2m,\bar{\Omega}} \leq \langle\langle L(u)\rangle\rangle_{k-2m,\Omega} + C_{I3}|u|_{k,\bar{\Omega}},$$

$$\langle\langle B_j(w_i u)\rangle\rangle_{k-m_j,\partial\Omega} \leq \langle\langle B_j(u)\rangle\rangle_{k-m_j,\partial\Omega} + C_{I3}|u|_{k,\Omega}, j=I,\dots,m$$

we obtain

$$|u|_{k,\alpha,\bar{\Omega}} \leq C(\langle\langle L(u)\rangle\rangle_{k-2m,\bar{\Omega}} + \sum_{j=I}^{m} \langle\langle B_j(u)\rangle\rangle_{k-m_j,\partial\Omega} + |u|_{k,\bar{\Omega}}).$$

The application of interpolation inequality (I.3) comletes the proof.

Remark I. Let d_ε be ε-neighbourhood of d in $\bar{\Omega} \times \bar{\Omega}$ and $\partial d_\varepsilon = d_\varepsilon \cap (\partial\Omega \times \partial\Omega)$. It is easy to check that the above arguments give a more general estimate than (2.IO):

$$|u|_{k,\alpha,\bar{\Omega}} \leq C(\langle\langle L(u)\rangle\rangle_{k-2m,d_\varepsilon} + \sum_{j=I}^{m} \langle\langle B_j u\rangle\rangle_{k-m_j,\partial d_\varepsilon} +$$

$$+ |u|_{0,\bar{\Omega}})$$

for any $\varepsilon > 0$, but the constant C depends on ε.

Remark 2. Statements of theorem 5 and remark I can be easily generalized for the case, when the boundary values are given only for the part of boundary $\partial\Omega$.

3. Properness of induced mappings.

In the second section we obtain the properness of the mapping F, introduced by equality (5), on the bounded subsets in $C^{k,\alpha}(\bar{\Omega})$. The main part of the statement is singled out as a certain theorem of convergence (theorem 6).

Let us consider non-linear elliptic mapping

$$F[u](x) = F(x, u(x), Du(x), \dots, D^{2m}u(x)), \quad x \in \bar{\Omega} \qquad (3.I)$$

and general non-linear boundary mappings

$$G_j[u](x) = G_j(x, u(x), \dots, D^{m_j}u(x)), \quad x \in \partial\Omega, \quad j=I,2,\dots,m. \quad (3.2)$$

Let, as in the previous sections, Ω be a bounded domain in R^n with the boundary $\partial\Omega$ of class $C^{k,\alpha}$ and Q be a bounded subset $\bar{\Omega} \times R^{M(2m)}$, $\partial Q = Q \cap (\partial\Omega \times R^{M(2m)})$. For the sake of simplicity we shall consider $m_j < 2m$, $j=I,2,\dots,m$.

Further in the main statements of the paper we shall use the following conditions on functions $F(x, \eta_0, \dots, \eta_{2m})$ and $G_j(x, {}_0, \dots, {}_{m_j})$, $j=I,2,\dots m$:

A_I) function F has continuous partial derivatives up to order $k-2m$ inclusively in all the variables; moreover, these derivatives on Q are functions of class $\Lambda^{0,\alpha}$ in variables $(x, \eta_0, \dots, \eta_{2m-I})$ and continuously differentiable in variables $\eta_{2m} = \{\eta_\beta : |\beta| = 2m\}$;

A_2) for any (x, η) from Q operator

$$L\, v(x) = \sum_{|\beta|=2m} \frac{\partial F(x,\eta)}{\partial \eta_\beta} D^\beta v(x), \quad x \in \bar{\Omega}$$

satisfies the condition of uniform ellipticity.

A_3) functions G_j have continuous partial derivatives up to order $k-m_j$ inclusively in all the variables, moreover these derivatives on ∂Q are functions of class $\Lambda^{0,\alpha}$ in variables $(x, \eta_0, \dots, \eta_{m_j-I})$ and continuously differentiable in variables η_{m_j};

A_4) linear operators

$$B_j v(x) = \sum_{|\gamma|=m_j} \frac{\partial G_j(x,\eta)}{\partial \eta_\gamma} D^\gamma v(x), \quad x \in \partial\Omega, \quad j=I,\dots,m$$

satisfy a complimentary condition relative to operator L.

For $u \in C^{2m}(\bar{\Omega})$ we define

$$\Gamma_u = \{(x, u(x), D_u(x), \dots, D^{2m}u(x)) \in \bar{\Omega} \times R^{M(2m)}, \ x \in \bar{\Omega}\}.$$

<u>Theorem 6</u>. Let $\{u_s\}$, a sequence of functions from $C^{k,\alpha}(\bar{\Omega})$,

be such, that

$$u_s \xrightarrow[S \to \infty]{C^k(\bar{\Omega})} u_o, \qquad (3.3)$$

$$F[u_s] \xrightarrow[S \to \infty]{C^{k-2m,\alpha}(\bar{\Omega})} F[u_o], \; G_j[u_s] \xrightarrow[S \to \infty]{C^{k-m_j,\alpha}(\partial\Omega)} G_j[u_o] \qquad (3.4)$$

If functions F and G_j, $j=1,2,\ldots,m$ satisfy conditions $A_1)$-$A_4)$ in some neighbourhood of Γ_{u_o}, then u_o belongs to space $C^{k,\alpha}(\bar{\Omega})$ and $u_s \xrightarrow[S \to \infty]{C^{k,\alpha}(\bar{\Omega})} u_o$

Proof. Let us apply operator D_α to functions $F[u_s]$, $G_j[u_s]$, $j=1,\ldots,m$; $s=0,1,\ldots$. According to conditions (3.4) and lemma 5 we have

$$D_\alpha F[u_s] \xrightarrow[S \to \infty]{C^{k-2m}((\bar{\Omega} \times \bar{\Omega}) \setminus d)} D_\alpha F[u_o],$$

$$D_\alpha G_j[u_s] \xrightarrow[S \to \infty]{C^{k-m_j}((\partial\Omega \times \partial\Omega) \setminus d)} D_\alpha G_j[u_o], \; j=1,\ldots,m$$

Let us consider operators $D_\alpha F$ and $D_\alpha G_j$. For any function $u \in C^{k,\alpha}(\bar{\Omega})$ such that $\Gamma_u \subset Q$ we present operator $D_\alpha F$ in the following form

$$D_\alpha F[u](x,\tilde{x}) = \sum_{|\beta|=2m} a_\beta[u](x,\tilde{x}) \cdot D_\alpha D^\beta u(x,\tilde{x}) + H[u](x,\tilde{x}), \qquad (3.5)$$

where $x,\tilde{x} \in \bar{\Omega}$, $x \neq \tilde{x}$,

$$H[u](x,\tilde{x}) = \frac{F(x,u(x),\ldots,D^{2m-1}u(x),D^{2m}u(x)) - F(\tilde{x},u(\tilde{x}),\ldots,D^{2m-1}u(\tilde{x}),D^{2m}u(x))}{|x-\tilde{x}|^\alpha}$$

and

$$a_\beta[u](x,\tilde{x}) = \int_0^1 \frac{\partial F}{\partial \eta_\beta}(\tilde{x},\ldots,D^{2m-1}u(\tilde{x}),\frac{\partial^{2m}}{\partial x_1^{2m}}u(\tilde{x}),\ldots, tD^\beta u(x) + (1-t)\cdot D^\beta u(\tilde{x}),\ldots,\frac{\partial^{2m}}{\partial x_n^{2m}}u(x))\,dt. \qquad (3.6)$$

This representation is well defined for $(x,\tilde{x})$ from some neighbourhood d_ε of d; moreover the value ε can be chosen as a constant for all u_s with sufficiently large number s. We note that

$$a_\beta[u](x,x) = \frac{\partial F}{\partial \eta_\beta}(x,u(x),Du(x),\dots,D^{2m}u(x)). \qquad (3.7)$$

Analogously operator $D_\alpha G_j$ can be represented in the following form

$$D_\alpha G_j[u](x,\tilde{x}) = \sum_{|\gamma|=m_j} b_{j\gamma}[u](x,\tilde{x})\, D_\alpha D^\gamma u(x,\tilde{x}) + E_j[u](x,\tilde{x}) \qquad (3.8)$$

where $x,\tilde{x} \in \partial\Omega$, $x \neq \tilde{x}$, and

$$b_{j\gamma}[u](x,\tilde{x}) = (G_j(\tilde{x},u(\tilde{x}),\dots,D^{m_j-1}u(\tilde{x}),\frac{\partial^{m_j}}{\partial x_I^{m_j}}u(\tilde{x}),\dots,D^\gamma u(x),\dots,$$

$$\dots,\frac{\partial^{m_j}u(x)}{x_n^{m_j}})$$

$$-G_j(\tilde{x},u(\tilde{x}),\dots,D^{m_j-I}u(\tilde{x}),\frac{\partial^{m_j}}{\partial x_I^{m_j}}u(x),\dots,D^\gamma u(\tilde{x}),\dots,\frac{\partial^{m_j}}{\partial x_n^{m_j}}u(x)) \cdot$$

$$\cdot\ (D^\gamma u(x)-D^\gamma u(x))^{-I},$$

$$E_j[u](x,\tilde{x}) =$$

$$= \frac{G_j(x,u(x),\dots,D^{m_j-I}u(x),D^{m_j}u(x) - G_j(\tilde{x},u(\tilde{x}),\dots,D^{m_j-I}u(\tilde{x}),D^{m_j}u(x))}{|x-x|^\alpha}$$

It is easy to see that functions $b_{j\gamma}[u](x,\tilde{x})$ can be expanded to continuous functions on a neighbourhood of points $(x,\tilde{x}) \in \partial\Omega \times \partial\Omega$, where $D^\gamma u(x) = D^\gamma u(\tilde{x})$, by the values

$$b_{j\gamma}[u](x,\tilde{x}) = \frac{\partial}{\partial \eta_\gamma} G_j(\tilde{x},u(\tilde{x}),\dots,D^{m_j-I}u(\tilde{x}),\frac{\partial^{m_j}}{\partial x_I^{m_j}}u(\tilde{x}),\dots,D^\gamma u(x),\dots,$$

$$\dots,\frac{\partial^{m_j}}{\partial x_n^{m_j}}u(x))$$

We also have

$$b_{j\gamma}[u](x,x) = \frac{\partial}{\partial \eta_\gamma} G_j(x,u(x),\dots,D^{m_j-I}u(x),D^{m_j}u(x)) \qquad (3.9)$$

We shall investigate the properties of functions $H[u]$, $a_\beta[u]$, $E_j[u]$, $b_{j\gamma}[u]$ in the following lemmas.

Lemma 6. Let F satisfy condition A_I). Then for any function $u \in C^{k,\alpha}(\bar{\Omega})$ such that $\Gamma_u \subset Q$ there exists a neighbourhood U in which the mappings

$$H : U \subset C^k(\bar{\Omega}) \longrightarrow \tilde{C}^{k-2m}(d_\varepsilon), \quad u \longmapsto H[u],$$

$$a_\beta : U \subset C^k(\bar{\Omega}) \longrightarrow \tilde{C}^{k-2m}(d_\varepsilon), \quad u \longmapsto a_\beta[u]$$

are well-defined and continuous for small $\varepsilon > 0$.

Proof. Let us show the continuity of operators at the point u. Let k=2m. Let us present the mapping H in the following form

$$H[u](x,\tilde{x}) = \frac{F(x,u(x),\dots,D^{2m-I}u(x),D^{2m}u(x)) - F(\tilde{x},u(\tilde{x}),\dots,D^{2m-I}u(\tilde{x}),D^{2m}u(x))}{\left(|x-x|^2 + \sum_{k=0}^{2m-I}\sum_{|\beta|=k}(D^\beta u(x)-D^\beta u(\tilde{x}))^2\right)^{\alpha/2}} \times$$

$$\times \left(I + \left(\frac{u(x)-u(\tilde{x})}{|x-\tilde{x}|}\right)^2 + \sum_{k=I}^{2m-I}\sum_{|\beta|=k}\left(\frac{D^\beta u(x) - D^\beta u(\tilde{x})}{|x-\tilde{x}|}\right)^2\right)^{\alpha/2}$$

All factors are bounded, so $H[u] \in \tilde{C}^0(d_\varepsilon)$. The mapping, defined by the second factor, is continuous. Let us show that the first factor also determines a continuous mapping. It is easy to see that this mapping is an operator of superposition with the function

$$h(x,\tilde{x},\eta_0,\tilde{\eta}_0,\dots,\eta_{2m-I},\tilde{\eta}_{2m-I},\eta_{2m}) =$$

$$= \frac{F(x,\eta_0,\dots,\eta_{2m-I},\eta_{2m}) - F(\tilde{x},\tilde{\eta}_0,\dots,\tilde{\eta}_{2m-I},\eta_{2m})}{\left(|x-\tilde{x}|^2 + \sum_{k=0}^{2m-I}|\eta_k - \tilde{\eta}_k|^2\right)^{\alpha/2}}$$

Set h be equal to zero for points $|x-x|^2 + \sum_{k=0}^{2m-I}|\eta_k - \eta_k|^2 = 0$

Function h obtained above is continuous due to condition A_I. Then the operator of superposition determined by h is also continuous. So, it is shown that mapping H is continuous.

Let us present mapping a_β in the form of sequential mappings, the first of which is the following operator of superposition

$$u \longmapsto \frac{\partial}{\partial \eta_\beta} F(\tilde{x},\dots,D^{2m-I}u(\tilde{x}),\frac{\partial^{2m}}{\partial x_I^{2m}}u(\tilde{x}),\dots,tD^\beta u(x)+(I-t)D^\beta u(\tilde{x}),\dots,$$

$$\dots,\frac{\partial^{2m}u(x)}{\partial x_n^{2m}})$$

and the second mapping is an integral operator.

Each of these mappings is continuous, so mapping a_β is also continuous.

Now, let k = 2m+I. To prove the continuity of the mapping H it is sufficient to show that every mapping

$$\frac{\tilde{\partial}}{\partial x_i}H : C^{2m+I}(\bar{\Omega}) \longrightarrow \tilde{C}^0(d_\varepsilon),\ i=I,2,\dots,n$$

is continuous at the point u. Elementary calculations show that

$$\frac{\tilde{\partial}}{\partial x_i}H[u](x-\tilde{x}) =$$

$$\frac{\frac{\partial}{\partial x_i}F(x,u(x),\dots,D^{2m-I}u(x),D^{2m}u(x))-\frac{\partial}{\partial \tilde{x}_i}F(\tilde{x},u(\tilde{x}),\dots,D^{2m-I}u(\tilde{x}),D^{2m}u(x))}{|x-\tilde{x}|^\alpha}+$$

$$+\sum_{l=0}^{2m}\sum_{|\beta|=l}\frac{\frac{\partial}{\partial \eta_\beta}F(x,u(x),\dots,D^{2m-I}u(x),D^{2m}u(x))-\frac{\partial}{\partial \eta_\beta}F(\tilde{x},u(\tilde{x}),\dots,D^{2m-I}u(\tilde{x}).D^{2m}u(x))}{|x-\tilde{x}|^\alpha}\cdot$$

$$\cdot\frac{\partial}{\partial x_i}D^\beta u(x)+\sum_{l=0}^{2m-I}\sum_{|\beta|=l}\frac{\partial}{\partial \eta_\beta}F(\tilde{x},u(\tilde{x}),\dots,D^{2m-I}u(\tilde{x}),D^{2m}u(x))\cdot$$

$$\cdot\frac{\frac{\partial}{\partial x_i}D^\beta u(x)-\frac{\partial}{\partial \tilde{x}_i}D^\beta u(\tilde{x})}{|x-\tilde{x}|^\alpha}$$

Continuity of the mapping, determined by the third group of summands, is obvious. As for the case k=2m it is proved that all mappings

$$H^{\eta_\beta}[u](x,\tilde{x})=$$

$$=\frac{\frac{\partial}{\partial \eta_\beta}F(x,\dots,D^{2m-I}u(x),D^{2m}u(x))-\frac{\partial}{\partial \eta_\beta}F(\tilde{x},\dots,D^{2m-I}u(\tilde{x}),D^{2m}u(x))}{|x-\tilde{x}|^\alpha}$$

and mappings H^{x_i} are continuous. This shows continuity of all the summands and therefore continuity of $\frac{\tilde{\partial}}{\partial x_i}H$.

The proof of continuity of mapping a_β is reduced to the checking of continuity of every mapping

$$\frac{\tilde{\partial}}{\partial x_i} a_\beta : C^{2m+I}(\bar{\Omega}) \longrightarrow \tilde{C}^o(d_\varepsilon), \ i=I,2,\dots,n$$

As in the case k=2m different operators of superposition appear here, and their continuity is obvious.

Our arguments for k=2m+I can be easily generalized for the case of arbitrary $k > 2m+I$. Lemma 6 is proved.

Lemma 7. Let G_j satisfy condition A_3). Then for any function $u \in C^{k,\alpha}(\bar{\Omega})$ such that $\Gamma_u \subset Q$ there exists a neighbourhood U in which the mappings

$$E_j : U \subset C^k(\bar{\Omega}) \longrightarrow C^{k-m_j}(\partial d_\varepsilon), \quad u \longmapsto E_j[u].$$

$$b_{j\gamma} : U \subset C^k(\bar{\Omega}) \longrightarrow C^{k-m_j}(\partial d_\varepsilon), \quad u \longmapsto b_{j\gamma}[u].$$

are well-defined and continuous for small $\varepsilon > 0$, where $\partial d_\varepsilon = d_\varepsilon \cap (\partial\Omega \times \partial\Omega)$.

Proof. Local unbending of the boundary allows us to write the functions $b_{j\gamma}[u]$ in the form, which is analogous to (3.6) and to repeat completely the arguments of the proof of lemma 6. Q.E.D.

Continue the proof of theorem 6.

Let us introduce for functions u_s, s=0,I,2,... the operators

$$L_\gamma[u_s]v = \sum_{|\beta|=2m} a_\beta[u_s] D_\alpha D^\beta v \quad ,$$

$$B_j[u_s]v = \sum_{|\gamma|=m_j} b_{j\gamma}[u_s] D_\alpha D^\gamma v \quad , \quad j=I,2,\dots,m.$$

According to (3.3) the operators are well-defined if s is large enough. The condition of a uniform ellipticity of operators $L[u_s]$ and complimentary condition of mappings $B_j[u_s]$, j = I,2,..., relative to $L[u_s]$ are satisfied. From lemmas 6 and 7 it follows that

$$\begin{aligned} &a_\beta[u_s] \xrightarrow[S \to \infty]{\tilde{C}^{k-2m}(d_\varepsilon)} a_\beta[u_o] \text{ for all } \beta : |\beta| = 2m \\ &b_{j\gamma}[u_s] \xrightarrow[S \to \infty]{C^{k-m_j}(\partial d_\varepsilon)} b_{j\gamma}[u_o] \text{ for all } j=I,2,\dots,m \\ &\qquad \gamma : |\gamma| = m_j, \end{aligned} \tag{3.I0}$$

if ε is small enough. Then constant C in estimates

$$|v|_{k,\alpha,\bar{\Omega}} \leq C(\langle\langle L[u_s]v\rangle\rangle_{k-2m,d_\varepsilon} + \sum_{j=1}^{m} \langle\langle B_j[u_s]v\rangle\rangle_{k-m_j,\partial d_\varepsilon} + |v|_{0,\bar{\Omega}}) \quad (3.11)$$

can be chosen independently of s=0,1,... and functions $v \in C^{k,\alpha}(\bar{\Omega})$

According to (3.5) and (3.8)

$$L[u_s]\,u_s = D_\alpha F[u_s] - H[u_s] \equiv y_s,$$

$$B_j[u_s]\,u_s = D_\alpha G_j[u_s] - E[u_s] \equiv z_{j,s}.$$

Since the sequences in the right-hand sides of the equalities converge, then

$$\langle\langle L[u_s]u_s\rangle\rangle_{k-2m,d_\varepsilon},\ \langle\langle B_j[u_s]u_s\rangle\rangle_{k-m_j,\partial d}$$

are jointly bounded. Set $v=u_s$ in (3.11).

Then $|u_s|_{k,\alpha,\bar{\Omega}}$, s=1,2,... are joinly bounded. From (3.3) it follows that $u_0 \in C^{k,\alpha}(\bar{\Omega})$.

For s=0 in (3.11) we choose $v = u_s - u_0$,

$$|u_s-u_0|_{k,\alpha,\bar{\Omega}} \leq C(\langle\langle L[u_0](u_s-u_0)\rangle\rangle_{k-2m,d_\varepsilon} +$$

$$+ \sum_{j=1}^{m}\langle\langle B_j[u_0](u_s-u_0)\rangle\rangle_{k-m_j,\partial d_\varepsilon} + |u_s-u_0|_{0,\bar{\Omega}}) \quad (3.12)$$

We shall show that every summand from the right-hand side converges to zero when $s \to \infty$, which will complete the proof of the theorem.

Let us represent $L[u_0]\,(u_s-u_0)$ in the form

$$L[u_0](u_s-u_0) = L[u_s]\,u_s - L[u_0]u_0 - (L[u_s]-L[u_0])u_s = y_s - y_0 -$$

$$- \sum_{|\beta|=2m} (a_\beta[u_s] - a_\beta[u_0])\, D_\alpha D^\beta u_s.$$

As it was shown above y_s converges to y_0. Continuity of mappings a_β and joint boundedness of $\langle\langle D_\alpha D^\beta u_s\rangle\rangle_{k-2m,d_\varepsilon}$ guarantee the convergence to zero of the second group of summands.

Analogous arguments can be applied to expression

$$B_j[u_o](u_s-u_o)=z_s-z_o-\sum_{|\gamma|=m_j}(b_{j,\gamma}[u_s]-b_{j,\gamma}[u_o])D_\alpha D^\gamma u_s.$$

for $j=1,2,\dots,m$.

Thus, it is shown that $|u_s-u_o|_{k,\alpha,\bar\Omega} \xrightarrow[s\to\infty]{} 0$. The theorem is proved.

As a consequence of the above theorem we shall obtain the statement about the properness of mappings induced by the boundary problems

Theorem 7. Let D be a bounded closed subset of $C^{k,\alpha}(\bar\Omega)$, $k\geq 2m$, $\Omega\subset R^n$ be a bounded domain of class $C^{k,\alpha}$ and

$$\Gamma_D=\bigcup_{u\in D}\Gamma_u$$

If functions F, G_j, $j=1,2,\dots,m$ satisfy conditions $A_1)$-$A_4)$ in some neibourhood Q for Γ_D, then mapping

$$\mathbb{F}: D\subset C^{k,\alpha}(\bar\Omega)\longrightarrow C^{k-2m,\alpha}(\bar\Omega)\times\prod_{j=1}^m C^{k-2m_j,\alpha}(\partial\Omega)$$

$$\mathbb{F}[u]=(F[u],G_1[u],\dots,G_m[u])$$

is proper on D.

Proof. Let $K\subset C^{k-2m,\alpha}(\bar\Omega)\times\prod_{j=1}^m C^{k-m_j,\alpha}(\partial\Omega)$ be an arbitrary compact subset. Let us show that $\mathbb{F}^{-1}(K)\cap D$ is compact in $C^{k,\alpha}(\bar\Omega)$. Since the space $C^{k,\alpha}(\bar\Omega)$ is a Banach space, then to prove the compactness it is sufficient to show that every sequence $\{u_s\}$, $u_s\in\mathbb{F}^{-1}(K)\cap D$ contains a convergent subsequence.

Let $\{u_s\}$, $u_s\in\mathbb{F}^{-1}(K)\cap D$ be an arbitrary sequence. Boundedness of set D and compactness of inclusion map of $C^{k,\alpha}(\bar\Omega)$ in $C^k(\bar\Omega)$ without loss of generality allow us to assume that $u_s\xrightarrow[s\to\infty]{C^k(\bar\Omega)}u_o$ and $u_o\in C^{k,\alpha}(\bar\Omega)$. Besides, we assume that sequence $\mathbb{F}[u_s]$, contained in K, is convergent. Then the convergence $\mathbb{F}u_s\xrightarrow[s\to\infty]{}\mathbb{F}u_o$ with respect to the norm of space $C^{k-2m}(\bar\Omega)\times\prod_{j=1}^m C^{k-m_j}(\partial\Omega)$ implies also the convergence in the norm of space $C^{k-2m,\alpha}(\bar\Omega)\times\prod_{j=1}^m C^{k-m_j,\alpha}(\partial\Omega)$. It means that

$$F[u_s]\xrightarrow[s\to\infty]{C^{k-2m,\alpha}(\bar\Omega)}F[u_o],$$

$$G_j[u_s]\xrightarrow[s\to\infty]{C^{k-m_j,\alpha}(\partial\Omega)}G_j[u_o],\quad j=1,2,\dots,m.$$

Since $\Gamma_{u_o}\subset\Gamma_D$, then all conditions of theorem 6 are fulfilled. By theorem 6 we have

$$u_s \xrightarrow[s \to \infty]{C^{k,\alpha}(\bar{\Omega})} u_0$$

and $u_0 \in D$. The theorem is proved.

We note that the boundedness of the domain D is used in the proof only once:to obtain the existence of a subsequence of the sequence $\{u_s\}$ converging in the norm of space $C^k(\bar{\Omega})$. If the boundedness D is not assumed, we can formulate the following statement.

Theorem 8. Let D be a closed subset in $C^{k,\alpha}(\bar{\Omega})$, $k \geq 2m$, Ω be a bounded domain of class $C^{k,\alpha}$ and functions $F, G_j, j=1,2,\ldots,m$ satisfy cinditions $A_1)-A_4)$ in some neighbourhood Q of Γ_D. If for any compact $K \subset C^{k-2m,\alpha}(\bar{\Omega}) \prod_{j=1}^{m} C^{k-m_j,\alpha}(\partial\Omega)$ the set $F^{-1}(K) \cap D$ is pre-compact in $C^k(\bar{\Omega}) \prod_{j=1}^{m} C^{k-m_j,\alpha}(\partial\Omega)$, then the mapping F is proper on D.

Corollary I. Let functions F and $G_j, j=1,2,\ldots,m$ satisfy conditions $A_1)-A_4)$ on set $Q = \bar{\Omega} \times R^{M(2m)}$. If the set of solutions of the boundary-value problem

$$F(x,u(x),Du(x),\ldots,D^{2m}u(x)) = 0, \; x \in \bar{\Omega},$$

$$G_j(x,u(x),Du(x),\ldots,D^{m_j}u(x)) = 0, x \in \partial\Omega, j=1,2,\ldots,m.$$

is compact in $C^k(\bar{\Omega})$ for $k \geq 2m$, then it is bounded and compact in $C^{k,\alpha}(\bar{\Omega})$.

Remark. The suggested method for investigation of properness of mappings in Holder space can also be used for parabolic equations.

REFERENCES

1. Elworthy K.D.,Tromba A.J. Degree theory on Banach manifolds.Proc. Symp.Pure Math,vol.18.A.M.S.,1970,p.86-94.
2. Zvyagin V.G. The properness of elliptic and parabolic differential operator.Lect.Notes in Math.,vol.1453,1990.
3. Zvyagin V.G. Theory of Fredholm maps and nonlinear boundary-value problems (manual for students).Voronezh University,1983.(in Russian).
4. Zvyagin V.G. On the number of solutions for certain boundary-value problems.Lect.Notes in Math.,vol.1334,1988.
5. Agmon S.,Douglis A.,Nirenberg L. Estimates near the boundary for solutions of elliptic partial differential equations satisfying general boundary conditions.I.Comm.pure and appl.math.vol.XII, 623-727 (1959).
6. Ladyzhenskaya O.A.,Ural'tseva N.N. Linear and quasilinear equations of elliptic type. Moscow,1973 (in Russian).

Lecture Notes in Mathematics

For information about Vols. 1–1323
please contact your bookseller or Springer-Verlag

Vol. 1324: F. Cardoso, D.G. de Figueiredo, R. Iório, O. Lopes (Eds.), Partial Differential Equations. Proceedings, 1986. VIII, 433 pages. 1988.

Vol. 1325: A. Truman, I.M. Davies (Eds.), Stochastic Mechanics and Stochastic Processes. Proceedings, 1986. V, 220 pages. 1988.

Vol. 1326: P.S. Landweber (Ed.), Elliptic Curves and Modular Forms in Algebraic Topology. Proceedings, 1986. V, 224 pages. 1988.

Vol. 1327: W. Bruns, U. Vetter, Determinantal Rings. VII, 236 pages. 1988.

Vol. 1328: J.L. Bueso, P. Jara, B. Torrecillas (Eds.), Ring Theory. Proceedings, 1986. IX, 331 pages. 1988.

Vol. 1329: M. Alfaro, J.S. Dehesa, F.J. Marcellan, J.L. Rubio de Francia, J. Vinuesa (Eds.): Orthogonal Polynomials and their Applications. Proceedings, 1986. XV, 334 pages. 1988.

Vol. 1330: A. Ambrosetti, F. Gori, R. Lucchetti (Eds.), Mathematical Economics. Montecatini Terme 1986. Seminar. VII, 137 pages. 1988.

Vol. 1331: R. Bamón, R. Labarca, J. Palis Jr. (Eds.), Dynamical Systems, Valparaiso 1986. Proceedings. VI, 250 pages. 1988.

Vol. 1332: E. Odell, H. Rosenthal (Eds.), Functional Analysis. Proceedings. 1986–87. V, 202 pages. 1988.

Vol. 1333: A.S. Kechris, D.A. Martin, J.R. Steel (Eds.), Cabal Seminar 81–85. Proceedings, 1981–85. V, 224 pages. 1988.

Vol. 1334: Yu.G. Borisovich, Yu.E. Gliklikh (Eds.), Global Analysis – Studies and Applications III. V, 331 pages. 1988.

Vol. 1335: F. Guillén, V. Navarro Aznar, P. Pascual-Gainza, F. Puerta, Hyperrésolutions cubiques et descente cohomologique. XII, 192 pages. 1988.

Vol. 1336: B. Helffer, Semi-Classical Analysis for the Schrödinger Operator and Applications. V, 107 pages. 1988.

Vol. 1337: E. Sernesi (Ed.), Theory of Moduli. Seminar, 1985. VIII, 232 pages. 1988.

Vol. 1338: A.B. Mingarelli, S.G. Halvorsen. Non-Oscillation Domains of Differential Equations with Two Parameters. XI, 109 pages. 1988.

Vol. 1339: T. Sunada (Ed.), Geometry and Analysis of Manifolds. Proceedings, 1987. IX, 277 pages. 1988.

Vol. 1340: S. Hildebrandt, D.S. Kinderlehrer, M. Miranda (Eds.), Calculus of Variations and Partial Differential Equations. Proceedings. 1986. IX, 301 pages. 1988.

Vol. 1341: M. Dauge, Elliptic Boundary Value Problems on Corner Domains. VIII, 259 pages. 1988.

Vol. 1342: J.C. Alexander (Ed.), Dynamical Systems. Proceedings, 1986–87. VIII, 726 pages. 1988.

Vol. 1343: H. Ulrich, Fixed Point Theory of Parametrized Equivariant Maps. VII, 147 pages. 1988.

Vol. 1344: J. Král, J. Lukes, J. Netuka, J. Vesely´ (Eds.), Potential Theory – Surveys and Problems. Proceedings, 1987. VIII, 271 pages. 1988.

Vol. 1345: X. Gomez-Mont, J. Seade, A. Verjovski (Eds.), Holomorphic Dynamics. Proceedings. 1986. VII. 321 pages. 1988.

Vol. 1346: O.Ya. Viro (Ed.), Topology and Geometry – Rohlin Seminar. XI, 581 pages. 1988.

Vol. 1347: C. Preston, Iterates of Piecewise Monotone Mappings on an Interval. V, 166 pages. 1988.

Vol. 1348: F. Borceux (Ed.), Categorical Algebra and its Applications. Proceedings, 1987. VIII, 375 pages. 1988.

Vol. 1349: E. Novak, Deterministic and Stochastic Error Bounds in Numerical Analysis. V, 113 pages. 1988.

Vol. 1350: U. Koschorke (Ed.), Differential Topology Proceedings, 1987, VI, 269 pages. 1988.

Vol. 1351: I. Laine, S. Rickman, T. Sorvali (Eds.), Complex Analysis, Joensuu 1987. Proceedings. XV, 378 pages. 1988.

Vol. 1352: L.L. Avramov, K.B. Tchakerian (Eds.), Algebra – Some Current Trends. Proceedings. 1986. IX, 240 Seiten. 1988.

Vol. 1353: R.S. Palais, Ch.-l. Teng, Critical Point Theory and Submanifold Geometry. X, 272 pages. 1988.

Vol. 1354: A. Gómez, F. Guerra, M.A. Jiménez, G. López (Eds.), Approximation and Optimization. Proceedings, 1987. VI. 280 pages. 1988.

Vol. 1355: J. Bokowski, B. Sturmfels, Computational Synthetic Geometry. V, 168 pages. 1989.

Vol. 1356: H. Volkmer, Multiparameter Eigenvalue Problems and Expansion Theorems. VI, 157 pages. 1988.

Vol. 1357: S. Hildebrandt, R. Leis (Eds.), Partial Differential Equations and Calculus of Variations. VI, 423 pages. 1988.

Vol. 1358: D. Mumford, The Red Book of Varieties and Schemes. V, 309 pages. 1988.

Vol. 1359: P. Eymard, J.-P. Pier (Eds.) Harmonic Analysis. Proceedings, 1987. VIII, 287 pages. 1988.

Vol. 1360: G. Anderson, C. Greengard (Eds.), Vortex Methods. Proceedings, 1987. V, 141 pages. 1988.

Vol. 1361: T. tom Dieck (Ed.), Algebraic Topology and Transformation Groups. Proceedings. 1987. VI, 298 pages. 1988.

Vol. 1362: P. Diaconis, D. Elworthy, H. Föllmer, E. Nelson, G.C. Papanicolaou, S.R.S. Varadhan. École d´ Été de Probabilités de Saint-Flour XV–XVII. 1985–87 Editor: P.L. Hennequin. V, 459 pages. 1988.

Vol. 1363: P.G. Casazza, T.J. Shura, Tsirelson's Space. VIII, 204 pages. 1988.

Vol. 1364: R.R. Phelps, Convex Functions, Monotone Operators and Differentiability. IX, 115 pages. 1989.

Vol. 1365: M. Giaquinta (Ed.), Topics in Calculus of Variations. Seminar, 1987. X, 196 pages. 1989.

Vol. 1366: N. Levitt, Grassmannians and Gauss Maps in PL-Topology. V, 203 pages. 1989.

Vol. 1367: M. Knebusch, Weakly Semialgebraic Spaces. XX, 376 pages. 1989.

Vol. 1368: R. Hübl, Traces of Differential Forms and Hochschild Homology. III, 111 pages. 1989.

Vol. 1369: B. Jiang, Ch.-K. Peng, Z. Hou (Eds.), Differential Geometry and Topology. Proceedings, 1986–87. VI, 366 pages. 1989.

Vol. 1370: G. Carlsson, R.L. Cohen, H.R. Miller, D.C. Ravenel (Eds.), Algebraic Topology. Proceedings, 1986. IX, 456 pages. 1989.

Vol. 1371: S. Glaz, Commutative Coherent Rings. XI, 347 pages. 1989.

Vol. 1372: J. Azéma, P.A. Meyer, M. Yor (Eds.), Séminaire de Probabilités XXIII. Proceedings. IV, 583 pages. 1989.

Vol. 1373: G. Benkart, J.M. Osborn (Eds.), Lie Algebras. Madison 1987. Proceedings. V, 145 pages. 1989.

Vol. 1374: R.C. Kirby, The Topology of 4-Manifolds. VI, 108 pages. 1989.

Vol. 1375: K. Kawakubo (Ed.), Transformation Groups. Proceedings, 1987. VIII, 394 pages, 1989.

Vol. 1376: J. Lindenstrauss, V.D. Milman (Eds.), Geometric Aspects of Functional Analysis. Seminar (GAFA) 1987–88. VII, 288 pages. 1989.

Vol. 1377: J.F. Pierce, Singularity Theory, Rod Theory, and Symmetry-Breaking Loads. IV, 177 pages. 1989.

Vol. 1378: R.S. Rumely, Capacity Theory on Algebraic Curves. III, 437 pages. 1989.

Vol. 1379: H. Heyer (Ed.), Probability Measures on Groups IX. Proceedings, 1988. VIII, 437 pages. 1989.

Vol. 1380: H.P. Schlickewei, E. Wirsing (Eds.), Number Theory, Ulm 1987. Proceedings. V, 266 pages. 1989.

Vol. 1381: J.-O. Strömberg, A. Torchinsky, Weighted Hardy Spaces. V, 193 pages. 1989.

Vol. 1382: H. Reiter, Metaplectic Groups and Segal Algebras. XI, 128 pages. 1989.

Vol. 1383: D.V. Chudnovsky, G.V. Chudnovsky, H. Cohn, M.B. Nathanson (Eds.), Number Theory, New York 1985–88. Seminar. V, 256 pages. 1989.

Vol. 1384: J. Garcia-Cuerva (Ed.), Harmonic Analysis and Partial Differential Equations. Proceedings, 1987. VII, 213 pages. 1989.

Vol. 1385: A.M. Anile, Y. Choquet-Bruhat (Eds.), Relativistic Fluid Dynamics. Seminar, 1987. V, 308 pages. 1989.

Vol. 1386: A. Bellen, C.W. Gear, E. Russo (Eds.), Numerical Methods for Ordinary Differential Equations. Proceedings, 1987. VII, 136 pages. 1989.

Vol. 1387: M. Petkovi´c, Iterative Methods for Simultaneous Inclusion of Polynomial Zeros. X, 263 pages. 1989.

Vol. 1388: J. Shinoda, T.A. Slaman, T. Tugué (Eds.), Mathematical Logic and Applications. Proceedings, 1987. V, 223 pages. 1989.

Vol. 1000: Second Edition. H. Hopf, Differential Geometry in the Large. VII, 184 pages. 1989.

Vol. 1389: E. Ballico, C. Ciliberto (Eds.), Algebraic Curves and Projective Geometry. Proceedings, 1988. V, 288 pages. 1989.

Vol. 1390: G. Da Prato, L. Tubaro (Eds.), Stochastic Partial Differential Equations and Applications II. Proceedings, 1988. VI, 258 pages. 1989.

Vol. 1391: S. Cambanis, A. Weron (Eds.), Probability Theory on Vector Spaces IV. Proceedings, 1987. VIII, 424 pages. 1989.

Vol. 1392: R. Silhol, Real Algebraic Surfaces. X, 215 pages. 1989.

Vol. 1393: N. Bouleau, D. Feyel, F. Hirsch, G. Mokobodzki (Eds.), Séminaire de Théorie du Potentiel Paris, No. 9. Proceedings. VI, 265 pages. 1989.

Vol. 1394: T.L. Gill, W.W. Zachary (Eds.), Nonlinear Semigroups, Partial Differential Equations and Attractors. Proceedings, 1987. IX, 233 pages. 1989.

Vol. 1395: K. Alladi (Ed.), Number Theory, Madras 1987. Proceedings. VII, 234 pages. 1989.

Vol. 1396: L. Accardi, W. von Waldenfels (Eds.), Quantum Probability and Applications IV. Proceedings, 1987. VI, 355 pages. 1989.

Vol. 1397: P.R. Turner (Ed.), Numerical Analysis and Parallel Processing. Seminar, 1987. VI, 264 pages. 1989.

Vol. 1398: A.C. Kim, B.H. Neumann (Eds.), Groups – Korea 1988. Proceedings. V, 189 pages. 1989.

Vol. 1399: W.-P. Barth, H. Lange (Eds.), Arithmetic of Complex Manifolds. Proceedings, 1988. V, 171 pages. 1989.

Vol. 1400: U. Jannsen. Mixed Motives and Algebraic K-Theory. XIII, 246 pages. 1990.

Vol. 1401: J. Steprans, S. Watson (Eds.), Set Theory and its Applications. Proceedings, 1987. V, 227 pages. 1989.

Vol. 1402: C. Carasso, P. Charrier, B. Hanouzet, J.-L. Joly (Eds.), Nonlinear Hyperbolic Problems. Proceedings, 1988. V, 249 pages. 1989.

Vol. 1403: B. Simeone (Ed.), Combinatorial Optimization. Seminar, 1986. V, 314 pages. 1989.

Vol. 1404: M.-P. Malliavin (Ed.), Séminaire d´Algèbre Paul Dubreil et Marie-Paul Malliavin. Proceedings, 1987–1988. IV, 410 pages. 1989.

Vol. 1405: S. Dolecki (Ed.), Optimization. Proceedings, 1988. V, 223 pages. 1989. Vol. 1406: L. Jacobsen (Ed.), Analytic Theory of Continued Fractions III. Proceedings, 1988. VI, 142 pages. 1989.

Vol. 1407: W. Pohlers, Proof Theory. VI, 213 pages. 1989.

Vol. 1408: W. Lück, Transformation Groups and Algebraic K-Theory. XII, 443 pages. 1989.

Vol. 1409: E. Hairer, Ch. Lubich, M. Roche. The Numerical Solution of Differential-Algebraic Systems by Runge-Kutta Methods. VII, 139 pages. 1989.

Vol. 1410: F.J. Carreras, O. Gil-Medrano, A.M. Naveira (Eds.), Differential Geometry. Proceedings, 1988. V, 308 pages. 1989.

Vol. 1411: B. Jiang (Ed.), Topological Fixed Point Theory and Applications. Proceedings. 1988. VI, 203 pages. 1989.

Vol. 1412: V.V. Kalashnikov, V.M. Zolotarev (Eds.), Stability Problems for Stochastic Models. Proceedings, 1987. X, 380 pages. 1989.

Vol. 1413: S. Wright, Uniqueness of the Injective III_1 Factor. III, 108 pages. 1989.

Vol. 1414: E. Ramirez de Arellano (Ed.), Algebraic Geometry and Complex Analysis. Proceedings, 1987. VI, 180 pages. 1989.

Vol. 1415: M. Langevin, M. Waldschmidt (Eds.), Cinquante Ans de Polynômes. Fifty Years of Polynomials. Proceedings, 1988. IX, 235 pages. 1990.

Vol. 1416: C. Albert (Ed.), Géométrie Symplectique et Mécanique. Proceedings, 1988. V, 289 pages. 1990.

Vol. 1417: A.J. Sommese, A. Biancofiore, E.L. Livorni (Eds.), Algebraic Geometry. Proceedings, 1988. V, 320 pages. 1990.

Vol. 1418: M. Mimura (Ed.), Homotopy Theory and Related Topics. Proceedings, 1988. V, 241 pages. 1990.

Vol. 1419: P.S. Bullen, P.Y. Lee, J.L. Mawhin, P. Muldowney, W.F. Pfeffer (Eds.), New Integrals. Proceedings, 1988. V, 202 pages. 1990.

Vol. 1420: M. Galbiati, A. Tognoli (Eds.), Real Analytic Geometry. Proceedings, 1988. IV, 366 pages. 1990.

Vol. 1421: H.A. Biagioni, A Nonlinear Theory of Generalized Functions, XII, 214 pages. 1990.

Vol. 1422: V. Villani (Ed.), Complex Geometry and Analysis. Proceedings, 1988. V, 109 pages. 1990.

Vol. 1423: S.O. Kochman, Stable Homotopy Groups of Spheres: A Computer-Assisted Approach. VIII, 330 pages. 1990.

Vol. 1424: F.E. Burstall, J.H. Rawnsley, Twistor Theory for Riemannian Symmetric Spaces. III, 112 pages. 1990.

Vol. 1425: R.A. Piccinini (Ed.), Groups of Self-Equivalences and Related Topics. Proceedings, 1988. V, 214 pages. 1990.

Vol. 1426: J. Azéma, P.A. Meyer, M. Yor (Eds.), Séminaire de Probabilités XXIV, 1988/89. V, 490 pages. 1990.

Vol. 1427: A. Ancona, D. Geman, N. Ikeda, École d'Eté de Probabilités de Saint Flour XVIII, 1988. Ed.: P.L. Hennequin. VII, 330 pages. 1990.

Vol. 1428: K. Erdmann, Blocks of Tame Representation Type and Related Algebras. XV. 312 pages. 1990.

Vol. 1429: S. Homer, A. Nerode, R.A. Platek, G.E. Sacks, A. Scedrov, Logic and Computer Science. Seminar, 1988. Editor: P. Odifreddi. V, 162 pages. 1990.

Vol. 1430: W. Bruns, A. Simis (Eds.), Commutative Algebra. Proceedings. 1988. V, 160 pages. 1990.

Vol. 1431: J.G. Heywood, K. Masuda, R. Rautmann, V.A. Solonnikov (Eds.), The Navier-Stokes Equations – Theory and Numerical Methods. Proceedings, 1988. VII, 238 pages. 1990.

Vol. 1432: K. Ambos-Spies, G.H. Müller, G.E. Sacks (Eds.), Recursion Theory Week. Proceedings, 1989. VI, 393 pages. 1990.

Vol. 1433: S. Lang, W. Cherry, Topics in Nevanlinna Theory. II, 174 pages.1990.

Vol. 1434: K. Nagasaka, E. Fouvry (Eds.), Analytic Number Theory. Proceedings, 1988. VI, 218 pages. 1990.

Vol. 1435: St. Ruscheweyh, E.B. Saff, L.C. Salinas, R.S. Varga (Eds.), Computational Methods and Function Theory. Proceedings, 1989. VI, 211 pages. 1990.

Vol. 1436: S. Xambó-Descamps (Ed.), Enumerative Geometry. Proceedings, 1987. V, 303 pages. 1990.

Vol. 1437: H. Inassaridze (Ed.), K-theory and Homological Algebra. Seminar, 1987 88. V, 313 pages. 1990.

Vol. 1438: P.G. Lemarié (Ed.) Les Ondelettes en 1989. Seminar. IV, 212 pages. 1990.

Vol. 1439: E. Bujalance, J.J. Etayo, J.M. Gamboa, G. Gromadzki. Automorphism Groups of Compact Bordered Klein Surfaces: A Combinatorial Approach. XIII, 201 pages. 1990.

Vol. 1440: P. Latiolais (Ed.), Topology and Combinatorial Groups Theory. Seminar, 1985–1988. VI, 207 pages. 1990.

Vol. 1441: M. Coornaert, T. Delzant, A. Papadopoulos. Géométrie et théorie des groupes. X, 165 pages. 1990.

Vol. 1442: L. Accardi, M. von Waldenfels (Eds.), Quantum Probability and Applications V. Proceedings, 1988. VI, 413 pages. 1990.

Vol. 1443: K.H. Dovermann, R. Schultz, Equivariant Surgery Theories and Their Periodicity Properties. VI, 227 pages. 1990.

Vol. 1444: H. Korezlioglu, A.S. Ustunel (Eds.), Stochastic Analysis and Related Topics VI. Proceedings, 1988. V, 268 pages. 1990.

Vol. 1445: F. Schulz, Regularity Theory for Quasilinear Elliptic Systems and – Monge Ampère Equations in Two Dimensions. XV, 123 pages. 1990.

Vol. 1446: Methods of Nonconvex Analysis. Seminar, 1989. Editor: A. Cellina. V, 206 pages. 1990.

Vol. 1447: J.-G. Labesse, J. Schwermer (Eds), Cohomology of Arithmetic Groups and Automorphic Forms. Proceedings, 1989. V, 358 pages. 1990.

Vol. 1448: S.K. Jain, S.R. López-Permouth (Eds.), Non-Commutative Ring Theory. Proceedings, 1989. V, 166 pages. 1990.

Vol. 1449: W. Odyniec, G. Lewicki, Minimal Projections in Banach Spaces. VIII, 168 pages. 1990.

Vol. 1450: H. Fujita, T. Ikebe, S.T. Kuroda (Eds.), Functional-Analytic Methods for Partial Differential Equations. Proceedings, 1989. VII, 252 pages. 1990.

Vol. 1451: L. Alvarez-Gaumé, E. Arbarello, C. De Concini, N.J. Hitchin, Global Geometry and Mathematical Physics. Montecatini Terme 1988. Seminar. Editors: M. Francaviglia, F. Gherardelli. IX, 197 pages. 1990.

Vol. 1452: E. Hlawka, R.F. Tichy (Eds.), Number-Theoretic Analysis. Seminar, 1988–89. V, 220 pages. 1990.

Vol. 1453: Yu.G. Borisovich, Yu.E. Gliklikh (Eds.), Global Analysis – Studies and Applications IV. V, 320 pages. 1990.

Vol. 1454: F. Baldassari, S. Bosch, B. Dwork (Eds.), p-adic Analysis. Proceedings, 1989. V, 382 pages. 1990.

Vol. 1455: J.-P. Françoise, R. Roussarie (Eds.), Bifurcations of Planar Vector Fields. Proceedings, 1989. VI, 396 pages. 1990.

Vol. 1456: L.G. Kovács (Ed.), Groups – Canberra 1989. Proceedings. XII, 198 pages. 1990.

Vol. 1457: O. Axelsson, L.Yu. Kolotilina (Eds.), Preconditioned Conjugate Gradient Methods. Proceedings, 1989. V, 196 pages. 1990.

Vol. 1458: R. Schaaf, Global Solution Branches of Two Point Boundary Value Problems. XIX, 141 pages. 1990.

Vol. 1459: D. Tiba, Optimal Control of Nonsmooth Distributed Parameter Systems. VII, 159 pages. 1990.

Vol. 1460: G. Toscani, V. Boffi, S. Rionero (Eds.), Mathematical Aspects of Fluid Plasma Dynamics. Proceedings, 1988. V, 221 pages. 1991.

Vol. 1461: R. Gorenflo, S. Vessella, Abel Integral Equations. VII, 215 pages. 1991.

Vol. 1462: D. Mond, J. Montaldi (Eds.), Singularity Theory and its Applications. Warwick 1989, Part I. VIII, 405 pages. 1991.

Vol. 1463: R. Roberts, I. Stewart (Eds.), Singularity Theory and its Applications. Warwick 1989, Part II. VIII, 322 pages. 1991.

Vol. 1464: D. L. Burkholder, E. Pardoux, A. Sznitman, Ecole d'Eté de Probabilités de Saint- Flour XIX-1989. Editor: P. L. Hennequin. VI, 256 pages. 1991.

Vol. 1465: G. David, Wavelets and Singular Integrals on Curves and Surfaces. X, 107 pages. 1991.

Vol. 1466: W. Banaszczyk, Additive Subgroups of Topological Vector Spaces. VII, 178 pages. 1991.

Vol. 1467: W. M. Schmidt, Diophantine Approximations and Diophantine Equations. VIII, 217 pages. 1991.

Vol. 1468: J. Noguchi, T. Ohsawa (Eds.), Prospects in Complex Geometry. Proceedings. 1989. VII, 421 pages. 1991.

Vol. 1469: J. Lindenstrauss, V. D. Milman (Eds.), Geometric Aspects of Functional Analysis. Seminar 1989-90. XI, 191 pages. 1991.

Vol. 1470: E. Odell, H. Rosenthal (Eds.), Functional Analysis. Proceedings, 1987-89. VII, 199 pages. 1991.

Vol. 1471: A. A. Panchishkin, Non-Archimedean L-Functions of Siegel and Hilbert Modular Forms. VII, 157 pages. 1991.

Vol. 1472: T. T. Nielsen, Bose Algebras: The Complex and Real Wave Representations. V, 132 pages. 1991.

Vol. 1473: Y. Hino, S. Murakami, T. Naito, Functional Differential Equations with Infinite Delay. X, 317 pages. 1991.

Vol. 1474: S. Jackowski, B. Oliver, K. Pawałowski (Eds.), Algebraic Topology, Poznań 1989. Proceedings. VIII, 397 pages. 1991.

Vol. 1475: S. Busenberg, M. Martelli (Eds.), Delay Differential Equations and Dynamical Systems. Proceedings, 1990. VIII, 249 pages. 1991.

Vol. 1476: M. Bekkali, Topics in Set Theory. VII, 120 pages. 1991.

Vol. 1477: R. Jajte, Strong Limit Theorems in Noncommutative L_2-Spaces. X, 113 pages. 1991.

Vol. 1478: M.-P. Malliavin (Ed.), Topics in Invariant Theory. Seminar 1989-1990. VI, 272 pages. 1991.

Vol. 1479: S. Bloch, I. Dolgachev, W. Fulton (Eds.), Algebraic Geometry. Proceedings, 1989. VII, 300 pages. 1991.

Vol. 1480: F. Dumortier, R. Roussarie, J. Sotomayor, H. Żoładek, Bifurcations of Planar Vector Fields: Nilpotent Singularities and Abelian Integrals. VIII, 226 pages. 1991.

Vol. 1481: D. Ferus, U. Pinkall, U. Simon, B. Wegner (Eds.), Global Differential Geometry and Global Analysis. Proceedings, 1991. VIII, 283 pages. 1991.

Vol. 1482: J. Chabrowski, The Dirichlet Problem with L^2-Boundary Data for Elliptic Linear Equations. VI, 173 pages. 1991.

Vol. 1483: E. Reithmeier, Periodic Solutions of Nonlinear Dynamical Systems. VI, 171 pages. 1991.

Vol. 1484: H. Delfs, Homology of Locally Semialgebraic Spaces. IX, 136 pages. 1991.

Vol. 1485: J. Azéma, P. A. Meyer, M. Yor (Eds.), Séminaire de Probabilités XXV. VIII, 440 pages. 1991.

Vol. 1486: L. Arnold, H. Crauel, J.-P. Eckmann (Eds.), Lyapunov Exponents. Proceedings, 1990. VIII, 365 pages. 1991.

Vol. 1487: E. Freitag, Singular Modular Forms and Theta Relations. VI, 172 pages. 1991.

Vol. 1488: A. Carboni, M. C. Pedicchio, G. Rosolini (Eds.), Category Theory. Proceedings, 1990. VII, 494 pages. 1991.

Vol. 1489: A. Mielke, Hamiltonian and Lagrangian Flows on Center Manifolds. X, 140 pages. 1991.

Vol. 1490: K. Metsch, Linear Spaces with Few Lines. XIII, 196 pages. 1991.

Vol. 1491: E. Lluis-Puebla, J.-L. Loday, H. Gillet, C. Soulé, V. Snaith, Higher Algebraic K-Theory: an overview. IX, 164 pages. 1992.

Vol. 1492: K. R. Wicks, Fractals and Hyperspaces. VIII, 168 pages. 1991.

Vol. 1493: E. Benoît (Ed.), Dynamic Bifurcations. Proceedings, Luminy 1990. VII, 219 pages. 1991.

Vol. 1494: M.-T. Cheng, X.-W. Zhou, D.-G. Deng (Eds.), Harmonic Analysis. Proceedings, 1988. IX, 226 pages. 1991.

Vol. 1495: J. M. Bony, G. Grubb, L. Hörmander, H. Komatsu, J. Sjöstrand, Microlocal Analysis and Applications. Montecatini Terme, 1989. Editors: L. Cattabriga, L. Rodino. VII, 349 pages. 1991.

Vol. 1496: C. Foias, B. Francis, J. W. Helton, H. Kwakernaak, J. B. Pearson, H_∞-Control Theory. Como, 1990. Editors: E. Mosca, L. Pandolfi. VII, 336 pages. 1991.

Vol. 1497: G. T. Herman, A. K. Louis, F. Natterer (Eds.), Mathematical Methods in Tomography. Proceedings 1990. X, 268 pages. 1991.

Vol. 1498: R. Lang, Spectral Theory of Random Schrödinger Operators. X, 125 pages. 1991.

Vol. 1499: K. Taira, Boundary Value Problems and Markov Processes. IX, 132 pages. 1991.

Vol. 1500: J.-P. Serre, Lie Algebras and Lie Groups. VII, 168 pages. 1992.

Vol. 1501: A. De Masi, E. Presutti, Mathematical Methods for Hydrodynamic Limits. IX, 196 pages. 1991.

Vol. 1502: C. Simpson, Asymptotic Behavior of Monodromy. V, 139 pages. 1991.

Vol. 1503: S. Shokranian, The Selberg-Arthur Trace Formula (Lectures by J. Arthur). VII, 97 pages. 1991.

Vol. 1504: J. Cheeger, M. Gromov, C. Okonek, P. Pansu, Geometric Topology: Recent Developments. Editors: P. de Bartolomeis, F. Tricerri. VII, 197 pages. 1991.

Vol. 1505: K. Kajitani, T. Nishitani, The Hyperbolic Cauchy Problem. VII, 168 pages. 1991.

Vol. 1506: A. Buium, Differential Algebraic Groups of Finite Dimension. XV, 145 pages. 1992.

Vol. 1507: K. Hulek, T. Peternell, M. Schneider, F.-O. Schreyer (Eds.), Complex Algebraic Varieties. Proceedings, 1990. VII, 179 pages. 1992.

Vol. 1508: M. Vuorinen (Ed.), Quasiconformal Space Mappings. A Collection of Surveys 1960-1990. IX, 148 pages. 1992.

Vol. 1509: J. Aguadé, M. Castellet, F. R. Cohen (Eds.), Algebraic Topology - Homotopy and Group Cohomology. Proceedings, 1990. X, 330 pages. 1992.

Vol. 1510: P. P. Kulish (Ed.), Quantum Groups. Proceedings, 1990. XII, 398 pages. 1992.

Vol. 1511: B. S. Yadav, D. Singh (Eds.), Functional Analysis and Operator Theory. Proceedings, 1990. VIII, 223 pages. 1992.

Vol. 1512: L. M. Adleman, M.-D. A. Huang, Primality Testing and Abelian Varieties Over Finite Fields. VII, 142 pages. 1992.

Vol. 1513: L. S. Block, W. A. Coppel, Dynamics in One Dimension. VIII, 249 pages. 1992.

Vol. 1514: U. Krengel, K. Richter, V. Warstat (Eds.), Ergodic Theory and Related Topics III, Proceedings, 1990. VIII, 236 pages. 1992.

Vol. 1515: E. Ballico, F. Catanese, C. Ciliberto (Eds.), Classification of Irregular Varieties. Proceedings, 1990. VII, 149 pages. 1992.

Vol. 1517: K. Keimel, W. Roth, Ordered Cones and Approximation. VI, 134 pages. 1992.

Vol. 1518: H. Stichtenoth, M. A. Tsfasman (Eds.), Coding Theory and Algebraic Geometry. Proceedings, 1991. VIII, 223 pages. 1992.

Vol. 1519: M. W. Short, The Primitive Soluble Permutation Groups of Degree less than 256. IX, 145 pages. 1992.

Vol. 1520: Yu. G. Borisovich, Yu. E. Gliklikh (Eds.), Global Analysis – Studies and Applications V. VII, 284 pages. 1992.